THE EVOLUTION OF THE BIOSPHERE

# ATMOSPHERIC SCIENCES LIBRARY

M. I. BUDYKO

*State Hydrological Institute*
*Leningrad, U.S.S.R.*

# The Evolution of the Biosphere

*Translated from the Russian by*
M. I. BUDYKO, S. F. LEMESHKO and V. G. YANUTA

D. Reidel Publishing Company

A MEMBER OF THE KLUWER ACADEMIC PUBLISHERS GROUP

Dordrecht / Boston / Lancaster / Tokyo

**Library of Congress Cataloging-in-Publication Data**

CIP

Budyko, M. I. (Mikhail Ivanovich)
The evolution of the biosphere.

(Atmospheric sciences library)
Translation of: Évoli͡utsii͡a biosfery.
Bibliography: p.
Includes index.
1. Evolution. 2. Biosphere. I. Title. II. Series.
QH371.B7713 1986 574.5 86–17799
ISBN 90–277–2140–8



Translated from the 1984 edition of
*Evolyutsia Biosfery*
published by Gidrometeoizdat

Published by D. Reidel Publishing Company,
P.O. Box 17, 3300 AA Dordrecht, Holland.

Sold and distributed in the U.S.A. and Canada
by Kluwer Academic Publishers,
101 Philip Drive, Assinippi Park, Norwell, MA 02061, U.S.A.

In all other countries, sold and distributed
by Kluwer Academic Publishers Group,
P.O. Box 322, 3300 AH Dordrecht, Holland.



Printed in The Netherlands.

# CONTENTS

# INTRODUCTION

## THE STUDY OF THE BIOSPHERE

The term 'biosphere' first appeared in the works of the French biologist J.-B. Lamarck and the Austrian geologist E. Suess in the 19th century. In the 20th century, the study of the biosphere attracted considerable attention, largely due to the research of V. I. Vernadsky (1863—1945). The results of Vernadsky's investigations have appeared in a number of publications, including the monograph *The Biosphere* published in 1926. This work consists of two parts, 'The Biosphere in Cosmos' and 'The Zone of Life', written in a form of speculation and reflection that is rarely used in modern studies. This work concerns the distinguishing properties of the space occupied by organisms and the exceptional importance of the activities of these organisms in the formation of their environment.

In this and subsequent studies, Vernadsky has laid the foundations of the science of the biosphere, which today plays an important role in the many branches of science concerned with the Earth. Several terms have been suggested for the science of the biosphere, including global ecology (a discipline studying the global ecological system, whose meaning is close to that of the biosphere).

One of the most prominent predecessors of Vernadsky was his teacher V. V. Dokuchaev (1846—1903), the founder of scientific pedology. Dokuchaev was the first to study the interrelationships between living organisms and their environment in different geographical zones. In a series of articles, which first appeared in 1898—1900 and which were published later in a small volume (Dokuchaev, 1948), he considered a system of geographical zones corresponding to different types of soil formation. Dokuchaev identified a relationship between geographical zones and climatic factors and constructed a table describing soils, vegetation, fauna and other components of the biosphere for each geographical zone.

Unlike Dokuchaev, Vernadsky was more interested in the global aspects of the biosphere rather than in the zonal ones. In his monograph *The Biosphere* he wrote: ". . . the author . . . attempts to describe a geologic development of life, to portray the planetary process that takes place around us". In this book, Vernadsky introduced the concept of a 'film of living matter' close to the Earth's surface, both on land and in the ocean. Pointing out that the zone in which living organisms exist comprises a considerable part of the atmosphere and the hydrosphere, as well as the upper layers of the lithosphere, Vernadsky went on to show that the life zone of the overwhelming majority of organisms is limited to a comparatively thin film, including the lower layer of the atmosphere, the upper layers of the oceans and soil. He also emphasized that the transformation of solar radiation into chemical energy, vital for the activities of organisms, largely takes place within this film.

Vernadsky indicated that, in the course of an organism's life, energy and various kinds of matter are continually being transferred from the Earth's surface to the lithosphere, atmosphere and hydrosphere. As a result, these activities appear to be the most essential factor in the geochemical evolution of geographical envelopes of the Earth (including the atmosphere, hydrosphere and upper layers of the lithosphere). Vernadsky paid much attention to the impact of human activities on the biosphere. Under this impact the biosphere is being converted into the noosphere (a word of Greek origin translated as a sphere of mind).

The great value of Vernadsky's study of the biosphere became apparent, particularly in the second half of our century, when, because of the rapidly growing impact of man's activities on the environment, ecological problems were given much greater attention.

Much research work of great importance in the study of the global ecological system (the biosphere) has been undertaken during the last few decades. This includes the studies of Sukachev (1945, 1964) on biogeocenose. The concept of biogeocenose is similar to that of the ecological system and has been essential in developing modern ideas of interrelations between organisms and their environment. This line of research was also developed in the studies of Lavrenko (1949, 1971) who examined the properties of the phytogeosphere, the sphere of plant life; and in the studies of Bazilevich and Rodin (1967) who presented detailed information on the biomass and productivity of different vegetation types.

Dokuchaev's ideas in the field of pedology have been further developed by Gerasimov (1945, 1960) who treated soils as an element of the biosphere closely connected with its other components. The works of Volobuev (1953) and Kovda (1973) also deal with this problem and examine the process of soil formation and its dependence on external factors.

A comprehensive description of interactions among natural processes that lead to the formation of geographical zones is given in the studies of Grigoriev (1966). In these he shows the dependence of geographical zones on such climatic elements as solar radiation and precipitation. Information on the biosphere's resources in the territory of the U.S.S.R. appeared in the monograph edited by Gerasimov (1971). This study also contains an analysis of human impact on the biosphere.

For examining global ecology it is necessary to have data on the transformations of solar energy in the biosphere and on the water balance of the Earth. Detailed information on the energy balance can be derived from a number of monographic studies and atlases, including *Atlas of the Heat Balance of the Earth* (1963). Comprehensive information on the water balance of land and ocean is presented in a collective monograph *The World Water Balance and Water Resources of the Earth* (1974) and in the studies of Lvovich (1974). There are a number of books in which data on the energy and water balances have been used in studying global ecology (Budyko, 1956, 1971, 1977a; H. T. Odum, 1971). A number of general studies of ecology also contain data that are relevant for the study of global ecology, among which are the monographs by E. Odum (1971) and Whittaker (1975).

The attention of many scientists has been drawn to changes in the biosphere resulting from anthropogenic influences. This subject has been treated in the works of Armand (1966), Duvigneaud and Tanghe (1967) Watt (1968), Commoner (1971), Fyodorov (1972), Izrael (1979b), amongst others. The important practical problem of developing an optimum system of observations of the state of the environment (the monitoring system) has been considered by Izrael (1974, 1979a).

The rapid change which the biosphere is undergoing at present, under the impact of anthropogenic factors, has been compared with the evolution of the biosphere in the geological past (Budyko, 1974, 1980). A number of studies have been concerned with the interrelationship between changes in the abiotic components of the biosphere and the evolution of organisms (Krasilov, 1977; Pearson, 1978, Kamshilov, 1979; Budyko, 1982). It might be thought that further development of this line of research would help in clearing up the general features of the biosphere's evolution.

The major reasons for the study of the biosphere's evolution have been discussed by Sokolov (1981) who considered the palaeontological and geological aspects of this problem and who pointed out that the study of the biosphere's history involves an extensive synthesis of material from both Earth and biological scientific disciplines.

In recent years attention has been drawn to the question, raised by Vernadsky, that the biosphere is turning into the noosphere. A modern interpretation of this question is given in the work of Yanshin (1981) who underlined the vital importance of the formation of the noosphere in our epoch.

## EVOLUTION OF THE BIOSPHERE

The evolution of the biosphere presents two aspects, one of which concerns the history of the biosphere since the time when life first emerged on the Earth to the time when human activity started influencing the biosphere, and the other concerns the study of the biosphere in the present epoch, which is experiencing an ever-growing impact of anthropogenic factors.

The available information relating to past changes in the biosphere does not provide much information on the evolution of the biosphere's major components. As a result of recent advances in palaeontological research, we now have a considerable body of information on changes in successive floras and faunas. However, information is scarce on changes in the abiotic environment of organisms and particularly variations in the atmosphere. As a result, there are great difficulties in understanding the mechanism of the biosphere's evolution, particularly concerning interactions between variations in the biosphere's biotic and abiotic components.

An opportunity to determine the general patterns of the biosphere's evolution during the geologic past has only recently arisen, due largely to progress in studying the development of abiotic components of the biosphere. The examination of these patterns, which are of importance in perceiving the history of the biosphere, is also necessary in order to understand the development of the biosphere in our epoch and to estimate possible future changes in the biosphere.

Although at present, anthropogenic factors often produce a greater impact on the biosphere than natural factors, evaluation of anthropogenic change in the biosphere requires in many cases a knowledge of the history of its natural change. At the same time, information on past natural changes in the biosphere is of great importance in confirming and checking models of present biosphere variations.

As mentioned above, anthropogenic changes in the biosphere have attracted considerable attention in the second half of the present century when the impact of human activities on the environment became particularly pronounced. The pollution of the atmosphere, of the continental waters and of the ocean, the destruction of natural vegetation cover, the extinction of a great many animal species, the deterioration of soils, and other consequences of the economic activities of man, have made it urgent to carry out ecological studies on a much greater scale. The overwhelming majority of these studies have been devoted to local ecological problems i.e. to examining interactions between organisms and their environment within regional ecological systems. Studies of global ecological problems concerning the biosphere as a whole, or its major parts, have been undertaken comparatively rarely.

The global problems of ecology have attracted much more attention in recent years when it appeared that human economic activities were beginning to affect large-scale natural processes. It should be mentioned that, although natural conditions throughout a considerable part of the globe have already changed because of anthropogenic impact, until recently these local effects resulting from the expansion of man's activities (which may individually affect vast areas) have, nevertheless, occurred independently of one another. For instance, the clearing of the forests on one of the continents did not affect the forests of the other continents. Similarly, the construction of dams on individual rivers did not change the run-off of the rivers that were not connected with them. A different situation arises when man starts to influence nature's global processes. In this case, the impact on the environment in one region might affect natural conditions in other regions, even if they are situated far from the initial impact.

Some examples of human influence on large-scale natural processes are known from earlier times. For instance, the annihilation of migratory birds in a number of middle latitude countries has altered the fauna of tropical countries to which these birds migrated in certain seasons of the year. A similar situation occurs with some migratory marine animals and fishes. These examples, however, refer to the biosphere's components, whose influence on the biosphere as a whole is comparatively slight.

It has been established in recent years that, under the impact of human activities, the chemical composition and physical state of the atmosphere and ocean have begun to alter and, as a result, there arises a possibility of anthropogenic changes in other components of the biosphere. Moreover, due to intensive horizontal mixing of the atmosphere and upper oceanic layers, man-made effects on the ocean and particularly on the atmosphere might influence areas at great distances from the region in which these effects originated.

Since atmospheric processes produce a significant influence on all the components of the biosphere, including living organisms, it is evident that large-scale changes in the atmosphere–ocean system will inevitably lead to changes in the biosphere as a whole, which might be unfavourable and even disastrous for human society.

In view of the uncontrolled human impact on large-scale processes in the atmosphere and ocean, which might lead in the long run to a global ecological crisis, it is necessary to study the possible large-scale anthropogenic changes in the biosphere, evaluating these changes quantitatively given different future scenarios of economic development. These forecasts are indispensable in finding the optimum technological progress' ways which will result in minimal harm to the environment.

To solve this and other problems of ecology, it is necessary to overcome a principal difficulty which does not emerge in many other scientific disciplines. The study of the interrelations of living organisms (man included) and their environment involves information and methods from many scientific disciplines including biology, geography, geophysics, geochemistry, geology, meteorology, climatology, economics and technology.

At present, with the volume of information rapidly increasing in every field of science and thus promoting an ever-growing specialization within specific scientific branches, the possibility of developing a synthesis of data from many lines of research becomes more and more limited. However, without such a synthesis, progress in modern ecology will be severely restricted. This synthesis is particularly necessary if we are to properly understand the effects of anthropogenic factors on the gobal ecological system (the biosphere), the structure of which is much more complicated than that of regional and local ecological systems.

In recent years a number of attempts have been made to comprehensively study anthropogenic changes in the biosphere. Some work in this direction has been devoted to examining recent climatic change. It has been established that, as a result of the combustion of ever-increasing amounts of coal, oil and other kinds of fossil fuels in the modern epoch, the chemical composition of the atmosphere has changed, mainly due to increasing carbon dioxide content. This has already caused some modification in global climate, and, in a few decades, will lead to changes as large as have occurred previously only over the course of many thousands or millions of years (Budyko, 1972, 1980; *Energy and Climate*, 1977; *Climatic Effects of Increased Atmospheric Carbon Dioxide*, 1982).

This change in global climate will introduce essential modifications in all basic components of the biosphere, including the hydrosphere, cryosphere, soils, vegetation cover and animal world. Because of these modifications it is important to try to forecast possible future changes in the biosphere under the influence of man's activities, using different scenarios of economic development. Since such forecasts determine to a great extent the long-term planning of economies associated with large investments, it is necessary that they should be highly reliable.

The reliability of forecasts of changes in the biosphere can be improved by further developments of global ecology, which require unfolding the following major lines of research:

1. Examination of the biosphere's components, undertaken to provide detailed qualitative descriptions for all regions of the world.
2. Study of the cycling of energy, basic types of mineral and organic matter in the biosphere for different geographic zones.
3. Construction of numerical models for each component of the biosphere in order to examine the processes occurring over large regions. As a result of these studies, a complex numerical model should be developed for the biosphere as a whole, incorporating the interactions among all biospheric components.
4. Collection of empirical data describing the state of the biosphere in the geological past for the purpose of understanding the general principles behind the biosphere's evolution.
5. Application of numerical models to estimate past changes in the biosphere, which will increase the possibility for explaining the mechanisms behind the biosphere's evolution. Comparison of results of calculated changes in the biosphere with empirical evidence is required in order to determine the reliability of the numerical models in making forecasts of future biosphere changes.
6. Application of numerical models for forecasting both anthropogenic modifications of the biosphere and changes arising from natural causes in order to make use of these forecasts while substantiating the optimum scenarios of economic development.
7. Devising methods for influencing large-scale processes in the biosphere in order to create a global system for controlling the biosphere for the sake of human society.

All these problems are still very far from being fully solved, but the available data for each of these lines of research make it possible to tackle the central problems of global ecology. The nature of the available data, however, restricts the possibility of obtaining detailed and accurate information on the past and future changes in the biosphere. That is why it is necessary to promote each of the indicated lines of research.

We would like to underline the point that, to obtain a correct estimate of the impending global changes in the biosphere, great value should be attached to the study of the biosphere's evolution in the geologic past, particularly in the epochs of the swiftest changes in the biosphere, namely, of global ecological crises. It should be recognized that these epochs, and the general principles governing the evolution of the biosphere as a whole, are still incompletely understood.

This book is concerned with three major problems.

The first five chapters give a description of the present state of the biosphere with an emphasis on energy exchange, and the cycles of water, carbon and oxygen in the biosphere.

In the next three chapters (6, 7 and 8) the history of the biosphere during the geologic past is considered. The possibility of clarifying general principles of the evolution of the biosphere, through examining interrelationships between variations in the biotic and abiotic components of the biosphere is discussed.

The concluding chapters (9 and 10) provide some information concerning human impacts on the biosphere, which appear today as a decisive factor in the biosphere's evolution. Chapter 10 presents considerations about the most likely future of the biosphere.

CHAPTER 1

# THE STRUCTURE OF THE BIOSPHERE

## 1.1. The Boundaries of the Biosphere

Definitions of the biosphere's boundaries have differed. Sometimes the biosphere is defined as a zone where both living organisms and the products of their activities are found. In this case the boundaries of the biosphere are quite vague: the oxygen molecules created by autotrophic plants reach the upper limits of the atmosphere, the position of which can only roughly be determined. Moreover, another product of autotrophic plants, organic carbon, can be carried by vertical tectonic movements from the Earth's surface to considerable depths which are difficult to ascertain precisely. An alternative definition is one that reduces the biosphere to the totality of living organisms. In this case biosphere becomes a synonym of the term 'biota'.

In this book, as well as in the author's previous works, the biosphere is defined as the zone in which modern living organisms exist. Thus, the biosphere includes the organisms themselves and embraces the troposphere, the hydrosphere and the upper, comparatively thin layers of the lithosphere. This definition is commonly used by many scientists.

There are different ways of dividing the biosphere into its major components. For instance, we can regard the biosphere as consisting of the principal abiotic spheres inhabited by living organisms (atmosphere, hydrosphere and lithosphere) and the organisms themselves (biota). At the same time the biosphere is often treated as a global ecological system incorporating a complex hierarchy of regional and local ecological systems, right down to the smallest systems.

The ecological system can be considered as a group of mutually related living organisms, together with those elements of the environment which produce the strongest influence on the organisms and are themselves dependent to some extent on the activities of these organisms.

It should be kept in mind that the boundaries and content of the ecological systems treated in ecological studies are purely conventional. Each group of living organisms has a great number of relations with other organisms and elements of the environment, among which one can select a wider or more limited set of relationships when distinguishing an ecological system. This will determine the extent of the ecological system and the number of components it contains. At the same time we should remember that, in some cases, ecological systems have more or less clearly defined natural boundaries (e.g. small closed water bodies or islands with comparatively homogeneous surfaces). In such cases it is easier to define the boundaries and components of the ecological system.

Depending on the purpose of the research, one can consider ecological systems of different sizes, including a system that embraces the entire sphere of existence of living organisms, i.e. the biosphere. Compared to the overall dimensions of the Earth, this is a thin envelope, whose thickness is no more than a few tens of kilometres.

As a whole, the global ecological system possesses more distinctive features than ecological systems of smaller sizes. This is because the matter and energy cycles in the global ecological system are largely closed processes (radiative exchange of energy between the biosphere and cosmic space excluded).

The biosphere takes a certain amount of matter from outer space (mainly as a result of falling meteorites) and releases a small portion of molecules of gases outwards, from the upper layers of the atmosphere. However, the exchange of matter between the biosphere and outer space is insignificant compared to the cycling of matter within the biosphere itself. Much more substantial is the exchange of some types of matter between the biosphere and lithosphere. Although this exchange is of great importance in the evolution of the biosphere (which is discussed in detail later), it is less intensive than the cycles of the same matter in the biosphere itself.

It is important that the cycles, formation and destruction, of living matter in the biosphere as a whole are completely closed. These cycles are not closed in individual ecological systems nor in the individual spheres of existence of the organisms (i.e. in the atmosphere, hydrosphere and lithosphere). This is further evidence that the biosphere is more integrated than its component parts. This absence of any exchange of living matter between the biosphere and outer space needs to be reconsidered since the epoch of space flights, in which first animals and then man took part. However, it is only in the remote future that this exchange might become important for the biosphere as a whole.

The possibility of organisms being transported from outer space to our planet in the epoch of the biosphere's formation is important for understanding the mechanism of the origin of life on Earth. This will be discussed further in a later part of this work.

## 1.2. The Atmosphere

The atmosphere of the Earth consists mainly of nitrogen and oxygen, which in dry air make up 78 and 21% of the air volume respectively. Of the minor permanent air constituents, carbon dioxide and ozone have considerable influence on the physical state of the atmosphere and on biological processes. The atmosphere also contains water vapour, the quantity of which varies over a wide range. It is generally present in concentrations of 0.1—1% and occasionally reaches as much as a few per cent of the atmospheric volume. The average water vapour content is about 2.4 g $cm^{-2}$.

The relative amounts of nitrogen, oxygen and carbon dioxide vary little with altitude in the troposphere and the lower layers of the stratosphere. The water vapour concentration, in contrast, diminishes rapidly above the Earth's surface towards the upper boundary of the troposphere. The concentration of water vapour in the stratosphere is low and does not substantially depend on the altitude. The relative amount of ozone ($O_3$) in the troposphere is not large. Ozone reaches a maximum in the lower stratosphere where it is formed from $O_2$ as a result of a photochemical reaction under the influence of solar ultraviolet radiation.

The total mass of the atmosphere is about $5.2 \times 10^{21}$ g (less than $10^{-6}$ of the Earth's mass). This is equivalent to about 1000 g for each square centimetre of the Earth's surface.

Air density and pressure both decrease with altitude; the atmospheric pressure decreases to half its surface value at a height of about 5 km. The troposphere, the lower layer of the atmosphere, extends up to approximately 8—10 km in polar regions and 16—18 km in the tropics. In the troposphere air temperature generally decreases with altitude at an average rate of 6 °C $km^{-1}$. In the stratosphere, which extends above the troposphere, temperature increases up to a height of about 50 km. More than two-thirds of the atmosphere's mass is concentrated within the troposphere.

Physical processes occurring in the troposphere determine changes in weather and fundamentally affect the climatic conditions in different regions of our planet. These processes include absorption of solar radiation, formation of outgoing longwave radiation flux (which varies only slightly in the higher air layers), the general atmospheric circulation, and the hydrologic cycle associated with the formation of clouds and the initiation of precipitation.

Since the atmosphere is more or less transparent to shortwave radiation, the greatest part of the solar radiation which the Earth absorbs at the top of the atmosphere is received at the Earth's surface. This radiation is the principal source of heat for the atmosphere. Heat energy is transferred from the surface to the atmosphere through longwave radiation and turbulent heat exchange. The heat consumed by evaporation from the surfaces of land and water bodies is also released into the atmosphere in the process of water vapour condensation.

Because the Earth is a sphere, most of the solar energy is absorbed at low latitudes where the air temperature near the surface is generally higher than in middle and high latitudes. This temperature difference appears to be the main source of energy for atmospheric and oceanic circulation. The atmospheric circulation is an intricate system of horizontal and vertical motions of the air. In the stratosphere, vertical motion is generally weak in contrast to the troposphere, where this plays an important role. The general circulation of the atmosphere and of the oceans ensures the transfer of a considerable amount of heat from low to middle and high latitudes, leading to a decrease in the meridional air temperature gradient.

Climatic conditions in different regions of the Earth depend on the incoming solar radiation and on the redistribution of this energy via circulation processes. A distinctive feature of the climate of our planet is its seasonal variations, which are particularly pronounced in high and middle latitudes. These variations result from the tilt of the Earth's axis by about 23° from the perpendicular to the plane of the Earth's orbit. Owing to this, the amount of solar radiation reaching the top of the atmosphere in different latitudinal zones varies considerably throughout the year. This process is treated in more detail in Chapter 2.

Heating of the Earth's surface by solar energy trends to produce vertical motion. If the upward moving air has a sufficiently high humidity this leads to condensation of water vapour and the formation of clouds and precipitation.

Evaporation from the surface of the oceans is a major source of water vapour for the atmosphere. In addition to this, water vapour is also released into the atmosphere as a result of evaporation from the land surface, although in regions with arid climates this kind of evaporation is insignificant.

Of some importance for weather and climate are processes occurring in the stratosphere, particularly those which are associated with the dissipation and absorption of radiation fluxes on liquid and solid particles (atmospheric aerosols). These particles, ranging in size from several nanometres to tens of micrometres, are observed both in the troposphere and in the upper layers of the atmosphere. Aerosols are formed at the Earth's surface as a result of soil erosion and evaporation of sea water. Aerosols can also be released into the atmosphere during volcanic eruptions as ejected solid particles and sulphurous gas products, or produced by other natural processes. Aerosols also enter the upper atmospheric layers from outer space. In addition to these natural sources of aerosols, anthropogenic sources such as industrial effluents, intensification of soil erosion and forest fires are also important.

Aerosol concentrations rapidly decrease with height in general, but the distribution is affected by secondary maxima associated with the existence of aerosol layers.

The atmosphere exerts an enormous influence on biological processes occurring on land and in water bodies. Of greatest importance for biological processes are the atmospheric oxygen used in the respiration of organisms and in the mineralization of organic matter, the carbon dioxide consumed in the photosynthesis of autotrophic plants, and the ozone which absorbs some of the potentially harmful ultraviolet solar radiation. No complex living organisms can exist on Earth without the atmosphere. The Moon, which has neither atmosphere nor life, is an obvious example illustrating this point.

Information concerning physical processes in the atmosphere is obtained through meteorological observations covering all regions of the globe. These observations come from a world network of constantly operating meteorological stations, including more than 10 000 climate stations and several tens of thousands of meteorological posts scattered over all the continents and on many islands. Daily observations of air temperature and humidity, as well as of precipitation, cloudiness, wind, pressure and other meteorological elements are made at climate stations. The programme of observations at meteorological posts is more limited and often restricted to observations of precipitation.

An important component of this system of meteorological observations is a global network of actinometric stations. This includes about one thousand stations at which observations of solar radiation and its transformation are made. Data from the network of aerological stations are also valuable in studying atmospheric processes. There are approximately one thousand such stations which carry out observations of meteorological conditions in the free atmosphere. Measurements up to heights of 30—40 km are made using radiosondes and other instruments. A number of stations also carry out observations of atmospheric ozone, elements of atmospheric electricity, and the chemical composition of the air.

Data collected by meteorological stations on land are supplemented by information from weather ships which continuously make observations in different ocean regions. Since the number of weather ships is not great, meteorological information from merchant and other ships at sea is of great importance in studying the atmosphere over the oceans.

In recent years meteorological Earth-orbiting satellites have acquired importance in observing the physical state of the atmosphere. For instance, measurements of radiation fluxes made from meteorological satellites permit the study of the radiation energy balance at the top of the atmosphere. Through the existing system of observations it is possible to obtain a detailed understanding of the chemical composition and physical state of the present atmosphere.

## 1.3. The Hydrosphere

Water in its three different states is a major element in all the components of the biosphere and one of the basic factors which permit the existence of living organisms. Most of the biosphere's water lies in the oceans. These cover about 71% of the Earth's surface. The average depth of the oceans is 3.7 km and the total volume of water contained exceeds 1300 million $km^3$. A considerable amount of water, over 30 million $km^3$, is comprised in glaciers, the Antarctic ice sheets contributing the major portion of this. The amount of ground water within 2 km of the Earth's surface is similar to the amount held in glaciers.

The volume of water in the remaining components of the world water supply is much smaller. The amount of water in lakes is about 0.18 million

$km^3$, the atmospheric water content is 0.013 million $km^3$, the river channels carry 0.002 million $km^3$ of water, and the amount of biological water (i.e. water contained in living organisms) is about 0.001 million $km^3$.

Data on the physical state of the ocean at different depths are mainly provided by research expeditions taking ship-based measurements of temperature, salinity, optical properties and of other elements of the hydrological regime. A set of such measurements is usually called an oceanographical station. For the period of oceanic investigations about 200 000 oceanographical stations have been functioning in open oceanic regions. Approximately the same number of stations have been running in seas where they are distributed rather unevenly. Although observations in the oceans and seas give an immense bulk of information on the physical state of water, the total amount is much smaller than that of atmospheric observations. Therefore, the physical processes in the ocean, despite their comparatively high stability, have been studied to a lesser degree than those in the atmosphere. Because the majority of ocean measurements are taken at shallow depths, particularly little information is available on the state of deep-sea waters (depths of more than 1—2 km).

The ocean comprises about 95% of the total volume of free water. Oceanic water contains a considerable amount of salts (on the average about 35 $g\,l^{-1}$). Salinity is somewhat higher in the subtropical high-pressure belt, where evaporation from the oceans exceeds rainfall. Outside these areas, salinity decreases near the equator and in middle and high latitudes where rainfall is greater than evaporation.

Of the gases dissolved in sea water, oxygen and carbon dioxide are of prime importance for organisms. The amount of oxygen in the ocean varies over a wide range, depending on temperature, on the activities of living organisms, and on various other factors. The carbon dioxide concentration in the ocean is also variable; the total carbon dioxide content of the ocean is approximately 60 times greater than that of the atmosphere.

Carbon dioxide in the ocean is assimilated by autotrophic plants through photosynthesis. A part of the carbon dioxide that enters the cycle of organic matter is consumed in the production of carbonate skeletons and shells of various living organisms. Following the destruction of such organisms, the dissolution of their skeletons returns part of this carbonate to the ocean. However, some of the material settles to the sea bottom forming sedimentary carbonate rocks.

In addition to oxygen and carbon dioxide, the autotrophic plants in the ocean require a variety of mineral compounds, among which phosphates (salts of phosphoric acid) are of great importance. The quantity of these mineral substances is very small in many areas of the ocean, leading to considerable decreases in the productivity of autotrophic plants. The lowest concentration of phosphates occurs in low latitudes. Concentrations increase somewhat in middle latitudes and are highest in polar regions.

Sea water is in constant motion, the major energy source for which is the temperature differences in surface waters at different latitudes and the dynamic interaction between the air fluxes and the surface of the ocean.

Under the influence of wave motion and other dynamic processes the upper layer of the ocean is highly mixed. Ocean currents reach their greatest speed in this layer. At depths exceeding several hundreds of metres the motion of water is generally weak and the rate of mixing is not great. The temperature of the ocean in low latitudes decreases with depth reaching 1—3 °C at a depth below 2 km. In high latitudes the vertical temperature gradient is small and temperature at all depths only slightly exceeds the freezing point of salt water.

In the Arctic and Antarctic regions a considerable part of the ocean is covered with ice fields, whose average thickness reaches a few metres. The area occupied by ice varies throughout the year, being greatest in spring and smallest towards the end of summer.

The greatest part of continental waters lies within lakes, the Caspian Sea being the largest of these. About half of the lake water occurs in fresh water lakes, the other half is in salt-water lakes.

The most important continental hydrological process is the river run-off which compensates for the water vapour transfer from water bodies to the continents. Measurements of run-off are carried out by hydrometric stations and posts, the total number of which is, at present, about 60000. The hydrometric network is very uneven. In Europe and North America there is one observation post per 1000 to 2000 $km^2$. In many tropical countries it is only one per 10000 $km^2$ or more.

Of the general principles of the atmospheric water cycle, treated in detail in Chapter 3, we need only mention here the moisture content of the upper layers of soil, a result of precipitation falling on land, which is of vital importance in living nature. The existence of autotrophic vegetation cover would not have been possible without soil moisture, and the nature of vegetation varies according to the moisture content of soil. For instance, on soils with surplus moisture marshy vegetation prevails, with moderate and optimum moisture content forests grow, with insufficient moisture savannah and steppe develop, and, where moisture levels are extremely low, deserts appear. The type of vegetation cover determines many of the other properties of geographical zones.

Continental waters, being one of the major factors in physical and chemical weathering, have an enormous influence on the upper layers of the lithosphere. In many regions the Earth's surface bears the marks of the continental glaciations of the past. Having retreated, the glaciers have left behind characteristic glacial landforms.

In conclusion, we underline once more the fact that physical and chemical processes in the hydrosphere are closely connected with similar processes in the atmosphere. The transformation of energy, the water cycle, and the circulation of carbon dioxide and other air and water constituents are all involved in the hydrosphere and the atmosphere, and are inextricably linked into a single system.

## 1.4. The Lithosphere

In the atmosphere and hydrosphere organisms are distributed within layers of a considerable thickness extending to several kilometres. In the lithosphere, however, most organisms live in the soil layer whose depth generally does not exceed several metres.

Soils are formed from a mixture of mineral substances (produced as a result of the weathering of rocks) and organic matter, the product of the activities of organisms and of the decomposition of dead organisms (mostly plants). These products are partly generated on the soil surface (from leaves, dry branches, etc.), and partly at some depths (for example, from dead roots). Various microorganisms such as actinomycetes and soil fungi play a decisive role in the decomposition of organic material in the soil.

As a result of these soil processes, deposits of carbon, nitrogen, phosphorus, potassium, magnesium and other elements accumulate in the soil in a form which can be used by higher plants. In this way soils exert a considerble influence on the productivity of plants.

Another important result of the decomposition of organic matter in the soil is the release of carbon dioxide, which is added to the carbon dioxide already in the atmosphere and hydrosphere thus compensating for its consumption in photosynthesis.

Soil usually consists of several layers that often differ in colour. The upper layer (the A horizon) contains the decomposition products of organic matter from dead organisms as well as mineral matter. The next layer (the B horizon) includes mostly mineral components. Soluble inorganic material is carried from the A horizon to the B horizon by the downward flow of soil water. The third layer (the C horizon) consists of slightly altered debris of the original rock. Sometimes this material is produced in situ and sometimes it is carried in from distant sites by water, glaciers or wind and then redeposited. The transport of fine fragments of soil by wind results, for instance, in the formation of loess deposits, which in some cases can be quite thick and very fertile.

Different types of soil are formed depending on the climate, the nature of parent rock, and the types of vegetation cover and soil organisms. The first two factors can be considered as external factors, independent of the process of soil formation, while the next two are themselves substantially dependent on the soil formation process. Climatic conditions have the greatest effect in determining the types of soil occurring in different areas.

The major types of soil at high and middle latitudes are the soils of tundra, podzolic soils of coniferous forests, brown and grey soils of deciduous forests, black earths and chestnut soils of steppes, and brown semi-desert and grey-brown desert soils. In the subtropics and tropics the principal types of soil are similar to those of middle latitudes. Reddish-yellow or dark lateritic soils are characteristic of humid areas of low latitudes and red-brown soils are typical of drier regions.

All soil types correspond to certain types of vegetation cover, the nature of the soil exerting a profound effect on the plant's activities. Vegetation is also appreciably affected by local soil features which are dependent on the properties of the parent rock and the local moisture regime.

Not all organisms are restricted to the soil layer: some exist deeper in the lithosphere in the ground water region. For instance, different microorganisms have been discovered in waters associated with oil deposits at a depth of several kilometres. Nevertheless the greatest part of the lithosphere, lying below the ground water zone, is usually completely devoid of living organisms.

## 1.5. Organisms

Living organisms are characterized by a number of significant features; specific structure, metabolism, growth, development, reproduction and other features that may be specific to particular groups of organisms.

Vernadsky repeatedly made the point that an important characteristic of living matter is that it can affect the non-living chemical and geological environment. He emphasized that, although living matter represents an insignificant part of the total mass of the biosphere, it can have substantial effects on all the abiotic components of the biosphere. Interactions between living matter and the abiotic environment, which is of great importance in the evolution of the biosphere, are treated thoroughly below.

An organism is a whole system comprising tissues and organs of different complexity. There is a detailed biological systematics which describes the existing and extinct organisms, based on the idea that different organisms can be combined into groups (taxons), whose hierarchy shows to what extent the organisms belonging to different groups are similar in structure and how close or distant is their origin. The basic unit of this taxonomic classification is the species. Related species are combined into higher taxonomic groups, which include genera, families, orders, classes and phyla (for animals) or divisions (for plants). This sequence of groups corresponds to their hierarchical rank; each of the higher groups includes all the lower groups.

We shall not consider this classification system further, except to note that all organisms can be combined into a small number of kingdoms, among which are the animal and plant kingdoms. In recent years it has been suggested that the plant kingdom should be divided into the plant kingdom proper and the fungi kingdom. In some studies, the most primitive unicellular organisms, the procaryotes (organisms having neither a distinct nuclei nor chromosomal apparatus) are also identified as an individual kingdom. Procaryotes include bacteria, blue-green algae and some other organisms. Some of them, such as blue-green algae, are autotrophic organisms, others are heterotrophs. All other organisms have nuclei and chromosomal apparatus and are called eucaryotes.

According to their mode of nourishment, all organisms are divided into two large groups. The autotrophic organisms produce organic matter from inorganic substances in the environment, heterotrophs exist by using organic material from other living things.

The majority of plants are characterized by autotrophic nourishment, in the course of which they absorb solar energy and assimilate water and carbon dioxide from their environment to create organic matter. This process, which is called photosynthesis, is of paramount importance in the balance of living matter in the biosphere and its energetics. The general principles of photosynthesis are treated in more detail in Chapter 4. The autotrophs also include some kinds of bacteria which can assimilate carbon dioxide at the expense of energy released through oxidation of inorganic compounds (chemosynthesis).

Fungi are heterotrophic organisms, like some secondary heterotrophic plants, which have lost their ability to synthesize organic matter from mineral substances during the course of evolution. All animals are heterotrophic organisms.

Plants are divided into several large groups. Thallophyta (algae, fungi — sometimes identified as an individual group, or groups of higher rank), Bryophyta (mosses, liverworts), and Tracheophyta (vascular plants). Vascular plants include a broadly distributed group of Angiospermae, or flowering plants. The total number of species of plants is about 500 000. On land, plants form vegetation covers constituting the greatest part of the 'film of living matter'. In water bodies, this film includes the phytoplankton, plants of the upper water layer, the layer which is penetrated by solar radiation. The thickness of this layer in the ocean is usually no more than 100 m.

Animals are divided into a number of phyla. The most important phyla are the Porifera (sponges), Coelenterata (coelenterates), Mollusca (molluscs), Annelida (segmented worms), Arthropoda (arthropods), Echinodermata (echinoderms), Chordata (chordates). The vertebrata (vertebrates) form a subphylum of the Chordata. It is believed that the total number of animal species is about 1 500 000, the majority of which (more than 1 million) belong to the Arthropoda phylum.

This rough outline of the higher groups of plants and animals is not necessarily accepted by all authors and considerable divergences in classification are apparent in different works.

Furthermore, the data on the number of plant and animal species are not complete, and there are grounds to believe that this number would probably be much greater if full inventories of all contemporary flora and fauna were carried out.

The species that are known at present are distributed very unevenly among the higher taxonomic groups of organisms. For instance, some classes of animals include only a few species, whereas the class of insects comprises more than a million species. Noteworthy, too, is the coexistence in the modern biosphere of extremely primitive organisms, such as procaryotes, with very complex living things. The primitive organisms are often represented by a vast number of species. In this connection it can be concluded that, among large animals and plants, the representatives of more progressive groups prevail, whereas the smallest organisms are always very primitive. This conclusion is partly based on the fact that minute organisms cannot have complex organs and tissues.

Today, nearly the entire primary production of living organisms depends on the photosynthesis of autotrophic plants. (Chemosynthesis, by comparison, provides only a minor quantity of organic matter.)

The energy transformations in living organisms may be generally represented in the form of an 'energy pyramid' whose base is the energy assimilated by autotrophic plants via photosynthesis and which is effectively stored in the organic matter of plants. A large part of this organic matter is expended by plants themselves in the course of their life activities.

Another part is assimilated by animals feeding directly on these plants. This is the second stage of the energy pyramid; the use of the energy stored in autotrophic plants by animals and heterotrophic plants. Usually no more than 10% of the energy initially assimilated is transferred to living organisms through consumption of autotrophic plants. In the course of their own life these second stage organisms also lose energy through respiration and other functions.

The third stage of the energy pyramid corresponds to the energy intake of animals that subsist, not on the primary energy of autotrophic plants, but on the energy already assimilated by other animals. This level includes carnivorous animals. Many living organisms are located simultaneously at the second and third stages of the energy pyramid. This includes man, who consumes both energy directly produced by plants and energy created by animals.

The energy pyramid reflects a close relationship between animals and plants, for the autotrophic plants are the basic source of energy that supports the vital functions of all animal species. This trophic linkage greatly influences the structure of animal bodies and their ecology. The energy pyramid thus shows the transformation of energy in living organisms. The transformation of energy also continues in dead organisms in the process of decomposition. These energy transformations proceed until energy is fully expended following the mineralization of organic matter. A large number of living organisms contribute to this process including a variety of microorganisms, whose role is particularly important.

In reality, the energy transmission from organism to organism takes place in ways that are far more complex than this oversimplified scheme.

In describing the trophic interconnecting lines existing in nature, an important concept is the trophic or food chain. A trophic chain consists of a group of organisms which, in the course of feeding, transfer the energy of their biomass to other organisms in a certain sequence. There are usually more than three links in a food chain. However, the total number of links cannot be very numerous, since generally each successive link in the chain contains roughly one-tenth of the assimilated energy of the preceding link. Accordingly, if the number of links in a food chain is ten, for example, then the quantity of energy assimilated by the last link will be billions of times smaller than that in the organisms of the first link. This implies a corresponding reduction in the biomass of organisms consuming this energy, so that, for food chains of any significant length, the biomass of the last link becomes negligible. Since heterotrophic plants and animals often have varied diets, food chains are, in fact, parts of complicated webs interconnecting different organisms rather than isolated lines.

It has already been noted that the distribution of living organisms within the biosphere is uneven. This reflects partly a still more uneven distribution of the primary production of organic matter in space and time.

It is not in all parts of the biosphere that life exists as an autonomous system, i.e. living organic mater is not necessarily produced at all localities where organisms are found. As already mentioned, in our epoch nearly all primary production depends on the activity of autotrophic plants. These plants create organic matter only on the land surface and in the upper layers of water bodies. Photosynthesis does not take place in the free atmosphere nor in deep layers of the ocean or in the soil. Moreover, there are vast areas on the continents and in the oceans, in which the process of photosynthesis does not occur and in a number of regions it is absent for long periods of the year. Autotrophic plants cannot exist in many continental regions because of insufficient heat or moisture deficiencies. Photosynthetic activity of plants is also limited over a large part of the surface of the ocean by a shortage of mineral substances, and, as a result, these zones are practically barren. Thus the surface of the Earth is far from being entirely a zone accommodating the energy pyramid of life. Even after thousands of millions of years of evolution, autotrophic organisms have not been able to adapt to the conditions covering a considerable part of the outer envelope of our planet.

The decisive influence that the environment has on the activities of living organisms has become apparent in the phenomenon of geographical zones studied by Dokuchaev. Depending on the heat and water regimes, the natural vegetation cover forms a number of distinctive types, which include tundra, coniferous forest of high and middle latitudes (taiga), mixed and deciduous forest of middle latitudes, subtropical and equatorial rain forest, deciduous forest of the tropics, forest-steppe, savannah, steppe, semi-desert and desert. These types of vegetation are closely associated with fauna whose zonation is also determined by climatic factors.

It is interesting to see to what extent organisms can alter their environment in order to make it more convenient for their existence. Leaving aside the role of organisms in the biosphere's evolution, which will be discussed later, we shall note here that the ability of plants and animals to regulate their environment is very limited. The vegetation cover produces some influence on the microclimate, modifying temperature and humidity of the lower air layer and reducing wind speed near the Earth's surface. At the same time many types of vegetation cover influence both the moisture content of soil and the rate of evaporation as well as the components of the surface energy balance.

Changes in the microclimate of the vegetation cover are a direct result of the structure and physiological functions of plants. We cannot, therefore, regard the influence of the vegetation cover on microclimate as an adaptation to the environment that has emerged in the course of evolution. This may be seen from the fact that certain plant-induced changes in microclimate (such as an increase in the total evaporation) are sometimes unfavourable to the vegetation cover. But, generally, many characteristics of the microclimate within the vegetation layer contribute to the survival of plants, notwithstanding fluctuations in external conditions. This is illustrated by the existence and preservation of relict forests in regions where the present (changed) climate would prevent their renewal if they were felled.

It is interesting to note the role of the vegetation cover in sustaining soils. In many cases soils deteriorate rapidly under erosion processes following the destruction of the vegetation cover. It is known that, in a number of regions of insufficient humidity, vegetation of dry steppes or savannahs that has been destroyed by excessive grazing, fires and other anthropogenic factors, is renewed either very slowly or not at all. Finally, one should note that the influence of plants on their environment is largely of a local nature.

The correspondence between different types of vegetation cover and climatic conditions shows that the development of plants depends ultimately on the global climate and that plants are able to control their environment only to a limited extent. A similar conclusion may be drawn from the widely known sensitivity of plants to climatic changes.

The vegetation cover does influence the climate over large areas, but its effects are relatively modest. Neither the expenditure of solar energy on photosynthesis nor the generation of heat through oxidation of organic matter is large, generally less than a fraction of a per cent of the solar radiation. These two processes, in fact, almost compensate for each other.

Of all biological processes it is transpiration from the vegetation cover that produces the greatest influence on the energy balance of the Earth's surface. The expenditure of heat on transpiration may represent several tens of a per cent of the radiation balance of the Earth's surface. Yet the effect of transpiration on the thermal regime is often not large, since, in the absence of transpiration, evaporation from soil surface increases.

Vegetation, and especially forests, appear to exert some influence on the amount of precipitation. This happens because the water cycle above forests becomes more intensive owing to increased evaporation and increased surface roughness (the latter enhances the vertical air currents). As a result, the total precipitation is somewhat increased, but not significantly. This process is considered more closely in Chapter 3.

Plants may influence the rate of absorption of solar radiation. On land the albedo of a thick vegetation cover is usually smaller than that of a weakly developed vegetation. Similarly, the albedo of water bodies with highly concentrated aquatic plants is also reduced. Although all these factors exert some influence on climate, the extent of this influence is not great and no appreciable changes in the global climate are produced.

The effects of animals on the environment are even more limited than those of the plants. These effects are most perceptible for animals living in the soil, which, in many cases, can alter the soil structure. These changes, however, are only local.

Of considerable and growing influence on the biosphere is man's economic activity. In earlier times this affected only local ecological systems. Today it has begun to modify the global ecological system, the biosphere as a whole. The problem of human impact on the biosphere is discussed in Chapter 9 and 10.

CHAPTER 2

# SOLAR RADIATION AND ITS TRANSFORMATIONS

## 2.1. Solar Radiation

The Sun is the star closest to the Earth and belongs to the class of yellow dwarf stars. The diameter of the Sun is about 1.4 million km, and the average distance from the Earth is 149.5 million km. The temperature of the Sun's surface, about 6000 K, results from nuclear reactions that occur inside the Sun. The Sun therefore radiates a considerable amount of energy into space.

The radiation coming from the Sun to the Earth is the only form of incoming radiant energy that determines the energy balance and the thermal regime of the Earth. Radiation energy coming to the Earth from all other celestial bodies is so insignificant that it does not exert any noticeable influence on the Earth's heat exchange processes.

The spectrum of solar radiation corresponds to that of a black body described by Planck's formula. In accordance with the temperature of the Sun's radiating surface, the maximum radiation flux is observed at the wavelengths of about 0.50 μm, and most of the radiated energy falls in the wavelength range from 0.3 to 2.0 μm.

The intensity of radiation decreases in inverse proportion to the square of the distance from the Sun. As the Earth moves in its elliptical orbit around the Sun, the intensity of solar radiation falling on the top of the atmosphere changes during the year according to the change in the Earth—Sun distance.

The smallest distance between the Earth and the Sun, 147 million km, occurs at the beginning of January. The greatest distance, at the beginning of July, is 153 million km.

The flux of solar energy per unit of time through a unit square perpendicular to the solar rays and situated out of the atmosphere at the average distance between the Sun and the Earth is called the solar constant. As a result of changes in the Sun—Earth distance the actual flux of solar energy at the top of the atmosphere varies from the solar constant value by approximately 3.5%.

Numerous studies have attempted to determine the value of solar constant. For many years the solar constant was obtained from ground station data on radiation. Such estimates have noticeable errors because the effect of the attenuation of the solar flux by the atmosphere can only be taken into account approximately. Recently, observations of solar constant value have been made at great altitudes. These include satellite measurements. As a result the conclusion has been drawn that the solar constant is equal to 1368 W $m^{-2}$ (Willson *et al.*, 1981).

Of great interest is the question of stability of the solar constant in time, an issue discussed in a number of studies. Many authors believed that, even though variations in the solar constant take place, they do not exceed the accuracy of available measurements. Along with this it should be taken into account that small variations in the solar constant, e.g. 1% or even 0.1% of its value, can exert certain effects on climate.

From satellite observations it has been found that the solar constant can change for a short period of time in the range of 0.1—0.2%. Whether or not long-term variations, on time scales exceeding one year, occur is not yet known, so the influence of such variations on climate cannot be considered as proven.

Knowing the value of solar constant one may calculate how much energy would reach the Earth's surface at different latitudes if the atmosphere does not influence the radiation. Some results of this calculation, namely, daily sums of solar radiation at different latitudes for various seasons, are tabulated in Table 2.1.

As seen from Table 2.1, the greatest daily sums of radiation are observed over the poles during the summer solstices. It should be noted that at these times of the year, when moving to lower latitudes, the radiation flux first decreases and then increases to a second small maximum. In the Southern

TABLE 2.1

Daily totals of solar radiation incident upon the surface of the Earth under a completely transparent atmosphere (MJ $m^{-2}$)

| Latitude | Mar. 21 | May 6 | June 22 | Aug. 8 | Sept. 23 | Nov. 8 | Dec. 22 | Feb. 4 |
|---|---|---|---|---|---|---|---|---|
| 90°N | — | 32.8 | 45.6 | 32.4 | — | — | — | — |
| 80 | 6.6 | 32.0 | 44.8 | 32.0 | 6.6 | — | — | — |
| 70 | 13.1 | 31.6 | 42.7 | 31.2 | 12.7 | 0.8 | — | 0.8 |
| 60 | 18.9 | 34.1 | 41.5 | 34.1 | 18.9 | 6.2 | 2.1 | 6.2 |
| 50 | 24.2 | 36.5 | 41.9 | 36.5 | 24.2 | 12.3 | 7.4 | 12.3 |
| 40 | 29.2 | 38.6 | 41.9 | 38.2 | 28.7 | 18.1 | 13.5 | 18.5 |
| 30 | 32.8 | 39.4 | 41.1 | 39.0 | 32.4 | 23.8 | 19.7 | 24.2 |
| 20 | 35.7 | 39.0 | 39.4 | 38.6 | 35.3 | 29.2 | 25.5 | 29.2 |
| 10 | 37.4 | 37.8 | 37.0 | 37.4 | 37.0 | 33.3 | 31.2 | 33.7 |
| 0 | 37.8 | 35.3 | 33.3 | 35.3 | 37.4 | 37.0 | 35.7 | 37.0 |
| 10 | 37.4 | 32.0 | 29.2 | 32.0 | 37.0 | 39.4 | 39.4 | 39.4 |
| 20 | 35.7 | 27.9 | 23.8 | 27.5 | 35.3 | 40.6 | 42.3 | 41.1 |
| 30 | 32.8 | 23.0 | 18.5 | 23.0 | 32.4 | 40.6 | 43.9 | 41.1 |
| 40 | 29.2 | 17.7 | 12.7 | 17.2 | 28.7 | 39.8 | 44.8 | 40.2 |
| 50 | 24.2 | 11.5 | 7.0 | 11.5 | 24.2 | 38.2 | 44.8 | 38.6 |
| 60 | 18.9 | 5.7 | 2.1 | 5.7 | 18.9 | 35.7 | 44.3 | 35.7 |
| 70 | 13.1 | 0.8 | — | 0.8 | 12.7 | 32.8 | 45.6 | 33.3 |
| 80 | 6.6 | — | — | — | 6.6 | 33.3 | 48.0 | 33.7 |
| 90°S | — | — | — | — | — | 34.1 | 48.4 | 34.1 |

Hemisphere the radiation flux decreases again, reaching zero at some latitude. At the equinoxes, the radiation maximum occurs at the equator. With increasing latitude the radiation sums reduce at first slowly and then faster.

Table 2.2 gives the radiation sums at different latitudes for the summer and winter half-years. Note that the radiation sums in each half-year at a given latitude of the Northern Hemisphere equal the radiation sums at the same latitude of the Southern Hemisphere.

TABLE 2.2

Sums of solar radiation incident on the surface of the Earth under a completely transparent atmosphere (MJ $m^{-2}$)

| Term | Latitude | | | | | | | | | |
|---|---|---|---|---|---|---|---|---|---|---|
| | 0° | 10° | 20° | 30° | 40° | 50° | 60° | 70° | 80° | 90° |
| Summer half-year | 6610 | 6980 | 7180 | 7140 | 6980 | 6610 | 6120 | 5710 | 5540 | 5460 |
| Winter half-year | 6610 | 6040 | 5300 | 4430 | 3450 | 2420 | 1400 | 530 | 120 | 0 |
| Year | 13220 | 13020 | 12480 | 11570 | 10430 | 9030 | 7520 | 6240 | 5660 | 5460 |

In reality, the atmosphere is not a completely transparent medium for solar radiation. A noticeable part of the radiation coming from the Sun is absorbed and scattered in the atmosphere, and also is reflected into space. Solar radiation is especially influenced by clouds; but even under a cloudless sky solar radiation is changed greatly in the atmosphere. Solar radiation is absorbed in the atmosphere by water vapour, water droplets, ozone, carbon dioxide, and dust. Scattering of solar radiation is caused both by the molecules of air and by various admixtures — dust, water droplets and so forth. The flux of direct solar radiation that passes through the atmosphere depends on the transparency of the air and the elevation of the Sun which defines the length of the path of the solar rays in the atmosphere.

As indicated above, in addition to changes of atmospheric transparency, cloudiness greatly influences the amount of direct radiation reaching the Earth's surface. With more or less dense clouds covering the Sun, direct radiation is equal to zero.

The greatest amount of direct radiation flux is recorded under cloudless sky and high atmospheric transparency. In these conditions, the flux of direct radiation incident on a perpendicular surface at sea level can amount to 1000—1200 W $m^{-2}$. Average noon values of this flux usually equal 700—900 W $m^{-2}$. With decreasing solar elevation during the daily cycle, the direct solar radiation is noticeably reduced as the optical mass of the atmosphere increases.

The amount of scattered radiation coming to the Earth's surface varies within wide limits, mainly as a function of cloudiness conditions and the Sun's elevation. The theoretical assessment of this radiation flux is rather complicated and does not yield accurate results. Available observational data show that, in many cases, the flux of scattered radiation is comparable in value with the flux of direct radiation incident on a horizontal surface. The greatest amounts of scattered radiation are recorded under cloudy skies. Scattered radiation is greatly influenced by the reflectivity of the Earth's surface. In particular, scattered radiation increases noticeably when the Earth's surface is covered with snow which reflects a considerable amount of solar energy.

The spectral composition of solar radiation reaching the Earth's surface differs significantly from that of radiation outside the atmosphere. The energy distribution in the solar spectrum obtained from ground observation data (curve *b*) and outside the atmosphere (curve *a*) is presented in Fig. 2.1. Comparison of these two curves shows that ultraviolet radiation (wavelengths less than 0.4 μm) is, to a considerable extent, absorbed in the atmosphere and does not reach the Earth's surface. It is known that the absorption of the ultraviolet radiation is mainly caused by atmospheric ozone. As seen in Fig. 2.1, the absorption of solar radiation in the long wavelength part of the spectrum is very uneven, demonstrating the heterogeneity of the spectral transparency of the atmosphere.

The greatest amount of energy in the solar spectrum falls in its visible part (the wavelengths from 0.40 to 0.74 μm). Alongside this, a considerable

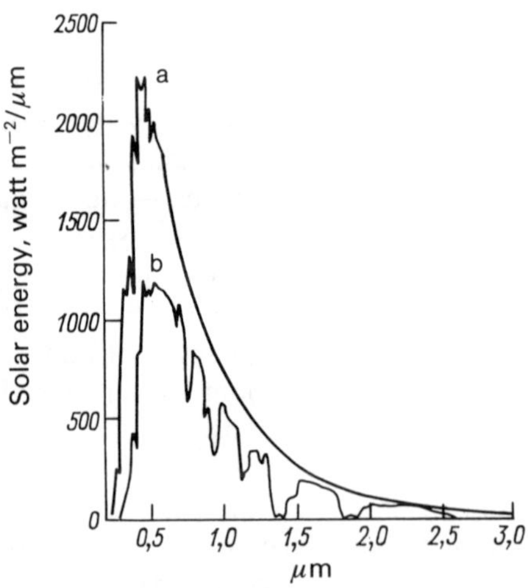


Fig. 2.1.

amount of energy is transmitted in the form of infrared radiation (wavelength exceeding 0.74 μm). Ultraviolet radiation contributes little to the energy flux, but it has a pronounced effect on various biological processes.

The general picture of the basic solar energy transformations in the geographic envelope of the Earth is as follows. The flux of solar energy at the mean distance between the Earth and the Sun is equal to the value of solar constant. Because of the spherical shape of the Earth, a unit area at the top of the atmosphere gets, on the average, one-fourth of the total flux — about 340 W $m^{-2}$, of which approximately 240 W $m^{-2}$ is absorbed by the Earth as a planet. The greater part of this amount is absorbed by the surface of the Earth and a considerably smaller fraction by the atmosphere.

The Earth's surface, since it is heated due to absorbing solar radiation, is a source of outgoing radiation that transmits heat into the atmosphere. Atmospheric water vapour, dust and various gases that absorb longwave radiation, detain the longwave emission from the Earth's surface. Thus, a considerable portion of the longwave radiation emitted from the Earth's surface is compensated for by the counter-radiation of the atmosphere.

The difference between the Earth's surface emission itself and the atmospheric counter-radiation is called the net longwave radiation. The value of net longwave radiation is usually several times smaller than the flux of longwave emission from the Earth's surface that would be observed if the atmosphere were completely transparent to outgoing radiation.

The algebraic sum of the radiative energy fluxes coming to and going from the surface of the Earth is called the radiation budget (balance) of the Earth's surface. It is evident that the value of the radiation budget is equal to the difference between the amount of direct and scattered shortwave radiation absorbed by the Earth's surface and the value of the net longwave radiation.

On the average for the whole surface of the Earth, the net longwave radiation is much less than the absorbed shortwave radiation. This is a consequence of the so-called greenhouse effect; that is, the result of the relatively large transparency of the atmosphere to shortwave radiation compared with that for outgoing emission. Therefore, the mean radiation balance of the Earth's surface is a positive value.

The energy in the radiation budget of the Earth's surface is expended in various ways: heating the atmosphere by means of turbulent heat conductivity, evaporation, heat exchange with the deeper layers of the hydrosphere or lithosphere, and so on. Terms describing all of the above forms of solar energy transformation must enter the equation of energy (heat) balance for the Earth's surface in addition to the algebraic sum of energy fluxes incident on and going from the Earth's surface. According to the law of energy conservation, this sum must equal zero.

The radiation and energy balances of the Earth's surface are definitely connected with the radiation and energy balances of the atmosphere. The

Earth as a planet receives energy from space and emits it back only through radiation. As the mean temperature of the Earth changes little in time, it is evident that the radiation balance of the Earth (the difference between radiation absorbed and emitted into space) is equal to zero.

The radiation budget of the atmosphere (equal to the difference between the radiation balances of the planet as a whole and that of the Earth's surface) must therefore, on average, be negative and equal in absolute value to the Earth's surface radiation budget. In the atmospheric energy balance, the negative radiation budget is compensated for by an influx of energy from condensation of water vapour when forming clouds and precipitation and by an influx of heat from the Earth's surface due to the turbulent heat conductivity of the lower layers of the atmosphere. In addition to the processes of redistributing solar energy in the vertical direction, powerful processes of horizontal heat redistribution are developed in the geographical envelope. Most important among them is the transport of thermal energy in the atmosphere and hydrosphere between high and low latitudes due to the heterogeneity of radiative heating of the spherical surface of the Earth. This transport is carried out through macroturbulent heat exchange and sensible heat transfer, as well as latent heat transfer through phase transformations of water.

The processes of solar energy transformation caused by radiative factors enumerated above, in turn significantly change the radiation regime, since this depends considerably on circulation in the atmosphere and hydrosphere, phase transformations of water and so on. Cloudiness and coverage of the Earth's surface by snow (or ice) produce an especially pronounced effect on the radiation regime.

In addition to the 'first order' processes of transformations of solar energy which substantially change the radiation and heat regimes, a number of solar energy transformation processes requiring comparatively small amounts of energy and, therefore, not affecting the radiation and heat regimes directly, are developing in the geographical envelope of the Earth. Some of these processes are very important for the formation of the biosphere. For example, the process of photosynthesis is based on the transformation of radiant energy into a comparatively stable form of chemical energy.

In studying all the forms of solar energy transformation in the geographical envelope, radiation and energy balance data are of prime importance. Data on the energy and radiation balances at the Earth's surface are especially important for investigating the biosphere since it is the main source of energy for biological processes.

## 2.2. The Energy Balance

The radiation balance of the Earth's surface, $R$, is equal to the difference between the absorbed solar radiation and the net longwave radiation:

$$R = Q(1 - \alpha) - I, \tag{2.1}$$

where $Q$ is the total shortwave radiation (the sum of direct and scattered radiation), $\alpha$ is the albedo (the reflectivity of the Earth's surface for total radiation) expressed in fractions of unity and $I$ is the net longwave radiation, i.e. the difference between the radiation emitted by the Earth's surface itself and the absorbed counter-radiation of the atmosphere.

The radiation balance of the Earth—atmosphere system, $R_s$, i.e. the radiation balance of the vertical column through the whole thickness of the atmosphere, may be determined in a similar way.

We find in this case that

$$R_s = Q_s(1 - \alpha_s) - I_s, \tag{2.2}$$

where $Q_s$ is the solar radiation incident on the outer boundary of the atmosphere, $\alpha_s$ is the albedo of the Earth—atmosphere system and $I_s$ is the longwave emission from the outer boundary of the atmosphere into space (outgoing emission).

The radiation balance or budget of the atmosphere, $R_a$, equals the difference between the values of $R_s$ and $R$. Using Equations (2.1) and (2.2), we get

$$R_a = Q_s(1 - \alpha_s) - Q(1 - \alpha) - (I_s - I). \tag{2.3}$$

Reflection of solar radiation from the Earth's surface is determined by the albedo, which is defined as the ratio of the amounts of reflected to incident radiation. Theoretically, the values of albedo might vary from unity for a completely white and reflective surface to zero for an absolutely black surface entirely absorbing the Sun's rays. Available observational data show that the albedo of natural surfaces does vary over a wide range, embracing almost the entire possible interval of values of reflective ability of various surfaces. In experimental investigations of solar radiation reflection, values of albedo for more or less all of the widespread natural surfaces have been found. These investigations have shown that the conditions of solar radiation absorption on land and water differ noticeably. Typical values of albedo for different kinds of land surfaces are given in Table 2.3.

The conditions of solar radiation absorption by water bodies differ from those at the land surface. Clear water is comparatively transparent to shortwave radiation, and so the Sun's rays penetrating the upper layers of a water body are widely scattered and then to a considerable extent absorbed. It is easily understood that the absorption of solar radiation in such conditions should depend on the solar elevation or zenith angle. If the Sun stands high, a great part of the incoming radiation penetrates the upper layers of water and is mainly absorbed there. At low solar elevations, the rays incident on the water surface at small angles are specularly reflected and do not penetrate

TABLE 2.3

Albedo of natural surfaces

| Surface | Albedo |
|---|---|
| *Snow and ice* | |
| Fresh, dry snow | 0.80—0.95 |
| Clean, moist snow | 0.60—0.70 |
| Dirty snow | 0.40—0.50 |
| Sea ice | 0.30—0.40 |
| *Bare soil* | |
| Dark soils | 0.05—0.15 |
| Moist, grey soils | 0.10—0.20 |
| Dry loam or grey soils | 0.20—0.35 |
| Dry light sandy soils | 0.35—0.45 |
| *Fields, meadows, tundra* | |
| Rye and wheat fields | 0.10—0.25 |
| Potato fields | 0.15—0.25 |
| Cotton fields | 0.20—0.25 |
| Meadows | 0.15—0.25 |
| Dry steppe | 0.20—0.30 |
| Tundra | 0.15—0.20 |
| *Arboreal vegetation* | |
| Coniferous forest | 0.10—0.15 |
| Deciduous forest | 0.15—0.20 |

into the depths of the water body. This leads to an abrupt increase of the albedo. Both theoretical calculations and experimental investigations show that the albedo of water surfaces for direct radiation varies over a wide range as a result of this dependence on solar elevation. The character of this dependence is seen from the graph in Fig. 2.2. At large solar elevations the albedo for direct radiation does not exceed several hundredths, while with

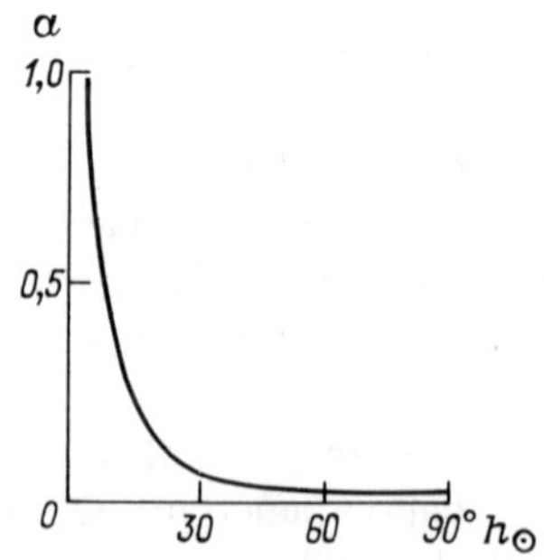


Fig. 2.2.

the Sun approaching the horizon the albedo reaches values of the order of several tenths. The albedo of a water surface for scattered radiation varies much less and is about 0.10.

The albedo of the Earth—atmosphere system is of a more complicated nature than that of the Earth's surface. Solar radiation received by the atmosphere is partially reflected as a result of backscattering. Under a cloudy sky, a considerable fraction of radiation can be reflected from the upper surfaces of the clouds. With little cloudiness or under clear sky, the albedo of the Earth—atmosphere system depends essentially on the albedo of the Earth's surface. The albedo of clouds depends on the thickness of their layers, its mean value being equal to 0.40—0.50. The albedo of the Earth—atmosphere system without clouds is usually greater than that of the Earth's surface, except in those cases when the surface is covered with more or less clean snow.

Both the reflection of shortwave radiation and the emission of longwave radiation are equally important forms of radiative energy loss. The basic features of the process of heat transfer by longwave radiation can be expressed using the Stefan—Boltzmann law which states that the flux of radiation from an absolutely black body equals $\sigma T^4$, where $T$ is the absolute temperature in the Kelvin scale and $\sigma$ is the Stefan constant, equal to $5.67 \times 10^{-8}$ W m$^{-2}$ K$^{-4}$. Available experimental data show that radiation from the Earth's natural surfaces is, in general, rather close to radiation from a black body at the corresponding temperature, the ratio of the observed values of radiation to black-body radiation in most cases amounting to 0.90—1.00.

The maximum intensity of longwave radiation from the Earth's surface is a function of wavelength and depends on the temperature of the radiating surface. The major part of the radiated energy occurs at wavelengths from five to several tens of micrometres.

A considerable part of the longwave radiation flux emitted by the Earth's surface is compensated for by the atmospheric counter-radiation. The longwave emission of the atmosphere under clear sky is basically determined by the content of water vapour and carbon dioxide. The effect of atmospheric ozone is usually relatively less important. Water vapour, as well as carbon dioxide, absorbs longwave radiation mainly in certain spectral bands so that the longwave emission from the atmosphere is rather spectrally heterogeneous. The most intensive absorption of radiation by water vapour takes place in the range of wavelengths from 5 to 7.5 μm, while in the 9—12 μm spectral range the absortion of radiation is relatively small. Carbon dioxide has several bands of absorption, of which the band with wavelengths from 13 to 17 μm is of greatest importance.

It should be mentioned that the atmospheric content of carbon dioxide varies comparatively little, while the amount of water vapour can be very different depending on meteorological conditions. Therefore, variations in air humidity are a substantial factor influencing radiation from the atmosphere.

Usually the atmospheric counter-radiation increases greatly with cloudiness. Low and middle level coulds are, as a rule, sufficiently dense so that they radiate like an absolutely black body at the corresponding temperature. Under such conditions, the net longwave radiation is mainly defined by the difference between the temperature of the Earth's surface and that of lower surface of the clouds. If this difference is small, the net radiation also approaches zero. High clouds, because of their low density, usually radiate less than a black body. Since the temperature of such clouds is comparatively low, their effects on net longwave radiation are insignificant compared with the influences of low and middle clouds.

Thus, the net longwave radiation from the Earth's surface depends mainly on the surface temperature, the water vapour content of the air, and cloudiness. Due to these factors, the net longwave radiation may vary from values close to zero up to several hundred watts per square metre.

Radiation from the Earth—atmosphere system is that portion of the surface emission that passes through the atmosphere, plus atmospheric radiation. Under clear sky, the Earth's surface emission at wavelengths from 9 to 12 μm plays a significant role in the radiation from the system. In conditions of continuous cloudiness, radiation from the top surface of clouds (which depends on the cloud-top temperature) is of principal importance. Since this temperature is usually much lower than the temperature of the Earth's surface, it is evident that cloudiness considerably reduces the output of heat to space through outgoing radiation.

Equations of the energy (heat) balance represent a particular formulation of one of the principal laws of physics — the energy conservation law. These equations can be derived for various volumes and surfaces in the atmosphere, the hydrosphere and the lithosphere. Of most frequent use are the energy balance equations for a part of the Earth's surface and for the Earth—atmosphere system, i.e. for the vertical column passing through the whole geographical envelope.

The surface energy balance equation includes the energy fluxes between the surface element and surrounding space. Among these fluxes are the radiative fluxes of heat, the sum of which is equal to the radiation balance. The radiation balance equal to the sum of several heat fluxes may be positive or negative. As the Earth's surface temperature is usually not equal to the air temperature, a heat flux arises between the underlying surface and the atmosphere due to turbulent heat conductivity. A similar heat flux is observed between the Earth's surface and deeper layers of the lithosphere or hydrosphere, if the surface temperature differs from that of the deeper layers. In the latter case, the heat flux in soil is determined by molecular heat conductivity, while the heat transport in water bodies is usually turbulent.

The heat expenditure for evaporation is very important in the energy balance of the Earth's surface. The rate of evaporation depends on the moistness of the surface, its temperature, atmospheric humidity, and the intensity of turbulent exchange in the lowest layer of air. This intensity determines the rate of water vapour transfer from the Earth's surface to the atmosphere. Because of the effects of the latter factor, the features of evaporative heat transport are similar in many respects to those of the turbulent exchange of sensible heat between the Earth's surface and the atmosphere.

The evaporative heat flux changes its sign when the direction of the water vapour flux changes. If the flux of water vapour is directed from the atmosphere to the surface, then the evaporative heat flux becomes an input to the surface of heat of water vapour condensation.

To derive the energy balance equation, the radiative heat flux is designated by $R$, the turbulent heat flux from the surface to the atmosphere by $P$, the heat flux from the surface to the deeper layers by $A$, and the heat loss by evaporation (or the gain of heat in condensation) by $LE$ ($L$ is the latent heat of vaporization; $E$ is the rate of evaporation). Since all other components of the energy balance are usually much smaller than the heat fluxes enumerated here, then, as a first approximation, the energy balance equation can be written in the form

$$R = LE + P + A. \tag{2.4}$$

The value of the radiation flux (the radiation balance) is considered to be positive if there is an energy income to the Earth's surface; all other values are assumed to be positive when they describe a loss of energy from the surface.

In a number of papers, a different system of signs for the energy balance components is used, according to which all the components have the same sign when they designate expenditure (or income) of energy. Although this system is more logical, it has some inconvenient features. For instance, evaporative and turbulent heat losses from the Earth's surface to the atmosphere would have negative values, which is at variance with established practice.

The scheme of heat fluxes included in the energy balance equation is shown in Fig. 2.3.

Among the energy balance components that do not enter Equation (2.4), the most significant is often the expenditure of heat for melting surface snow or ice (or the gain of heat from freezing water). Although, as a rule, this

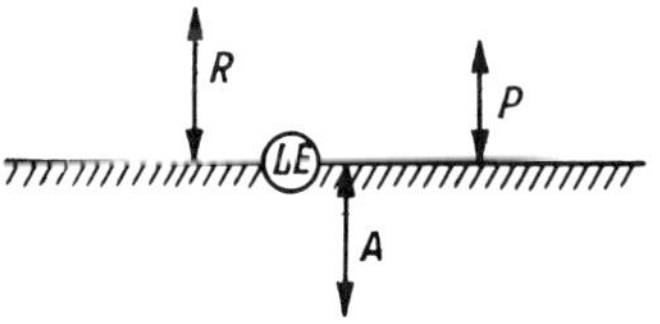


Fig. 2.3.

factor is relatively small compared with other terms in the energy balance when averaging periods of a year or more are considered, in some cases (for instance, periods of snow melting in middle and high latitudes) it should be included in Equation (2.4) as an additional term.

Other components of the energy balance (heat fluxes arising from energy dissipation due to friction of air current, wind-driven waves and tides; the heat flux (positive or negative) transferred by precipitation whose temperature differs from that of the Earth's surface; as well as the energy loss in photosynthesis and gain from biomass oxidation) are considerably less than the values of the chief components of the balance for averaging periods of any duration. Exceptions to this rule are possible, but these are relatively rare. As an example, in a forest fire, great amounts of energy previously accumulated in the process of photosynthesis are rapidly released.

Due to the linearity of Equation (2.4), the various heat flux components may be replaced by their sums over any given period of time, leaving the form of the equation unchanged.

The heat flux $A$ going from the surface to lower layers of soil or water can be determined through the other energy balance components of the upper layers of the lithosphere or hydrosphere. If we derive an energy balance equation for a vertical column with its upper boundary at the Earth's surface and its base at a sufficient depth, we obtain:

$$A = B + F_0. \tag{2.5}$$

Here $F_0$ is the heat gained by horizontal exchange between the column and the surrounding lithosphere or hydrosphere, and $B$ is the change in the amount of heat inside the column over a given period of time (Fig. 2.4). The vertical flux of heat through the base of the column can be assumed to be equal to zero, since the heat flow from the depth of the Earth's crust is usually negligibly small compared with the main components of the energy balance.

The $F_0$ component — the difference between the heat fluxes entering and leaving through the vertical sides of the column — may be of great importance for large reservoirs where currents exist and where strong horizontal heat conductivity takes place due to macroturbulence.

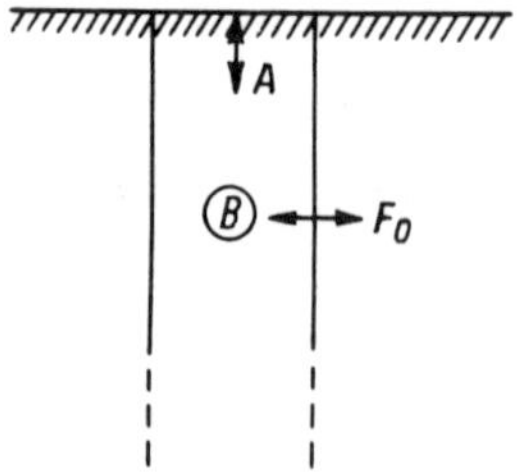


Fig. 2.4.

In the lithosphere, the value of $F_0$ is usually insignificant because of the small heat conductivity of soil. Therefore, for land, $A = B$. Since, when averaged over a year, the upper soil layer is neither cooled nor heated, both $A$ and $B$ must be zero when considering annual averages over a sufficiently long time period.

In the energy balance of closed reservoirs of significant area and depth and analysed as whole bodies, the values of $A$ and $B$ must be very close, since the heat exchange between water and ground is usually small in comparison with the basic components of the energy balance. However, for some portions of oceans and other water bodies (lakes and seas), the values of $A$ and $B$ might be very different. In this case, therefore, the mean annual value of heat exchange between the Earth's surface and the deeper layers is not zero, but equal to the amount of heat received or lost by the hydrospheric part of the vertical column, due to the effects of currents and macroturbulence (i.e. $A = F_0$).

In summary, then, the mean annual energy balance equation for land has the form

$$R = LE + P, \tag{2.6}$$

while, for the ocean, the equation is

$$R = LE + P + F_0. \tag{2.7}$$

In some cases, both of these equations can be simplified. For deserts, for example, where evaporation is close to zero, Equation (2.6) will be $R = P$. For the World Ocean as a whole, where total redistribution of heat by currents, due to self-compensation, is zero, Equation (2.7) is transformed into

$$R = LE + P. \tag{2.8}$$

To derive an energy balance equation for the Earth—atmosphere system, an analysis must be carried out of the gains and losses of thermal energy in a vertical column extending through the whole atmosphere and through the upper layers of the hydro- and lithosphere down to the levels where noticeable seasonal (or daily) temperature variations disappear. The heat exchange between the column and outer space is characterized by its radiation balance $R_s$, equal to the difference between the solar radiation absorbed in the entire volume of the column and the total outgoing emission from the column (Fig. 2.5). We shall consider the value of $R_s$ as being positive when it describes a heat gain to the Earth—atmosphere system. By extending the column deep into the lithosphere or the hydrosphere down to where the thermal regime does not depend on meteorological factors, we may assume that the heat flux through the base of the column is practically equal to zero.

The fluxes of heat through the lateral surfaces of the column are determined by the horizontal transfer of heat in the atmosphere and the hydro-

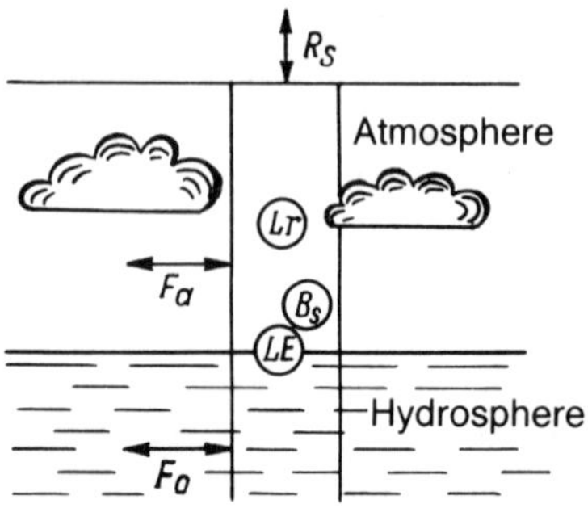


Fig. 2.5.

sphere. The difference between the gains and losses of heat due to heat transfer in the atmosphere is denoted in Fig. 2.5 by the arrow $F_a$ and for the hydrosphere by the arrow $F_0$. $F_a$ is similar in its physical meaning to $F_0$. It defines heat gains or losses in the air column due to the action of atmospheric advection and macroturbulence.

In addition to the heat exchange through the surface of the column, its energy balance is also affected by heat sources and sinks located inside the column. Among these, changes in heat content due to phase transformations of water, particularly evaporation and condensation, are of fundamental importance. The gain of heat from condensation in the atmosphere is equal to the difference between the gain and loss of heat due to condensation and evaporation of water droplets in clouds and fog. Over a sufficiently homogeneous surface for long periods of time, the average difference between the values of condensation and evaporation in the atmosphere is equal to the total precipitation, $r$. In this case, the heat income is equal to $Lr$. The corresponding component of the energy balance represents the difference between the heat income from condensation and the expenditure of heat in evaporating water droplets. It might, however, differ from the $Lr$ value over broken country and/or for individual short intervals of time.

The expenditure of heat for evaporation from the Earth's surface (the difference between heat loss for evaporation from the surfaces of reservoirs, soil and vegetation, and heat income from condensation on these surfaces) is equal to $LE$. The total effect of condensation and evaporation on the energy balance of the column can be approximated by $L(r-E)$. Among the other components in the energy balance of the column, account should be taken of the change in heat content of the column $B_s$. The other components of the balance (the gain of heat from dissipation of mechanical energy, the difference between expenditure of heat for photosynthesis, and income from oxidation of organic matter, and so on) are usually insignificant in the energy balance of the Earth–atmosphere system and need not be considered further.

From the above, the energy-balance of the Earth—atmosphere system may be written as

$$R_s = F_s + L(E - r) + B_s, \tag{2.9}$$

where $F_s = F_a + F_0$. All the terms on the right side are taken as positive if they show a loss of heat. For the annual average period the value of $B_s$ is close to zero and Equation (2.9) is transformed into

$$R_s = F_s + L(E - r). \tag{2.10}$$

For land, this equation is as follows:

$$R_s = F_a + L(E - r). \tag{2.11}$$

Since, for the Earth as a whole, over the year $E = r$, and the horizontal heat transfer in the atmosphere and the hydrosphere is equal to zero, the energy balance equation for the entire Earth—atmosphere system acquires the simple form

$$R_s = 0. \tag{2.12}$$

The energy balance equation of the atmosphere can be obtained as the sum of relevant heat fluxes, or simply derived from the energy balance equations of the Earth—atmosphere system and the Earth's surface. Since the radiation balance of the atmosphere, $R_a$, is equal to $R_s - R$, and since the change in heat content of the atmosphere, $B_a$, is equal to $B_s - B$, we find that

$$R_a = F_a - Lr - P + B_a, \tag{2.13}$$

and, for the average annual period,

$$R_a = F_a - Lr - P. \tag{2.14}$$

## 2.3. Distribution of the Energy Balance Components

Studies of the energy balance of the globe began in the 19th century, when instruments for measuring direct solar radiation were designed and calculations were carried out on the amount of solar radiation incident on the top of the atmosphere according to latitude and season. In the 1910—1930s, W. Schmidt, A. Ångström, F. Albrecht, and S. I. Savinov found the energy balance components of the Earth's surface for individual regions of the globe.

The present author and his colleagues have produced world maps of the energy balance components of the Earth's surface for every month and for average annual conditions, published as the *Atlas of the Heat Balance* (1955). In further studies these maps were improved and, together with several maps of the energy balance components of the Earth—atmosphere system, these improved maps were published as the *Atlas of the Heat Balance of the Earth* (1963).

Since 1963, when the second atlas of heat balance was published, a great body of radiation data has been obtained from observations at newly established radiation stations on the continents. Of particular significance is information obtained from oceanic radiation observations. This has led to clarification of various aspects of the radiation regime of reservoirs. Advances in studying turbulent diffusion allowed refinements to be made to previously used methods for calculating evaporative heat expenditure and turbulent heat exchange between the Earth's surface and the atmosphere over the oceans. These new observational materials and the development of calculation techniques for determining the energy balance components made it possible to construct new world maps of the energy-balance components that were an improvement over the earlier maps (Budyko *et al.*, 1978).

When constructing these maps, the energy balance equation of the Earth's surface was used in the above-mentioned form

$$R = LE + P + A, \tag{2.15}$$

where $R$ is the radiation balance of the Earth's surface, $LE$ the expenditure of heat for evaporation or the gain of heat from condensation at the surface ($L$ is the latent heat of vaporization and $E$ the rate of evaporation or condensation), $P$ the turbulent heat flux between the surface and the atmosphere, and $A$ the heat flux between the surface and the lower layers of reservoirs or soil.

In (2.15) the radiation balance $R$ is assumed to be equal to $Q(1 - \alpha) - I$ (where $Q$ is the total shortwave radiation incident on the Earth's surface, $\alpha$ the albedo of the surface, and $I$ the net longwave radiation), $R$ being positive for a radiation income to the Earth's surface. The other energy balance components are positive when they correspond to heat loss.

World maps of the energy balance components of the Earth's surface for mean annual conditions based on the above-mentioned maps of heat balance are shown below. Unlike the values in the afore-mentioned study, where the values of the balance components were expressed in kilocalories per square centimetre per year and in kilocalories per square centimetre per month, in this book they are presented in watts per square metre (1 $W\,m^{-2}$ is equal to 0.75 $kcal\,cm^{-2}\,yr^{-1}$ and 0.063 $kcal\,cm^{-2}\,month^{-1}$). Maps of the energy balance components for the oceans have been recently corrected by Strokina.

As seen from Fig. 2.6, the average annual values of total radiation on the globe vary from less than 80 to more than 280 $W\,m^{-2}$. The greatest values of total radiation over both land and oceans are observed in the high-pressure belts in the Northern and Southern Hemispheres. Total radiation decreases towards the poles. Some decrease in the total radiation takes place also at equatorial latitudes, as a result of the frequent occurrence of overcast sky.

The isopleths of total solar radiation mainly show a zonal character which is considerably disturbed by variations of cloudiness. Deviations from this

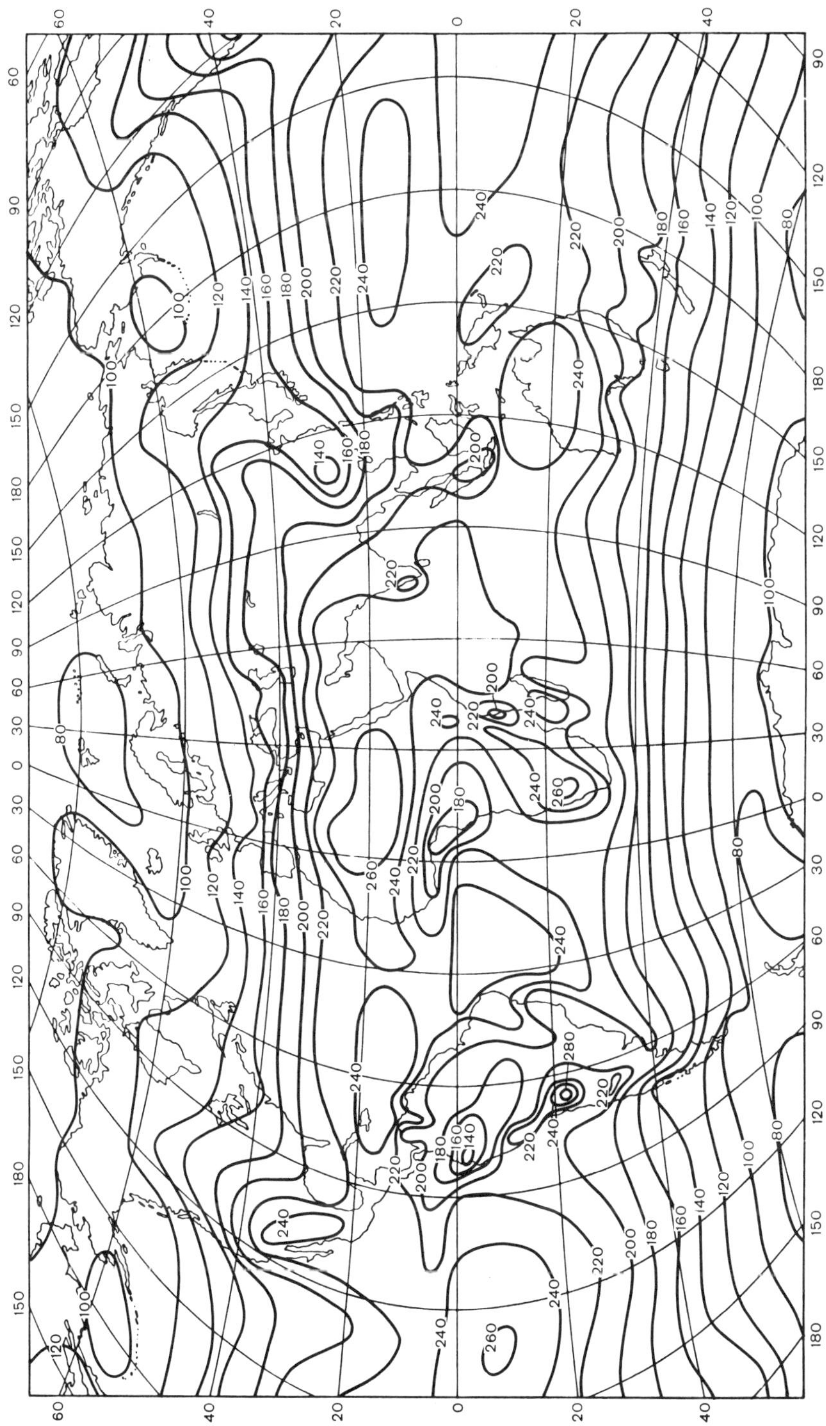

Fig. 2.6.

zonal structure take place in the mid-latitudes of both hemispheres where cyclonic activity is most intensive (the west coast of Canada, northern Europe, the southwest coast of South America, etc.), in the eastern regions of tropical zones, over the oceans under the influence of trade-wind inversions and cold marine currents, and in the regions of monsoon circulation (Hindustan, the eastern coast of Asia, the northwest of the Indian Ocean).

When analysing the data on the total radiation distribution for winter months, we should note that it decreases rapidly towards the poles. This is explained by the lower elevation of the Sun at noon and by the smaller length of the day. Along with this, in winter considerable interlatitudinal changes of total radiation are recorded: from about 200—220 W $m^{-2}$ in low latitudes to the zero values in polar regions. The greatest monthly mean radiation in low latitudes is observed in the regions of the equatorial monsoons, where cloudiness is insignificant in winter.

The summer distribution of total radiation is characterized by high values over the entire hemisphere with little geographical variability. The maximum solar radiation is received in tropical and subtropical deserts — above 300 W $m^{-2}$. A great amount of solar energy in summer also reaches the polar regions, where the effect of low solar elevation is compensated for by the longer days. The greatest solar radiation in summer months is recorded on the central plateau in the Antarctic. In January, monthly means vary from 250—300 W $m^{-2}$ on the coast to 450 W $m^{-2}$ in the interior of the continent. These values are noticeably above the values for tropical deserts (Marshunova, 1980).

The data for the radiation balance of the Earth's surface are presented in Fig. 2.7. Annual average values of the radiation balance of the land surface change from magnitudes less than −7 W $m^{-2}$ in the Antarctic and close to zero in the central Arctic, to 120—130 W $m^{-2}$ in tropical latitudes. The influence of astronomical factors produces a zonal pattern for annual and monthly means of the radiation balance over flat regions in high and middle latitudes of the Northern Hemisphere. This latitudinal distribution breaks down in the regions where circulation factors considerably change the cloudiness pattern.

Geographical zones in high and middle latitudes may be characterized by their annual radiation balance totals: Arctic tundra less than 15 W $m^{-2}$; tundra, forest-tundra, 15—25 W $m^{-2}$; northern and middle taiga, 25—40 W $m^{-2}$; southern taiga, 40—50 W $m^{-2}$; mixed, deciduous and broadleaf forests, forest-steppe and steppe in middle latitudes, 50—70 W $m^{-2}$. In subtropical, tropical and equatorial zones, the specific properties of the atmospheric circulation create drastic differences in cloudiness and moisture patterns, which result in variations in the annual values of the radiation balance from 70 to 130 W $m^{-2}$. At the same time, the lowest values of the radiation balance are observed in subtropical and tropical deserts because of both the high reflectivity of the desert surface and the great loss of heat through longwave radiation that occurs under conditions of little cloudiness, low air humidity and high temperature at the soil surface.

Data on the seasonsal distribution of the radiation balance show that the minimum monthly means are observed in high polar latitudes: from $-15$ to $-30$ W m$^{-2}$ in winter compared to about 65 W m$^{-2}$ in summer. In middle latitudes of the Northern Hemisphere, the monthly means of the radiation balance vary from $-15$ to $-30$ W m$^{-2}$ in January and from 110 to 145 W m$^{-2}$ in July. In tropical extra-equatorial latitudes around the time of the winter solstice the radiation balance values amount to 55—65 W m$^{-2}$. In these latitudes in summer the values range from 85—95 W m$^{-2}$ (in deserts and equatorial monsoon regions) to maximum values of 145—160 W m$^{-2}$.

The distribution of the surface radiation balance values over the oceans, as presented in Fig. 2.7, is similar to that for total radiation. The maximum average annual value of the radiation balance over the oceans is approximately 200 W m$^{-2}$. The minimum annual means for the ice-free ocean surface, recorded at the boundary of floating ice, are about 20—40 W m$^{-2}$. It should be noted that the annual means of the radiation balance over the entire ice-free ocean surface are positive.

During winter time, the radiation balance of the oceans varies from 140 to 160 W m$^{-2}$ in equatorial latitudes to negative values (of about $-30$ W m$^{-2}$) in middle latitudes. In this case, the radiation balance becomes negative in both hemispheres poleward of 45°N and 45°S.

In the summer months, the mean values of the radiation balance reach their maxima, exceeding 200 W m$^{-2}$ in low and 130—140 W m$^{-2}$ in high latitudes. At this time of the year the spatial distribution of the radiation balance, unlike winter, deviates noticeably from a zonal one. Regions with high and low values correspond to those with little and heavy cloudiness.

The annual map of heat expenditure for evaporation is presented in Fig. 2.8. Over land heat expenditure values were constructed using data for the long-term monthly evaporation means for stations located in flat regions. For mountainous regions the evaporative heat loss has been estimated using data on evaporation variations with altitude (*The World Water Balance and Water Resources of the Earth*, 1974).

Monthly means of evaporative heat loss (and turbulent heat exchange with the atmosphere) over the oceans were evaluated using data from long-term ship observations in the Atlantic, Indian and Pacific Oceans. From analyses of the latent heat flux distribution on land for the year, we can conclude that its value varies over a range of about 100 W m$^{-2}$. In regions of sufficient moisture, the mean annual heat loss by evaporation increases as the radiation balance rises from high latitudes towards the equator, varying from less than 10 W m$^{-2}$ on the northern coasts of continents to more than 80 W m$^{-2}$ in humid equatorial forests in South America, Africa and the Malay Archipelago. In regions with insufficient moisture, the value of heat expenditure for evaporation is determined by the aridity of the climate, decreasing with increasing aridity. The minimum values of the mean annual latent heat flux are recorded in tropical deserts where they are only a few watts per square metre.

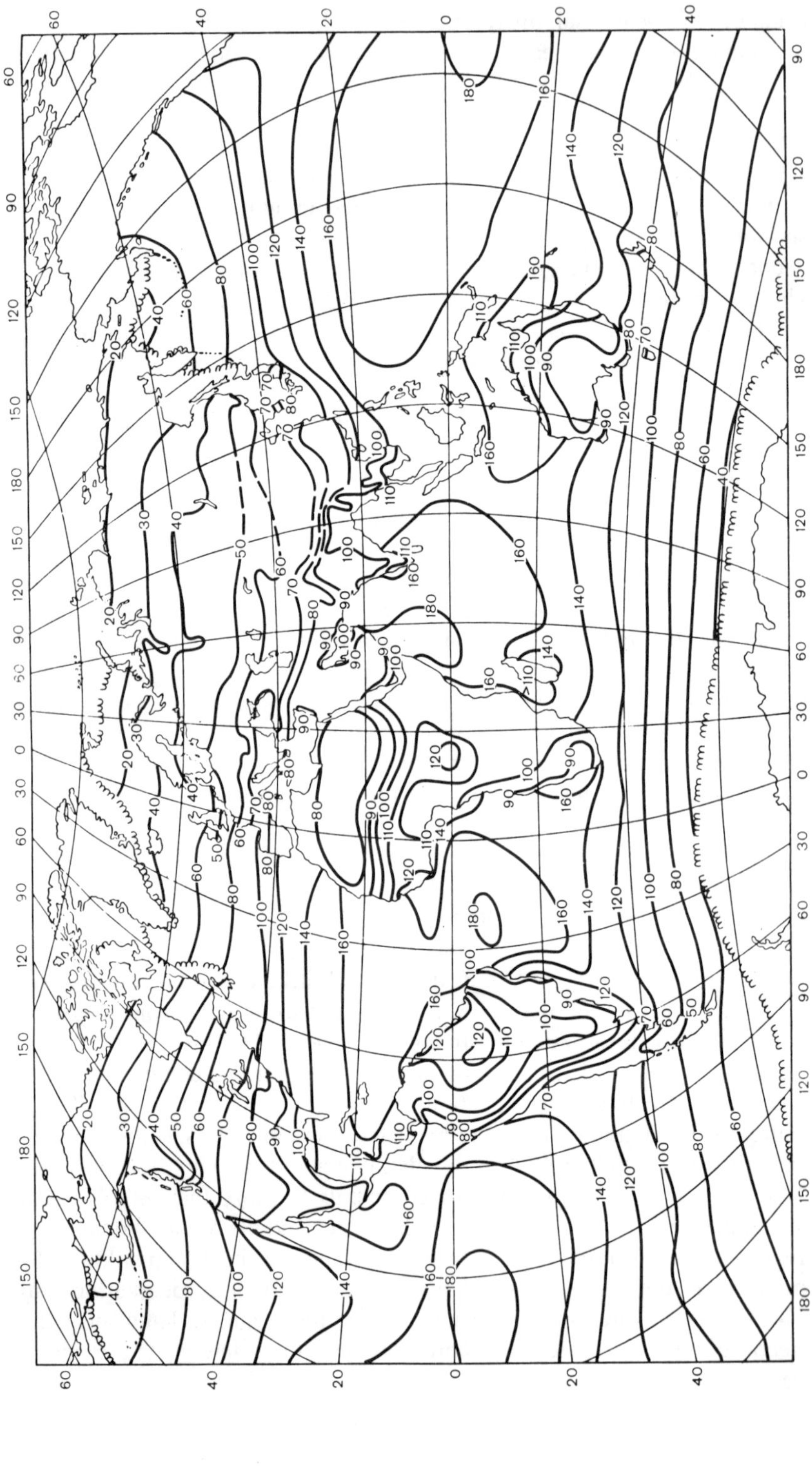

Fig. 2.7.

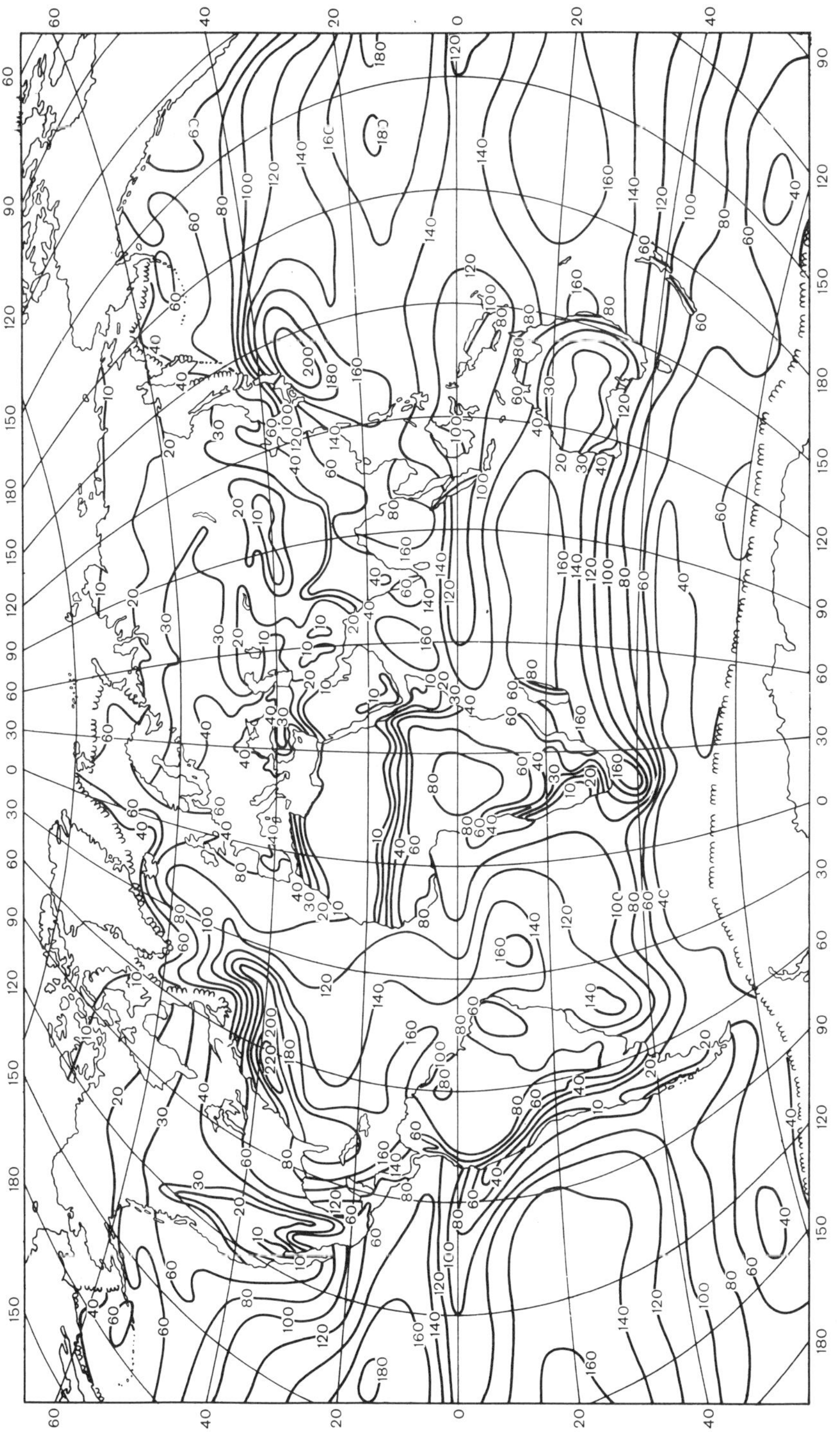

Fig. 2.8.

The annual course of the heat expenditure for evaporation is also determined by the availability of heat energy and water. In extratropical latitudes in conditions of sufficient moisture, maximum values of heat used in evaporation follow the annual variations in the radiation balance and take place in summer reaching 80—100 W m$^{-2}$. In winter the evaporative heat loss is small. In regions with insufficient moisture, the maximum heat expenditure for evaporation is also observed during the warm season, but the time at which the maximum is reached depends strongly on the moisture conditions. In deserts the annual course of $LE$ is determined by the yearly variation of precipitation.

In tropical latitudes with a humid climate, the expenditure of heat for evaporation remains high throughout the year and equals about 80 W m$^{-2}$. In regions with seasonal variations in precipitation latent heat losses are less in the drier seasons; however, the amplitude of the annual variation is comparatively small. In areas with a clearly defined arid period, maximum values of evaporative heat loss are recorded at the end of wet periods and minima occur at the end of arid periods. For the land area as a whole, including Antarctica, the mean annual evaporative heat loss is 38 W m$^{-2}$.

The spatial distribution of the annual means of evaporative heat loss from the oceans is, in general, similar to that of the radiation balance. As seen from Fig. 2.8, the range of values for the annual average loss of heat for evaporation is rather great: from more than 160 W m$^{-2}$ in tropical latitudes to about 40 W m$^{-2}$ at the edge of polar ice. In equatorial latitudes the average expenditure of heat in evaporation is somewhat less than in higher latitudes (less than 130 W m$^{-2}$), a consequence of greater cloudiness and air humidity.

In addition to radiation as an energy source for evaporation from the oceans, in a number of regions evaporation consumes heat transferred by ocean currents. The zonal character of the latent heat flux distribution is, therefore, noticeably disturbed in some areas by the activity of warm and cold currents. This is clearly seen in Fig. 2.8. The greatest latent heat fluxes are recorded in the Northern Hemisphere: more than 220 W m$^{-2}$ in the Gulf Stream and about 210 W m$^{-2}$ in the Kuro Shio areas. In these regions, the vapour pressure differential that provides the driving force for evaporation is larger, not only because of high water temperature, but also because of the comparatively low humidity of air masses incoming into the areas from North America and Asia during the cold season.

The annual means of evaporative heat loss from the oceans depend mainly on values prevailing during the autumn—winter period. The winter latent heat flux distribution is similar to the annual pattern. At this time of the year the influence of the warm currents is intensified so that the distinctive features of each ocean show up most vividly. The heat expenditure for evaporation from the surface of the North Atlantic in middle latitudes is twice as great as at the same latitudes in the Pacific Ocean. The lowest latent-heat fluxes are observed in middle latitudes in the Southern Hemisphere in the Atlantic and Indian Oceans. These areas have comparatively low water temperatures and are influenced by warmer air masses from tropical latitudes. These two factors tend to reduce heat loss in evaporation.

In spring the effect of warm currents on evaporative heat loss declines as a result of the decreasing energy resources of the currents. In summer the average wind speed decreases and the air—water temperature difference is lowered so that the expenditure of heat for evaporation drops noticeably. In addition, ocean-to-ocean differences in evaporative heat losses diminish.

A map of the annual mean turbulent flux of sensible heat is given in Fig. 2.9.

The greatest annual means of sensible heat flux between the land surface and the atmosphere are recorded in the tropical deserts, where they reach 70—80 W $m^{-2}$. In regions with higher humidity the sensible heat flux declines. For example, in humid tropical forest regions, the mean annual sensible heat flux is 15—40 W $m^{-2}$. Moving towards higher latitudes the turbulent flux reduces as the radiation balance decreases. Over the northern coasts of the Northern Hemisphere continents the sensible heat flux is less than 10 W $m^{-2}$. The same values are observed in several middle latitude regions with sufficient moisture.

In the annual course the same distinctive pattern can be seen: an increase in the sensible heat flux with rising radiation balance. Therefore, in extra-tropical latitudes, the greatest values of the sensible heat flux are recorded in summer, the smallest in winter. For areas north of 40°N and south of 40°S in both hemispheres the direction of sensible heat flux changes during the year. In winter in these regions the Earth's surface receives a small amount of heat from the atmosphere by means of turbulent heat exchange (even in high latitudes the amount is less than 10 W $m^{-2}$).

The surface of the continents between 40°N and 40°S loses heat by the turbulent heat flux throughout the year. In this area the annual course of the sensible heat flux is strongly dependent on moisture. The greatest monthly means of the turbulent flux occur during the driest months. In subtropical latitudes with Mediterranean-type climate, the maximum monthly means of the sensible heat flux, up to 100 W $m^{-2}$, are recorded in summer. In deserts (particularly coastal deserts) where the turbulent heat exchange is considerably affected by the processes of air mass transfer at the water—land boundary, the values of sensible heat flux exceed 100 W $m^{-2}$. In humid tropical regions, the sensible heat flux is small throughout the year, its monthly means being less than 30 W $m^{-2}$.

Overall, less than half of the annual value of the radiation balance of the continental surface is accounted for by turbulent heat exchange with the atmosphere.

The map of the turbulent heat exchange between the ocean surface and the atmosphere (Fig. 2.9) shows that almost everywhere the oceans lose heat.

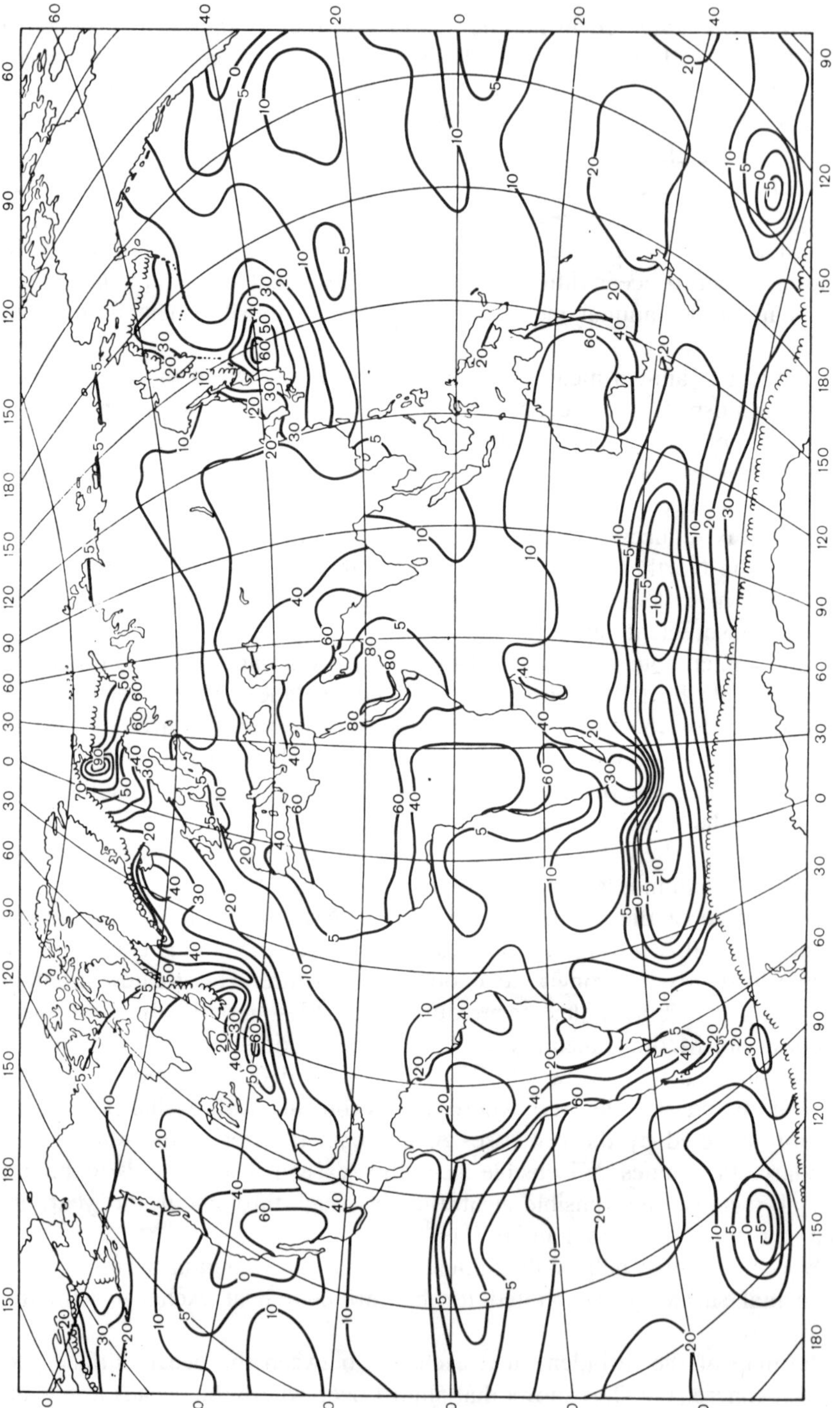

Fig. 2.9.

The greatest sensible heat loss (as is the case for heat loss in evaporation) takes place in western and northwestern regions of the Northern Hemisphere oceans. There the sensible heat flux may exceed 50 $W m^{-2}$. In southern regions, particularly near the equator where temperature differences between the water surface and the air masses over it are small throughout the year, the sensible heat flux from the ocean surface is less than 10 $W m^{-2}$. In the Southern Hemisphere because of comparatively small water—air temperature differences, the sensible heat flux is considerably less than that in the Northern Hemisphere and, as a rule, does not exceed 20 $W m^{-2}$.

Negative annual sums of the turbulent heat exchange with the atmosphere (corresponding to a flux from atmosphere to ocean) are observed in the zones of the cold California Current and the West Wind Current in the Southern Hemisphere. These negative values have small absolute magnitudes.

As for the heat expenditure for evaporation, the annual sums of ocean-atmosphere turbulent heat exchange depend mainly on their values during the autumn—winter period. The closest relationship between the heat exchange conditions and the effects of ocean currents and the atmospheric circulation is observed in winter. An average monthly heat loss by the oceans in the Northern Hemisphere during winter reaches 150—200 $W m^{-2}$. In the Southern Hemisphere the maximum is about 80 $W m^{-2}$. In summer, for both hemispheres, the turbulent heat exchange between the ocean surface and the atmosphere is close to zero everywhere: its values lie in the range from 15 to $-20$ $W m^{-2}$.

Heat exchange also occurs between the ocean surface and the deeper layers of the ocean due to the effects of currents. The value of this exchange is the algebraic sum of the radiative balance, the heat loss in evaporation, and the turbulent heat exchange between the ocean surface and the atmosphere. Figure 2.10 shows the annual map of this residual component in the energy balance equation — the heat gained or lost by the ocean surface by the activity of ocean currents. This map demonstrates the role of currents and atmospheric circulation in heat interaction between the surface and the deeper layers of the oceans. At the same time, it describes the redistribution of heat between different ocean regions.

In near-equatorial latitudes of both hemispheres, the ocean surface gains 40—80 $W m^{-2}$ per year as a result of large radiative heat income and decreased evaporative and turbulent heat exchange with the atmosphere. The higher the latitude in either hemisphere, the less the accumulation of heat by ocean waters. In high latitudes, the amount of radiative heat appears to be insufficient to compensate for evaporative and sensible heat losses, so the ocean surface must obtain heat from deeper layers transferred by currents. The areas with maximum heat loss from the ocean into the atmosphere are located in the regions influenced by warm currents — the Gulf Stream and Kuro Shio. In these regions during the winter half-year, the warm ocean waters interact with cold continental air from North America and Asia to

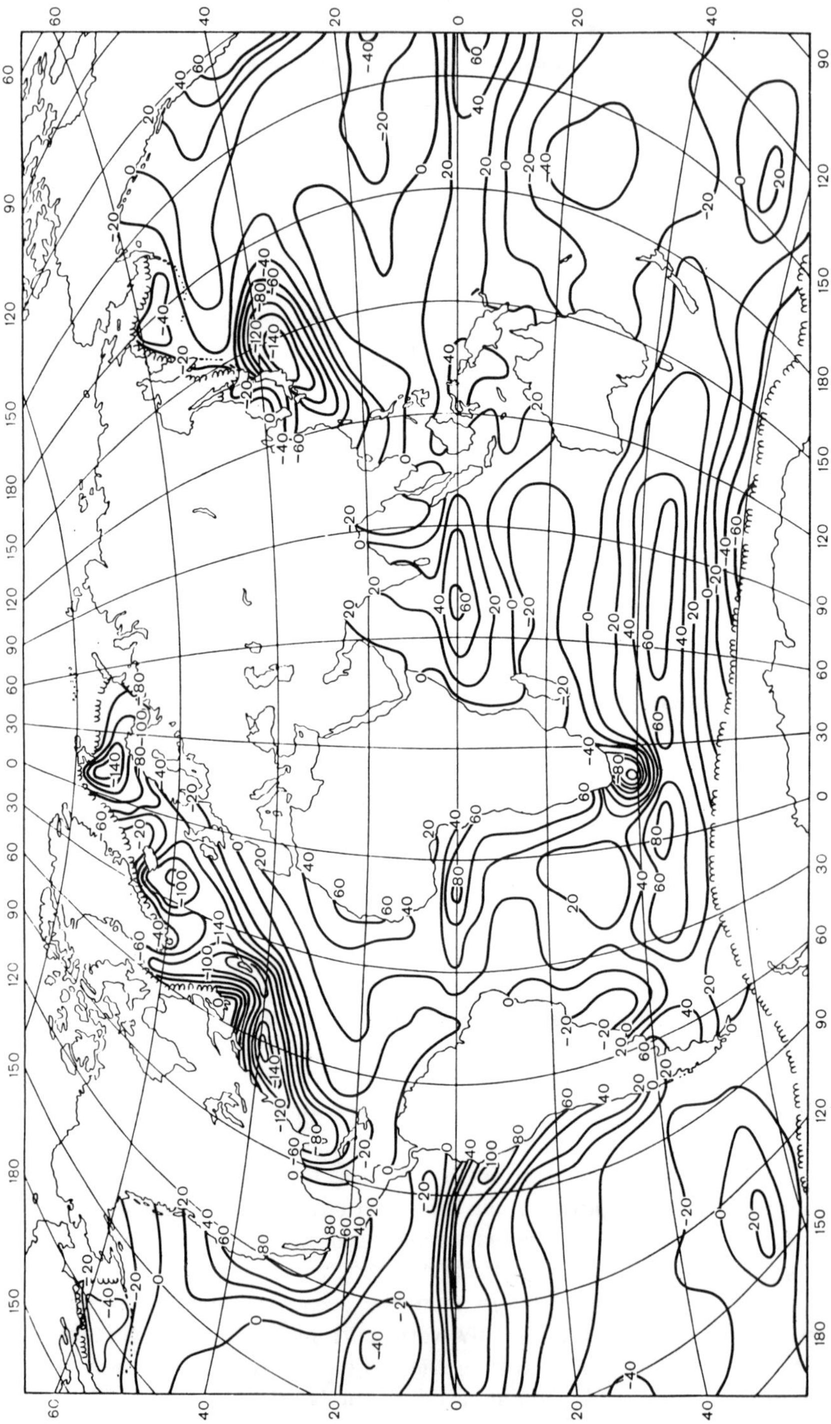

Fig. 2.10.

produce conditions for intensive heat transfer from the deeper ocean layers through the water surface, into the atmosphere, its value reaching in absolute magnitude 150 W $m^{-2}$.

It should be mentioned that, in high and middle latitudes of the Northern Hemisphere, because of large water—air temperature differences, an analogous process takes place in regions influenced not only by warm, but even by some cold currents. In the Southern Hemisphere, where the water—air temperature differences are considerably smaller, the heat loss from the ocean surface is generally less in absolute magnitude than 50 W $m^{-2}$.

In most regions of cold currents, the ocean surface gains heat from the Sun and the atmosphere and transfers a significant portion to deeper layers. The maximum amount of such heat transfer occurs in the areas of the California, Peru, Benguela, Falkland and West Wind currents.

How accurate are these determinations of the energy-balance components of the Earth's surface based on surface meteorological observations? This question has been discussed in a number of the author's works (Budyko, 1971, and so on). It is worth noting that Holloway and Manabe (1971) constructed world maps of the surface energy balance components using a general circulation model of the atmosphere. They estimated values of the radiation balance at the Earth's surface, evaporation (which is proportional to heat loss in evaporation), and the turbulent heat flux from the surface into the atmosphere. When their maps were compared with similar ones from *Atlas of the Heat Balance of the Earth* (1963), Holloway and Manabe found satisfactory agreement between them.

Oort and Vonder Haar (1976) suggested another method of determining some of the energy balance components of the Earth's surface. They used satellite data on the radiation balance components of the Earth—atmosphere system to calculate the heat flux from the surface to the atmosphere by using the atmospheric energy balance equation. They assumed this value to be equal to the algebraic sum of the Earth—atmosphere system radiation balance, the change in heat content of a vertical column passing through the atmosphere, and the horizontal heat flux into this column caused by atmospheric circulation. The first of these values was found from satellite data, the second and the third from aerological observations. In this way they estimated, for every month of the year, latitudinal means of the total heat flux (i.e. the sum of the radiation balance at the Earth's surface and the evaporative and sensible heat fluxes from the Earth's surface to the atmosphere). Given the limited accuracy in estimating, from aerological data, the heat flux into the air column due to atmospheric circulation, and the fact that the residual term of the energy balance equation obtained by Oort and Vonder Haar is frequently less in absolute magnitude than its main components, their results are fairly accurate. They found, in particular, that the total heat flux at different latitudes of the Northern Hemisphere varies from 40 to −140 W $m^{-2}$ in January, and from −30 to 90 W $m^{-2}$ in July. The corresponding values in *Atlas of the Heat Balance of the Earth* (1963) range from 40 to −110 W m$^{-2}$ in January and from 10 to 90 W m$^{-2}$ in July. Since the estimation of these quantities as a residual term in the energy balance equation from data of *Atlas of the Heat Balance of the Earth* is subject to noticeable errors, this agreement shows that the results obtained, even in approximate calculations of the energy balance components, are quite reliable.

Data on the spatial distribution of the energy balance components at the Earth's surface make it possible to determine their means for the entire globe. Table 2.4 shows annual values of the energy balance components for different latitudinal zones of land and oceans and for the Earth as a whole.

TABLE 2.4

Latitudinal means of the energy balance components at the Earth's surface (W m$^{-2}$)

| Latitudinal zone | Land | | | Ocean | | | | Earth | | | |
|---|---|---|---|---|---|---|---|---|---|---|---|
| | $R$ | $LE$ | $P$ | $R$ | $LE$ | $P$ | $F_0$ | $R$ | $LE$ | $P$ | $F_0$ |
| 70–60°N | 29 | 21 | 8 | 32 | 44 | 25 | −37 | 30 | 28 | 13 | −11 |
| 60–50 | 42 | 30 | 12 | 59 | 71 | 29 | −41 | 49 | 48 | 19 | −18 |
| 50–40 | 60 | 33 | 27 | 88 | 89 | 20 | −21 | 73 | 60 | 24 | −11 |
| 40–30 | 77 | 31 | 46 | 119 | 132 | 20 | −33 | 101 | 89 | 31 | −19 |
| 30–20 | 85 | 25 | 60 | 149 | 139 | 11 | −1 | 125 | 96 | 29 | 0 |
| 20–10 | 98 | 42 | 56 | 163 | 154 | 9 | 0 | 146 | 125 | 21 | 0 |
| 10–0 | 105 | 76 | 29 | 165 | 135 | 9 | 21 | 151 | 121 | 14 | 16 |
| 0–10 | 105 | 81 | 24 | 170 | 123 | 6 | 41 | 155 | 113 | 10 | 32 |
| 10–20 | 100 | 60 | 40 | 162 | 149 | 9 | 4 | 148 | 129 | 16 | 3 |
| 20–30 | 94 | 37 | 57 | 147 | 143 | 15 | −11 | 135 | 118 | 25 | −8 |
| 30–40 | 82 | 38 | 44 | 124 | 114 | 16 | −6 | 119 | 105 | 19 | −5 |
| 40–50 | 58 | 29 | 29 | 94 | 66 | 6 | 22 | 93 | 65 | 7 | 21 |
| 50–60°S | 46 | 29 | 17 | 59 | 48 | 12 | −1 | 59 | 48 | 12 | −1 |
| Earth as a whole | 66 | 38 | 28 | 121 | 108 | 13 | 0 | 105 | 88 | 17 | 0 |

These data can be compared with the results of previous studies of the surface energy balance. Table 2.5 shows relative values of the energy balance components (expressed as per cent of the solar constant value), obtained from a number of different studies. From the data given in the last three entries in this table we can see that the improvements in the global maps of the various energy-balance components have led to increases in the estimates of absorbed radiation, radiation balance and evaporative heat loss. The relative values derived in the most recent study (Budyko *et al.*, 1978) are close to those obtained by Houghton in the 1950s (except for the value of the turbulent heat flux).

TABLE 2.5

Energy balance at the Earth's surface (components of the balance are expressed as per cent of the amount of solar radiation coming to the top of the atmosphere)

| Investigation | Components of the balance | | | | |
|---|---|---|---|---|---|
| | Absorbed radiation | Net longwave radiation | Radiation balance | Heat loss in evaporation | Turbulent heat flux |
| Dines (1917) | 42 | 14 | 28 | 21 | 7 |
| Alt (1929) | 43 | 27 | 16 | 16 | 0 |
| Baur und Philipps (1934) | 43 | 24 | 19 | 23 | −4 |
| Houghton (1954) | 47 | 14 | 33 | 23 | 10 |
| Lettau (1954) | 51 | 27 | 24 | 20 | 4 |
| *Atlas of the Heat Balance* (1955) | 42 | 16 | 26 | 21 | 5 |
| *Atlas of the Heat Balance of the Earth* (1963) | 43 | 15 | 28 | 23 | 5 |
| Budyko *et al.* (1978) | 46 | 15 | 31 | 26 | 5 |

Calculations of the energy balance components of the Earth—atmosphere system were first made in the 1920—1930s by G. Simpson, F. Baur, H. Philipps and G. G. Trolle. World maps of the radiation balance of the Earth—atmosphere system, heat input from condensation in the atmosphere, and heat transfer by air currents constructed by Vinnikov were published in the *Atlas of the Heat Balance of the Earth* (1963). In the late 1960s, progress in satellite meteorology made it possible to construct maps of the radiation regime elements of the Earth—atmosphere system directly from satellite observation data. The first maps of this kind contained data for individual periods of time (e.g. Raschke *et al.*, 1968). Later, these data were used in constructing maps of the radiation regime elements averaged over a number of years (Vonder Haar and Suomi, 1969; Raschke *et al.*, 1973; Ellis and Vonder Haar, 1976; Stephens *et al.*, 1981; etc.).

The energy balance equation of the Earth—atmosphere system has been presented earlier in the form

$$R_s = F_s + L(E - r) + B_s, \tag{2.16}$$

where $R_s$ is the radiation balance of the Earth—atmosphere system; $F_s$ is the gain or loss of heat resulting from horizontal transfer in the atmosphere or the hydrosphere; $L(E - r)$ is the difference between the evaporative heat loss and the heat gain from condensation, which is considered to be proportional to the rate of precipitation $r$; and $B_s$ is the rate of change in the heat content of the vertical column passing through the atmosphere and the upper layers of the hydrosphere (or lithosphere). The radiation balance is assumed to be equal to $Q_s(1-\alpha_s)-I_s$ (where $Q_s$ is the incoming flux of solar radiation at the top of the atmosphere, $\alpha_s$ is the albedo of the Earth—atmosphere system, and $I_s$ is the net longwave radiation escaping into space). $F_s$ is considered to be equal to $F_a+F_0$, where $F_a$ is the redistribution of heat in the atmosphere and $F_0$ is the corresponding hydrosphere term. Latitudinal means of the energy balance components for the Earth—atmosphere system obtained from observational data (Budyko, 1974) are listed in Table 2.6.

In Table 2.6, $Q_a$ is the radiation absorbed by the Earth—atmosphere system. The difference between $Q_a$ and $I_s$ (defined above), namely, $R_s$, is the radiation balance of the Earth—atmosphere system. The total amount of heat redistribution due to atmospheric advection, equal to $F_a+L(E-r)$, is denoted by $C_a$. The values for $F_0$ in Tables 2.4 and 2.6, determined by independent methods, differ only slightly. The upper left and lower right parts in Table 2.6 refer to the warm seasons, whereas the lower left and upper right parts refer to the cold seasons.

This table shows that the absorbed radiation, $Q_a$, is not the only factor determining the values of outgoing emission, $I_s$. In high and middle latitudes

TABLE 2.6

Latitudinal means of the components of the heat balance of the Earth—atmosphere system for the summer and winter half-years (W m$^{-2}$)

| Latitude | April—September | | | | | October—March | | | | |
|---|---|---|---|---|---|---|---|---|---|---|
| | $Q_a$ | $C_a$ | $F_0$ | $B_s$ | $I_s$ | $Q_a$ | $C_a$ | $F_0$ | $B_s$ | $I_s$ |
| 80—90°N | 123 | −72 | 0 | 13 | 182 | 1 | −145 | 0 | −13 | 159 |
| 70—80 | 131 | −69 | 0 | 13 | 187 | 8 | −148 | 0 | −13 | 169 |
| 60—70 | 183 | −29 | −6 | 19 | 199 | 29 | −107 | −24 | −19 | 179 |
| 50—60 | 233 | 0 | −20 | 44 | 209 | 64 | −75 | −8 | −44 | 191 |
| 40—50 | 269 | 16 | −32 | 62 | 223 | 103 | −46 | 10 | −62 | 201 |
| 30—40 | 306 | 22 | −26 | 68 | 242 | 151 | −15 | 11 | −68 | 223 |
| 20—30 | 318 | 31 | −6 | 46 | 247 | 218 | 14 | 10 | −46 | 240 |
| 10—20 | 313 | 38 | 14 | 22 | 239 | 271 | 29 | 16 | −22 | 248 |
| 0—10 | 293 | 21 | 35 | −1 | 238 | 298 | 34 | 23 | 1 | 240 |
| 0—10°S | 287 | 37 | 35 | −24 | 239 | 313 | 38 | 12 | 24 | 239 |
| 10—20 | 258 | 41 | 15 | −40 | 242 | 329 | 55 | −5 | 40 | 239 |
| 20—30 | 206 | 21 | 5 | −54 | 234 | 328 | 55 | −19 | 54 | 238 |
| 30—40 | 142 | −6 | −6 | −67 | 221 | 301 | 32 | −23 | 67 | 225 |
| 40—50 | 94 | −31 | −25 | −55 | 205 | 253 | 0 | −10 | 55 | 208 |
| 50—60 | 53 | −62 | −41 | −40 | 196 | 204 | −41 | 10 | 40 | 195 |
| 60—70 | 16 | −151 | 0 | −13 | 180 | 130 | −68 | 0 | 13 | 185 |
| 70—80 | 4 | −156 | 0 | 0 | 160 | 70 | −103 | 0 | 0 | 173 |
| 80—90 | 0 | −140 | 0 | 0 | 140 | 54 | −110 | 0 | 0 | 164 |

during the cold season (and in high latitudes of the Southern Hemisphere during the whole year), the main source of heat is its transfer from lower latitudes by the atmospheric circulation.

Of the two processes of heat exchange in the oceans, seasonal heat accumulation and expenditure in oceanic waters is more important. In some latitude belts, the value of this energy balance component can reach 25—30% of the outgoing radiation. The redistribution of heat by ocean currents plays a smaller role, although in particular zones its value can amount to 15—20% of the longwave radiation. It should be noted that the values of the energy balance components in question are related to the total areas of the latitudinal zones. In most cases, this decreases their values compared with the energy balance components for the ocean surface areas.

As seen from Table 2.6, to determine the outgoing emission from other energy balance components one should take into account all the balance components incorporated in the table.

Table 2.7 presents latitudinal means of the radiation regime of the Earth—atmosphere system obtained in one of the recent generalizations of satellite observational data (Stephens *et al.*, 1981).

TABLE 2.7

Components of the radiation regime of the Earth—atmosphere system (W m$^{-2}$)

| Latitude | June | | | December | | | Year | | |
|---|---|---|---|---|---|---|---|---|---|
| | $\alpha_s$ | $I_s$ | $R_s$ | $\alpha_s$ | $I_s$ | $R_s$ | $\alpha_s$ | $I_s$ | $R_s$ |
| 90—80°N | 0.75 | 207 | −77 | — | 177 | −177 | 0.67 | 177 | −118 |
| 80—70 | 0.54 | 211 | 19 | — | 156 | −156 | 0.57 | 179 | −97 |
| 70—60 | 0.42 | 221 | 60 | 0.66 | 167 | −164 | 0.46 | 191 | −72 |
| 60—50 | 0.39 | 222 | 73 | 0.55 | 182 | −156 | 0.41 | 201 | −46 |
| 50—40 | 0.35 | 235 | 81 | 0.46 | 198 | −129 | 0.36 | 217 | −18 |
| 40—30 | 0.28 | 254 | 92 | 0.37 | 222 | −97 | 0.31 | 239 | 6 |
| 30—20 | 0.25 | 268 | 86 | 0.30 | 251 | −63 | 0.26 | 258 | 25 |
| 20—10 | 0.24 | 251 | 90 | 0.26 | 259 | −13 | 0.24 | 254 | 55 |
| 10—0 | 0.26 | 233 | 73 | 0.26 | 250 | 38 | 0.25 | 241 | 72 |
| 0—10°S | 0.22 | 261 | 24 | 0.25 | 246 | 85 | 0.23 | 251 | 70 |
| 10—20 | 0.21 | 273 | −28 | 0.24 | 253 | 108 | 0.23 | 262 | 52 |
| 20—30 | 0.24 | 260 | −70 | 0.23 | 262 | 124 | 0.24 | 259 | 33 |
| 30—40 | 0.30 | 229 | −100 | 0.27 | 249 | 126 | 0.29 | 239 | 12 |
| 40—50 | 0.39 | 209 | −137 | 0.33 | 229 | 116 | 0.35 | 218 | −16 |
| 50—60 | 0.47 | 195 | −166 | 0.42 | 216 | 83 | 0.42 | 203 | −51 |
| 60—70 | 0.72 | 172 | −169 | 0.49 | 209 | 52 | 0.51 | 185 | −77 |
| 70—80 | — | 146 | −146 | 0.64 | 200 | −7 | 0.64 | 159 | −92 |
| 80—90 | — | 110 | −110 | 0.75 | 190 | −54 | 0.70 | 135 | −83 |

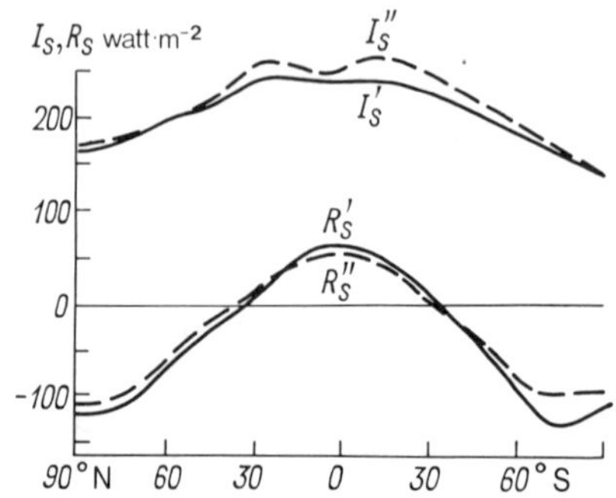


Fig. 2.11.

Figure 2.11 compares latitudinal means of the annual values of the Earth—atmosphere radiation balance $R_s'$ obtained from the *Atlas of the Heat Balance of the Earth* (1963) with $R_s''$ obtained from the data of Ellis and Vonder Haar (1976), as well as corresponding values of outgoing longwave emission ($I_s'$ and $I_s''$). As seen in this figure, the calculated values of the radiation balance and outgoing radiation differ somewhat from those obtained directly from satellite data due to errors in determining these values by each of these two methods.

However, the differences between the results of climatological calculations (Fig. 2.11, solid lines) and satellite observations (dashed lines) in many cases are so small that to speak of the advantage of one method over the other is impossible. For example, the differences at all latitudes, except the Antarctic, are less than the probable error of either method. Data on outgoing emission coincide completely in high and middle latitudes of the Northern Hemisphere, where more complete data from standard meteorological observations are available for the climatological calculations.

The differences in outgoing radiation in low and middle latitudes of the Southern Hemisphere seem to indicate that satellite data are more accurate in that region. Here the results of standard meteorological observations are insufficient for reliable estimations to be made. But even in this case, the relative discrepancy between these two kinds of data amounts to no more than 10%. Since the probable error in determining the outgoing longwave flux by either method amounts to not less than several per cent, whether or not a systematic discrepancy exists due to greater reliability of one of the methods of determining $I_s$, is not clear.

Using the available materials on the energy balance components, we can construct a diagram of the energy balance of the Earth as a whole. Considering the solar constant as 1368 W m$^{-2}$ and the Earth's albedo close to 0.30, we find that the shortwave radiation absorbed by the planet amounts to 237 W m$^{-2}$, i.e. $1368(1-0.307)/4$, the factor 4 arising from the relative areas of the circular cross-section of the Earth to the total surface area of the planet; this term is $Q_{sa}$ in Fig. 2.12. Of this quantity, 157 W m$^{-2}$ are absorbed at the Earth's surface ($Q_a$) and 80 W m$^{-2}$ are absorbed in the atmosphere. The radiation balance of the Earth's surface is equal to 105 W m$^{-2}$ and the net outgoing emission from the surface, corresponding to the difference between the absorbed radiation (157 W m$^{-2}$) and the radiation balance (105 W m$^{-2}$), is 52 W m$^{-2}$ ($I$). The total value of the net longwave emission from the Earth, equal to the amount of absorbed shortwave radiation, is designated by $I_s$. The ratio $I/I_s$ is considerably less than the ratio $Q_a/Q_{sa}$, which characterizes the influence of the greenhouse effect on the radiation balance of the Earth.

Another characteristic of the greenhouse effect is the value of the radiation balance of the Earth's surface (105 W m$^{-2}$). This radiation balance energy is used up in the evaporation of water (88 W m$^{-2}$, shown as $LE$ in Fig. 2.12) and in sensible heat exchange between the Earth's surface and atmosphere (17 W m$^{-2}$, arrow $P$).

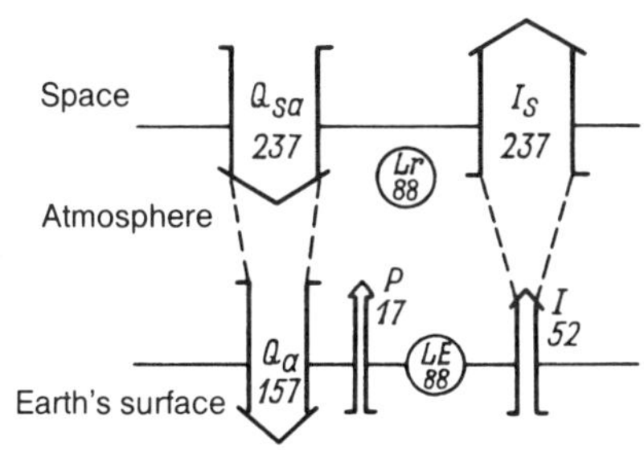


Fig. 2.12.

From Fig. 2.12 and the above figures it can be seen that the atmosphere gains heat energy from three sources:

(1) absorbed shortwave radiation (80 W m$^{-2}$; $Q_{sa} - Q_a$);
(2) condensation of water vapour (88 W m$^{-2}$, shown by the circle $Lr$);
(3) sensible heat flux from the Earth's surface (17 W m$^{-2}$).

The sum of these values is the output of heat as longwave radiation into space, which equals the difference between $I_s$ and $I$, i.e. 185 W m$^{-2}$.

CHAPTER 3

# CLIMATE AND GEOGRAPHICAL ZONES

## 3.1. Contemporary Climate

Climate has a profound influence on living organisms. The geographical distribution of plants and animals, and the nature and intensity of biological processes depend to a great extent on climatic conditions. Climatic change is an important factor in the evolution of the biosphere.

Let us consider the principal features of modern climate. Climatic conditions over the last century can be determined using instrumental meteorological observations carried out by the world network of climatological stations formed in the second half of the 19th century. These data show that elements of the meteorological regime vary noticeably in time. Alongside their periodic fluctuations (daily and yearly), there are non-periodic variations occurring over different time scales. For short time intervals (of the order of days or months) non-periodic variations in the meteorological regime represent fluctuations in weather; these variations are spatially inhomogeneous and largely explained by the unstable atmospheric circulation.

For longer time intervals (of the order of several years and more), in addition to irregular fluctuations in the elements of the meteorological regime, one often observes long-term changes that are similar over vast areas. It is these changes that characterize fluctuations in climate.

Since contemporary climatic fluctuations are relatively small, in order to describe the climatic features of our epoch it is expedient to make use of average values of meteorological elements over a period of several decades. This averaging allows us to eliminate the effects of short time scale instability of the atmospheric circulation. We give here some brief information on contemporary climate with emphasis on two meteorological elements: the air temperature near the Earth's surface and the sum of precipitation reaching the Earth's surface (total precipitation).

Table 3.1 presents latitudinal mean values of both surface air temperatures for January and July and annual precipitation. As can be seen in the table, the differences between mean surface air temperatures at different latitudes can be up to almost 80 °C, with the maximum temperatures near the equator and the minimum values at the South Pole. In high latitudes, where, for much of the year, the air temperature does not rise above freezing point, there is permanent ice cover. The causes of these latitudinal differences will be considered at length later. Here we mention only that, by producing latitudinal variations in total radiation reaching the top of the atmosphere, it is the spherical form of the Earth which determines, to a great extent, the distribution of these temperatures.

TABLE 3.1

Surface air temperature and precipitation

| | North Latitude | | | | | | | | |
|---|---|---|---|---|---|---|---|---|---|
| | 90–80 | 80–70 | 70–60 | 60–50 | 50–40 | 40–30 | 30–20 | 20–10 | 10–0 |
| January temperature (deg.) | −32 | −27 | −22 | −12 | − 3 | 8 | 17 | 24 | 26 |
| July temperature (deg.) | − 1 | 2 | 10 | 13 | 18 | 23 | 26 | 27 | 26 |
| Precipitation ($cm\ yr^{-1}$) | 19 | 26 | 52 | 80 | 75 | 77 | 73 | 114 | 201 |
| | South Latitude | | | | | | | | |
| | 0–10 | 10–20 | 20–30 | 30–40 | 40–50 | 50–60 | 60–70 | 70–80 | 80–90 |
| January temperature (deg.) | 26 | 26 | 24 | 19 | 12 | 5 | − 1 | −16 | −26 |
| July temperature (deg.) | 25 | 23 | 18 | 14 | 8 | 1 | −11 | −40 | −53 |
| Precipitation ($cm\ yr^{-1}$) | 150 | 122 | 91 | 103 | 108 | 101 | 67 | 25 | 11 |

In addition to a considerable change in the mean surface air temperature in the meridional direction, there are also noticeable temperature variations at different longitudes in most latitudinal zones. These are largely accounted for by the position of the continents and oceans.

The thermal regime of the ocean affects a substantial part of the continents, where, in middle and high latitudes, so-called maritime climates are observed with a relatively small annual amplitude of the air temperature. In extratropical continental regions, where the thermal influence of the oceans is less pronounced, and annual air temperature amplitudes rise sharply, corresponding to the conditions of a continental climate.

The distribution of latitudinal mean precipitation is characterized by a principal maximum in the equatorial zone, a decrease in high tropical and subtropical latitudes, two secondary maxima in middle latitudes, and a decrease in precipitation with increasing latitudes when moving to the polar regions. These variations are explained by the distribution of mean air temperature and by specific features of the atmospheric circulation.

Other conditions being equal, including the same relative humidity, total precipitation increases with increasing temperature, simply because the amount of water vapour available for the process of condensation also increases. It is evident that, if other factors did not affect precipitation, the distribution of its mean latitudinal values would show only a single maximum in low latitudes. This is not the case because the regime of vertical air currents that carry water vapour through the level of condensation and bring about the formation of clouds and precipitation is of great importance.

The general circulation of the atmosphere is closely connected with the geographical distribution of stable pressure systems, the most important being the low-pressure belt near the equator, the high-pressure regions in high tropical and subtropical latitudes and the areas of frequent cyclonic formations in middle latitudes. Since the high-pressure belt is mostly characterized by downward air motion, precipitation in this zone is decreased. This is manifested by two minima in the latitudinal distribution of precipitation. Strong upward motions of the air in equatorial latitudes and in a number of mid-latitude regions promote a heavier fall of precipitation there.

The largest deserts of the world, where precipitation is negligible, are found in the subtropical high-pressure zone. Total precipitation also decreases in continental regions at middle latitudes situated far off the ocean, because the amount of water vapour brought in by air currents from the ocean is small. This reduces the air's relative humidity and slows down the process of water vapour condensation.

Thus, on the continents, the zones of humid climate are found mainly in equatorial latitudes and in maritime climate regions at middle and high latitudes. High tropical and subtropical latitudes as well as regions of continental climate are characterized mostly by conditions of insufficient moisture.

## 3.2. Polar Ice

In the contemporary epoch, ice surrounds both poles of the globe. In the Northern Hemisphere, permanent ice cover takes up more than two-thirds of the surface of the Arctic Ocean, and also covers Greenland and several other high-latitude islands and continental regions. The total area of Arctic ice is about 12 million km$^2$, sea ice occupying approximately 10 million km$^2$ and land ice 2 million km$^2$.

Arctic sea ice resembles an enormous lens, in the centre of which mean ice thickness reaches 3—4 m, gradually decreasing towards its periphery. As this ice consists of a great many individual ice fields, it is constantly moving under the action of air and sea currents, which leads to fluctuations in the boundaries of the ice cover. As the ice moves, contraction of the ice fields often occurs, resulting in the formation of hummocks — concentrations of ice, the thickness of which far exceeds the mean thickness of the ice cover. Such hummocks, however, amount to a comparatively small part of the total area of the Arctic sea ice.

In summer months, the ice thickness diminishes as a result of thawing, which occurs mainly at the upper surface of the ice. In the cold season, ice becomes thicker as a result of freezing of sea water at the lower surface of the ice. These seasonal variations in the thickness of sea ice are accompanied by variations in the total area of ice cover. Ice area in the Northern Hemisphere reaches its maximum in March (solid line) and its minimum in September (broken line). The mean boundaries of sea ice in these months are given in Fig. 3.1 (from *Atlas of Oceans*, vol. 1 (1974), vol. 2 (1977), and vol. 3 (1980)).

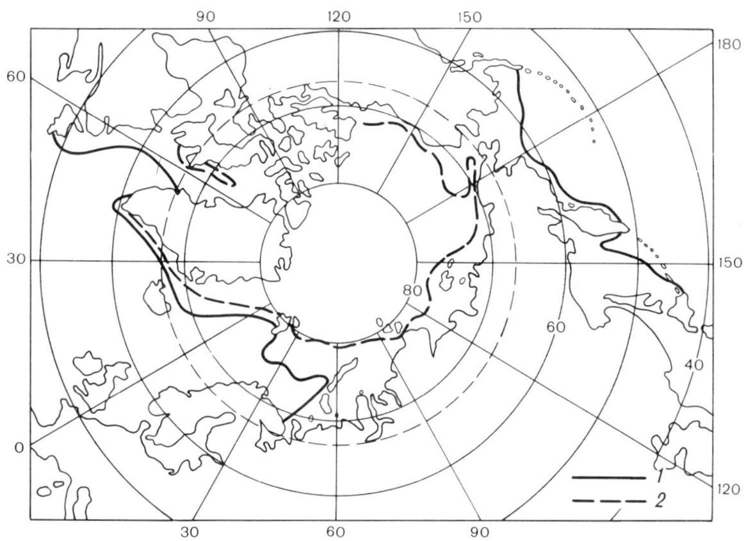


Fig. 3.1.

In the Southern Hemisphere, sea ice is in the form of a broad ring encompassing the Antarctic continent. Its thickness diminishes with distance from the pole. Variations in the sea ice area in the Southern Hemisphere are similar to changes in the area of Arctic ice, although with a time lag of six months.

Figure 3.2 shows a graph of the seasonal variations in the polar sea ice area for the World Ocean (curve 1), the Northern Hemisphere (curve 2) and the Southern Hemisphere (curve 3) constructed by Zakharov and Strokina (1978). As can be seen, because of the larger sea ice area in the Southern Hemisphere, the global extent of sea ice varies in accordance with fluctuations in the area of Antarctic sea ice and in counterphase with variations in the sea ice area in the Northern Hemisphere.

High-latitude ice cover, with its high albedo, has a fundamental influence on the climate, decreasing the air temperature in high latitudes, and leading

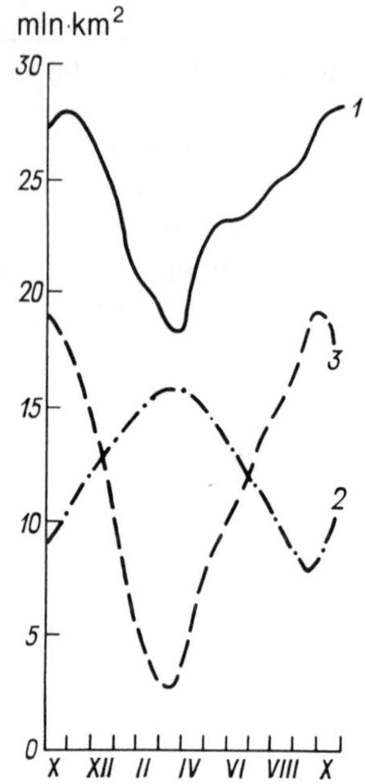


Fig. 3.2.

to an increase in the meridional temperature gradient. Studies of the interrelationship between polar ice and climate are, therefore, of great importance in the study of the evolution of climate.

Data on the thermal regime of the central Arctic are presented in Fig. 3.3. This regime is characterized by the annual march of mean latitudinal air temperature at 80°N ($T_{80}$ in the figure). For comparison, this figure shows also annual variations in the mean latitudinal air temperature near the equator (curve $T_0$). As seen in Fig. 3.3, the air temperature near the Earth's surface in the central Arctic is lower than the air temperature near the

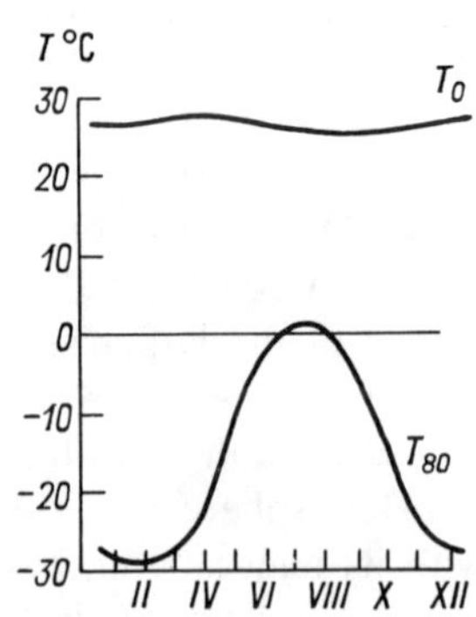


Fig. 3.3.

equator by 26 °C in summer, by 55 °C in winter and by 42 °C for the annual mean.

It is quite obvious that the low air temperatures in the polar zone are the main cause of the formation of polar ice. Less clear is the question of polar ice effects on the thermal regime in high latitudes. The idea that the ice cover is a substantial factor influencing the thermal regime was set forth in the works of Brooks (1950), who pointed out that the high albedo of ice cover significantly diminishes the amount of solar radiation absorbed by a surface covered by snow and ice. He suggested that the ice cover in high latitudes lowers the Arctic temperature by several tens of degrees. This suggestion, however, could not be substantiated by calculations based on the methods of physical climatology, which were only poorly developed at the time when Brooks was carrying out his research. For this reason, Brooks' views on the genesis of the Arctic climate were not generally accepted, and, until recently, the influence of polar ice on the climate of high latitudes was not taken into consideration in works on the theory of climate. However, ice cover *is* an important factor in the formation of climate, as can be shown using the following simple considerations (Budyko, 1969).

Let us consider the energy balance of the Earth—atmosphere system (i.e. of the vertical column embracing the atmosphere, the hydrosphere and the upper layers of the lithosphere). The radiation balance of this system, $R_s$, is equal to the difference between the absorbed solar radiation, $Q_s(1 - \alpha_s)$, and the longwave radiation into outer space, $I_s$:

$$R_s = Q_s(1 - \alpha_s) - I_s. \tag{3.1}$$

The radiation balance must also be equal to the sum of the change in heat content of matter in the indicated column ($B_s$), and the heat gained as a result of horizontal motions in the atmosphere and hydrosphere ($C$): hence

$$Q_s(1 - \alpha_s) - I_s = B_s + C. \tag{3.2}$$

Assuming a solar constant equal to 1340 W m$^{-2}$,[1] we find that, in June at 80°N latitude, 494 W m$^{-2}$ of solar radiation energy, and at the equator 374 W m$^{-2}$, reach the upper boundary of the denser layers of the atmosphere.

At the equator as well as in the central Arctic, changes in the heat content of the Earth—atmosphere system in June are small in comparison with the income of solar radiation. The cause of this is the fact that this component of the energy balance reaches appreciable values only with substantial changes in ocean temperature in the annual cycle. In addition, the temperature of ocean water changes little throughout the year both near the equator (because of the lack of significant annual variation in the mean elevation of the Sun) and in the central region of the Arctic Ocean (because of the isolating influence of sea ice on the heat exchange between the ocean and the atmosphere). Thus $B_s$ in Equation (3.2) must be relatively small.

[1] This is the conditional solar constant, which corresponds to the flux of shortwave radiation at the outer boundary of the troposphere.

Taking this conclusion into account and considering the fact that the horizontal flux of heat in the atmosphere and hydrosphere is directed from equator to pole, we can easily see that, if the albedo at all latitudes were the same, the longwave radiation into outer space at 80°N latitude in June would have to be much greater than it is near the equator.

Now, modern investigations have shown (see Section 3.3) that radiation emission into space, which is closely connected with mean temperature near the Earth's surface, increases with increasing temperature. Thus, if the albedo of the Earth—atmosphere system were the same at the equator and in high latitudes, the air temperature near the Earth's surface in June in high latitudes would have to be higher than the temperature near the equator. This is contrary to what is actually observed.

The only explanation of this paradox lies in the significant difference in the albedo of the Earth—atmosphere system with and without ice cover. This difference was determined with the help of material from direct experimental observations in the works of Raschke *et al.* (1968) and other authors, who constructed maps of albedo using satellite data. From these works, the mean value of albedo at 80°N turns out to be 0.62, while the mean albedo for regions free of ice cover is about 0.30. Taking these albedo values into account, we find that in June the absorbed radiation is 260 W m$^{-2}$ at the equator and 190 W m$^{-2}$ at 80°N. The air temperature at the equator in summer is, therefore, higher than it is at the pole because the albedo of the Earth—atmosphere system in the central Arctic is much greater than that at lower latitudes.

One might ask the question: how would the thermal regime of the Arctic have changed without polar ice, if the albedo in high latitudes did not differ much from that of low latitudes? It should be pointed out that the thermal regime of the Arctic in this case ought to depend, not only on the quantity of radiation absorbed (which is determined by the new value of albedo), but also on other factors connected with the disappearance of the ice cover. In particular, the meridional heat exchange in the atmosphere and hydrosphere would change significantly if there were no Arctic ice. The water temperature in an ice-free Arctic Ocean would change noticeably during the annual cycle, and this in turn would lead to a substantial redistribution of heat in the Earth—atmosphere system throughout the year.

To answer this question more fully we need to determine the components of the energy balance of the Arctic with and without ice, and to estimate the effect of changes in these components on the thermal regime. In this way, the importance of the ice cover as a factor in the genesis of the Arctic climate can be elucidated.

Let us consider the energy balance of the central Arctic (assumed to be represented by the mean conditions for 80°N). Figure 3.4 presents values of the elements of the radiation regime at the ice surface in this region in different months of the year. Curve $Q$ defines variations in the total radia-

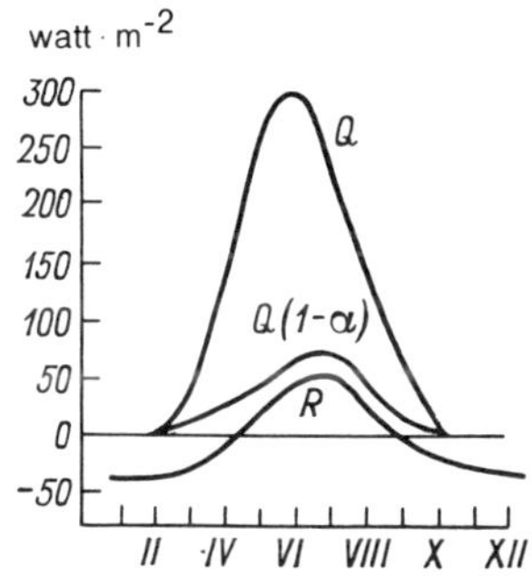


Fig. 3.4.

tion, $Q(1-\alpha)$ absorbed radiation, and $R$ the radiation balance. It can be seen that, in summer months, a considerable amount of solar radiation reaches the ice surface. This is explained by the long duration of the polar day and also by downward scattering of a fraction of the radiation reflected from the ice and snow surface.

The amount of absorbed radiation $Q(1-\alpha)$ depends on the albedo value $\alpha$. According to the available observational data (from drifting polar stations and from aircraft), we may consider the albedo of the central Arctic in spring and autumn to be about 0.80; in summer it decreases to about 0.70 in the period of the most active melting. These high surface albedos, which far exceed albedos in the absence of ice and snow, cause a substantial reduction in absorbed radiation compared with the total radiation.

The radiation balance $R$, equal to the difference between the absorbed shortwave radiation and the net longwave emission, has a small positive value in summer months. At this time, it is noticeably less than the radiation balance observed in lower latitudes, where ice cover is absent.

In winter, with solar radiation absent, the radiation balance is determined by the value of the net longwave radiation and is negative. The annual sum of the radiation balance in this region turns out to be negative.

This result confirms the earlier conclusion (Budyko, 1956) that the radiation regime of polar regions with permanent ice cover differs radically from that of the regions without ice cover. One of the chief peculiarities of this regime is that the annual sums of the radiation balance are small in absolute magnitude and often negative, as opposed to comparatively large positive values in all other regions of the globe.

The quantity of heat brought in to this zone by currents is relatively small. For instance, according to the data of Leonov (1947), its annual amount is approximately 3 W m$^{-2}$; Bedgley (1966) considers it to be about 2 W m$^{-2}$. To this value, one should add the latent heat of ice melting which is carried away from the Arctic every year by currents. Antonov (1968) points out that estimates of this ice flux made by different authors vary widely, from

900 to 3000 $km^3 yr^{-1}$. The most probable value is close to 1900 $km^3 yr^{-1}$, corresponding to an additional income of heat to the central Arctic that is about 2 $W m^{-2}$. Thus, owing to the oceanic circulation, the central Arctic receives on average 4—5 $W m^{-2} yr^{-1}$. Almost all of this is off-set by the negative value of the radiation balance of the ice surface.

As a result, the sum of the remaining components of the energy balance of the ocean, including the sensible heat flux and the conversion of heat for evaporation, must be close to zero. This result can be compared with the conclusion from the calculation of the components of the energy balance of ice fields. According to Doronin's data (1963), the annual average sum of the heat consumed for evaporation on the ice fields amounts to about 4 $W m^{-2}$, and the income of heat owing to turbulent sensible heat exchange is approximately 3.5 $W m^{-2} yr^{-1}$. The algebraic sum of these values is close to zero, confirming the estimate given above based on the heat budget for the central Arctic as a whole. In particular, as a consequence of these calculations, heat exchange at the surface of ice leads (i.e. open water regions) does not play a great role in the energy balance of the central regions of the Arctic Ocean. It follows that, in the cold season, ice leads probably account for only a very small part of the total area of the central Arctic.

The annual march of the components of the energy balance of an ice field is given in Fig. 3.5. This figure presents, along with values of the radiation balance $R$, the turbulent heat flux $P$ and the amount of heat consumed for evaporation $LE$, the latter values based on Doronin's data referred to above. The values of the radiation balance in this figure are considered positive when they express an income of heat, and the values of turbulent heat flux and evaporative heat consumption are considered positive when they represent heat expenditure at surface of the ice.

The sum of the annual values of the given components of the energy balance corresponds to a heat loss. This is compensated for by the inflow of heat through the lower boundary of the ice, which in turn is equal to the

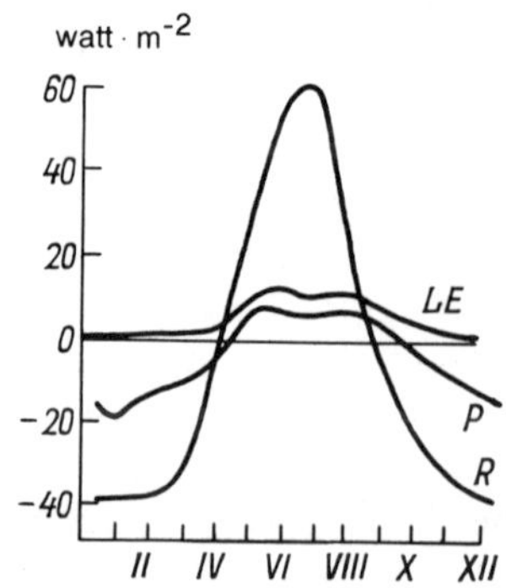


Fig. 3.5.

income of heat resulting from ocean current activity (without taking into account removal of ice).

The formation and destruction of ice on the ocean surface is closely connected with the conditions of the energy balance. Ice cover appears when, as a result of cooling, the temperature of the upper layer of water reaches the freezing point. For average salinity conditions, this temperature is −1.8 °C. The thickness of ice increases when the temperature of its upper surface is lower than the water temperature at the lower surface of the ice. Under these conditions, freezing occurs at the lower surface of the ice and its rate is determined by the energy balance of this surface.

If the flux of heat from the ocean to the lower surface of the ice is small, then the rate of freezing is equal to the value of the heat flux from the lower ice surface to its upper surface divided by the latent heat of melting. This vertical heat flux in the ice depends on the temperature difference between the upper and lower surfaces of the ice, its heat conductivity and thickness (together with the snow cover on it). The temperature at the surface of the ice (or at the surface of snow, if there is any) depends on the conditions of the energy balance. For small values of the vertical heat flux between ice and atmosphere, this temperature is usually close to that of the lower air layer. As the temperature of the lower ice surface is usually constant, and the thermal conductivity of ice varies comparatively little, the rate of freezing turns out to depend largely on two factors, the air temperature and the ice thickness. In these circumstances, the rate of freezing increases with a fall in air temperature and decreases with a growth in ice thickness.

The dependence of the thickness of ice cover on the factors mentioned above was established nearly a century ago by Weinprecht on the basis of observations made in the Arctic. This problem was then theoretically investigated by Stefan, who used the equation of heat conduction in solids for this purpose. Stefan found out that, at a constant difference ($\Delta T$) between the temperatures at the upper and lower surfaces of ice, the ice thickness is proportional to the square root of the product of $\Delta T$ and the length of time of ice accretion. Later, different empirical formulas were offered for calculating the rates of freezing. These allowed more accurate determinations of ice thickness than the schematic solution of Stefan.

One such formula, suggested by Zubov (1945), was used to construct a table showing the relationship between the rate of ice accretion and the initial thickness of ice and the mean daily temperature of the air (Table 3.2).

The rate of ice accretion is influenced, not only by air temperature, but also by circulation conditions in the ocean. An additional heat flux to the lower surface of the ice, associated with horizontal heat transport from low to high latitudes, can substantially retard ice accretion. The rate of growth also depends on the thickness of snow cover on the ice surface, because the low heat conductivity of snow decreases the heat flux between the lower surface of the ice and the atmosphere. The accuracy of empirical formulas

TABLE 3.2

Ice growth (cm $day^{-1}$)

| Initial thickness (cm) | Temperature (°C) | | | | | | | |
|---|---|---|---|---|---|---|---|---|
| | −5 | −10 | −15 | −20 | −25 | −30 | −35 | −40 |
| 0 | 0.8 | 1.6 | 2.4 | 3.2 | 3.8 | 4.7 | 5.5 | 6.3 |
| 10 | 0.6 | 1.1 | 1.7 | 2.3 | 2.9 | 3.4 | 4.0 | 4.6 |
| 20 | 0.4 | 0.9 | 1.3 | 1.8 | 2.2 | 2.6 | 3.1 | 3.5 |
| 30 | 0.4 | 0.7 | 1.1 | 1.5 | 1.8 | 2.2 | 2.6 | 3.0 |
| 40 | 0.3 | 0.6 | 0.9 | 1.2 | 1.5 | 1.8 | 2.1 | 2.4 |
| 50 | 0.3 | 0.5 | 0.8 | 1.1 | 1.3 | 1.6 | 1.9 | 2.1 |
| 75 | 0.2 | 0.4 | 0.6 | 0.8 | 1.2 | 1.4 | 1.6 | 1.8 |
| 100 | 0.2 | 0.3 | 0.5 | 0.6 | 0.8 | 1.0 | 1.1 | 1.0 |
| 150 | 0.1 | 0.2 | 0.3 | 0.5 | 0.6 | 0.7 | 0.8 | 0.9 |
| 200 | 0.1 | 0.2 | 0.3 | 0.4 | 0.4 | 0.5 | 0.6 | 0.7 |

for calculating the freezing rate is limited by the fact that they usually take into account only the mean effect of the factors mentioned.

In the mass balance of polar sea ice, increases of the mass of water in its solid state are determined by the freezing process and, to a lesser extent, by the fall of solid precipitation. In this mass balance, the loss of mass depends on the melting process, on evaporation from the ice surface, and on the outflow of ice into lower latitudes. Melting of ice generally occurs on its surface, when the temperature of this surface reaches the freezing point. Because the salinity of the upper layers of ice decreases during the process of melting, the temperature at which melting takes place is approximately zero.

The rate of ice melting is determined by energy balance conditions at the ice surface (Budyko, 1962b). The equation of this balance can be expressed in the form:

$$R = LE + P + A + lh_{\rho_1}, \tag{3.3}$$

where $R$ is the radiation balance, $L$ the latent heat of vaporization, $E$ evaporation from the ice surface, $P$ the turbulent flux of sensible heat between the ice surface and the atmosphere, $A$ the heat flux from the ice surface to the lower layers of the ice, $l$ the latent heat of melting of ice, $h$ the change in ice thickness due to melting and $\rho_1$ the ice density.

The dependence of the first three components of the energy balance on meteorological factors can be presented in the following form:

$$R = Q(1 - \alpha) - I_0 - 4\delta\sigma T^3(T_w - T), \tag{3.4}$$

$$LE = L\rho\varkappa u(q_s - q), \tag{3.5}$$

$$P = \rho c_P \varkappa u(T_w - T), \tag{3.6}$$

where $Q$ is the total shortwave radiation, $\alpha$ the albedo of ice, $I_0$ the net longwave radiation (a function of the air temperature), $\delta$ the coefficient describing the difference between the properties of the radiating surface and those of a black body, $\sigma$ the Stefan constant, $T$ the air temperature, $T_w$ the ice surface temperature, $\varkappa$ the coefficient of heat exchange, $\rho$ the air density, $u$ the wind speed, $q_s$ the specific humidity of saturated air at temperature $T_w$; $q$ the specific humidity of the air, and $c_p$ the heat capacity of the air at constant pressure.

From Equation (3.3)–(3.6), we obtain

$$h = \frac{1}{\rho_1 l}[Q(1-\alpha) - I_0 - L\rho\varkappa u(q_s - q) - \\ - (\rho\varkappa c_p u + 4\delta\sigma T^3)(T_w - T) - A]. \tag{3.7}$$

In applying this formula to calculate ice melting, one may use some simplifications. Since the relative humidity of the air above the ice varies only slightly on average, and the difference between temperatures of the ice surface and air is usually small, we may, without material damage to the accuracy of the calculation, consider that $L\rho\varkappa u(q_s - q) = L\rho\varkappa u q_s \eta$, where the coefficient $\eta$ is about 0.1. Considering the value of $A$ in the melting period to be relatively small, it is possible to employ an approximate estimate of it in accordance with the change in the mean temperature of the ice, taking into account the values of its effective heat capacity.

Using Equation (3.7), it is possible to evaluate which factors have the greatest influence on ice melting. Such calculations show that the rate of melting depends primarily on total radiation $Q$ and the air temperature $T$. It follows from these calculations that the total amount of melting in the 'central Arctic must be several tens of centimetres per year. This value, which agrees well with observational data, is markedly greater than the loss of ice mass by evaporation, which in this region, according to the available data, is only a few centimetres a year. Compared with evaporation, the transport of ice into lower latitudes has a more important influence on the ice balance in the central Arctic. As noted above, approximately 2000 km$^3$ of ice per year is carried by ocean currents away from high latitudes, equivalent to an annual decrease in thickness of the ice cover of about 20 cm.

Using the above methods, we can calculate, for any region, the annual march of ice thickness corresponding to the climatic conditions of the region. Let us assume in this case that changes in ice thickness are mainly determined by freezing and melting processes, and use the method of successive approximations (Budyko, 1962b). For the beginning of the calculation period an arbitrary ice thickness is assumed, and then, for every subsequent month, the rate of freezing (or melting) is calculated in accordance with the mean climatic conditions. This process is continued for a number of successive years until the total annual values of melting and freezing are equal. The ice thickness obtained in this way for each month will correspond to the climatic regime of the given region.

This calculation method gives satisfactory results in determining the *changes* in ice thickness over the annual cycle. However, the absolute values of ice thickness computed by this method do not always correspond to observations and, particularly for the central Arctic, often prove to be larger than the observed values of the ice thickness. The cause of this discrepancy lies in the effect of drift of ice fields on the ice balance in high latitudes. As a result of this effect, ice is carried out of the central Arctic area until it reaches a region where its thickness corresponds to the prevailing climatic conditions.

Zubenok (1963) used this method to calculate the mean boundaries of stable polar ice in different months of the year. The result of this calculation were found to agree well with the observational data.

This method has been also used for studying the effect of anomalies of meteorological elements on the thickness and area of ice cover. In this way, changes of ice thickness resulting from various air temperature anomalies have been calculated (Budyko, 1966). Figure 3.6 presents the results of just such a calculation for ice in the central Arctic with a positive temperature anomaly in the three summer months (June—August) equal to 2 °C (Fig. 3.6(a)), and with a positive anomaly in the six cold months (November—April) equal to 10 °C (Fig. 3.6(b)). Curves 1 and 2 show the changes in the maximum and minimum values of ice thickness in the annual cycle, respectively. Comparing Figs 3.6(a) and (b) shows that a comparatively small anomaly in summer temperature reduces ice thickness much more than the far greater temperature anomaly in the cold months.

A third anomaly scenario is shown in Fig. 3.6(c), the effect of a positive anomaly of summer temperature equal to 4 °C. In this case ice that is 4 m thick under mean climatic conditions for the central Arctic disappears completely in four years. Thus, for an increase of 4 °C in the mean summer temperature in the Arctic, the multi-year ice that covers most of the Arctic Ocean turns into one-year ice.

It is clear, however, that this evaluation of the effect of a given temperature anomaly could underestimate the amount of ice loss. With a reduction in

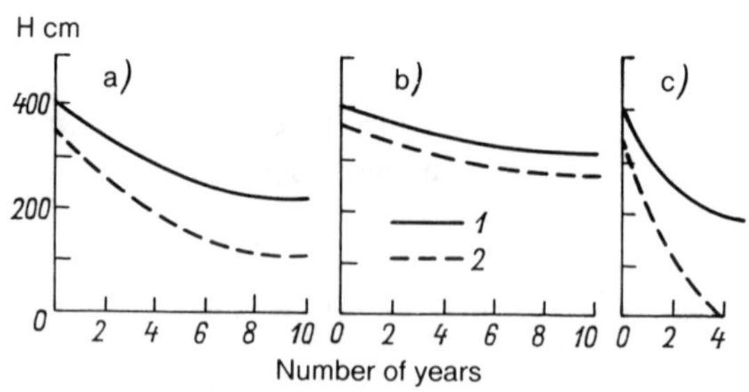


Fig. 3.6.

ice area, the amount of radiation absorbed at the open ocean surface would increase considerably. Partial destruction of the ice would therefore lead to a further rise in temperature, which in turn would intensify the melting of the ice. Since this positive feedback effect is not easily incorporated into the simple analysis given above, in order to clarify what anomaly of the meteorological regime would be needed for complete destruction of multi-year ice, it is necessary to apply more complex climatic models.

## 3.3. The Theory of Climate

The purpose of the theory of climate is to determine, using rigorous physical principles, the mean distribution of meteorological elements in space and time and their responses to forcing by external factors.

One of the first numerical climate models was developed by Milankovich (1930), who calculated the mean latitudinal distribution of air temperature using data on solar radiation at the outer boundary of the atmosphere. Milankovich assumed that the thermal regime was determined solely by radiative heat exchange in each individual latitudinal belt. He did not take account of the atmospheric greenhouse effect, nor of albedo variations with latitude. Milankovich also neglected the influences of meridional heat transfer in the atmosphere, heat exchange attributable to changes in the water state, and interactions between the atmosphere and the ocean.

The neglect of these factors produced quite large errors in the results, but these were compensated for by selecting particular values of the model parameters. As a result, Milankovich obtained mean latitudinal air temperature profiles that were quite close to observed values. Such an agreement was, however, rather fortuitous.

The work of Kochin (1936) was the first attempt to apply the methods of dynamical meteorology to the development of climatic theory. Kochin calculated the mean meridional profiles of pressure and wind velocity in the lower layers of the atmosphere for a given distribution of air temperature and pressure at the Earth's surface. Later, Dorodnitsin *et al.* (1939), using a similar method, constructed a model of the zonal circulation for summer in the Northern Hemisphere. This line of research was further developed by Blinova (1947) who calculated the mean distribution of temperature, pressure and wind velocity for the Northern Hemisphere.

A great contribution to the theory of climate was made by Shuleikin (1941), who established the significant and often decisive influence on climate of the interactions among physical processes in the ocean and in the atmosphere. Subsequently, many new climate models were developed that allowed the average fields of meteorological elements to be calculated. Advances in computer technology promoted the development of numerical models describing the time-varying fields of meteorological elements. By averaging such results over sufficiently long intervals of time average fields characterizing climatic conditions could be estimated. Numerous investigations in this line were carried out by Smagorinsky, Manabe, Mintz and other authors (Smagorinsky, 1963; Smagorinsky *et al.*, 1965; Manabe and Bryan, 1969; Holloway and Manabe, 1971; Mintz, 1965).

The work of Manabe and Bryan presented a numerical climatic model incorporating the effects on climate of the ocean circulation. Holloway and Manabe constructed, from theory, global maps of the basic components of the energy and water balances of the Earth's surface. Their results were similar to the maps based on empirical methods.

Climate theory can be used to estimate climatic variations produced by both natural and anthropogenic factors. In order to calculate climatic changes using numerical models, these models should satisfy several requirements. These requirements are more stringent than those for models used in investigating the steady-state climate regime. The first is that the model should not include empirical data on the distribution of individual climate elements, particularly those that vary considerably in the process of climatic change. Second, the model should consider realistically all types of heat flux that may noticeably influence the temperature field. Third, the model should incorporate the principal feedbacks among the various climatic elements. This third requirement will now be treated more comprehensively.

Of great importance for climatic change are positive feedbacks which amplify any increase in the anomalies of meteorological elements and which, as a result, decrease the climate stability. The dependence of absolute humidity on air temperature is one such positive feedback. With an increase in temperature, the rate of evaporation from open water or from humid surfaces usually increases, leading to a comparatively constant relative humidity of the air in the majority of climatic regions (except for dry continental regions). Warming, therefore, tends to cause an increase in absolute humidity, a fact which is confirmed by numerous empirical data. Since longwave radiation decreases as the absolute humidity increases, this effect partly compensates for the increase of longwave radiation caused directly by rising temperature.

This relationship has been examined by Manabe and Wetherald (1967). They found that the effect of variations in the solar constant on the air temperature near the Earth's surface is almost twice as much for the case of constant relative humidity as for the case of constant absolute humidity. It is evident that numerical models of the Earth's thermal regime used to study climatic change must always incorporate this feedback.

Another positive feedback, the result of the influence of snow and ice cover on the albedo of the Earth's surface, is of equal importance.

Expedition data from the Arctic and Antarctic, together with satellite observations, allow us to evaluate the albedos of the Earth's surface and of the Earth–atmosphere system in high latitudes and to compare these values with albedos in regions free from snow and ice. The data show that, in summer months, the surface albedo of the ice in the central Arctic is about 0.70 and in the Antarctic it is about 0.80—0.85. Since the average albedo of the Earth's surface in regions free from snow and ice does not exceed 0.15, it might be concluded that, other conditions being equal, snow and ice cover reduces the amount of shortwave solar radiation absorbed by the surface by several times.

Snow and ice also have a significant influence on the albedo of the Earth—atmosphere system. Satellite data (Raschke *et al.*, 1973) show that, in summer, the albedo of the Earth—atmosphere system in the central Arctic is about 0.55 and in the Antarctic, about 0.60. These values are approximately twice the albedo of the Earth as a whole obtained as 0.28 in the same work. It is evident that such great differences in the values of both the Earth's surface and the Earth—atmosphere albedo must have profound effects on the thermal regime of the atmosphere.

If an air temperature decrease leads to the formation of a greater snow and ice cover on the Earth's surface, the amount of the absorbed radiation must decrease. This will cause a further decrease in the air temperature and yet further increases in the area covered by snow and ice. The opposite result will be obtained for an air temperature increase, if it brings about the melting of snow and ice. The incorporation of this feedback mechanism into a numerical model of the thermal regime of the atmosphere has shown that it strongly affects the distribution of air temperatures near the Earth's surface (Budyko, 1968).

In order to estimate this effect, let us consider a simple example showing the change in global mean air temperature for the case of a cloud-free Earth entirely covered with snow and ice (Budyko, 1962b). Under such conditions, the albedo of the Earth is much greater than it is at present, and this will influence the temperature of the air. The 'effective' temperature of the Earth corresponding to its longwave emission is proportional to $\sqrt[4]{1-\alpha_s}$ (where $\alpha_s$ is the albedo). Therefore, as the albedo's value changes from $\alpha_s'$ to $\alpha_s''$ the absolute magnitude of the 'effective' temperature changes as $\sqrt[4]{(1-\alpha_s'')/(1-\alpha_s')}$. Since the present albedo of the Earth is approximately 0.30 and the albedo of the dry snow cover is about 0.80, we find that for a snow-covered Earth the mean 'effective' temperature must be reduced by approximately 75 °C.

One might think that the reduction of the mean air temperature near the Earth's surface will be even greater than that. At present, the mean temperature of the lower air layers is considerably increased all over the Earth's surface by the greenhouse effect, which is associated with the absorption of longwave radiation by water vapour and carbon dioxide in the atmosphere. At very low temperatures, this effect is of no importance. Furthermore, at such temperatures, dense clouds that might considerably change the radiation fluxes cannot form. Under these conditions, the atmosphere becomes more or less transparent for both longwave and shortwave radiation.

The mean temperature of the Earth's surface with a transparent atmosphere is determined by the simple formula $T = \sqrt[4]{S_0(1-\alpha)/4\sigma}$, where $S_0$ is the solar constant and $\sigma$ is the Stefan constant. It follows from this formula that, at $\alpha = 0.80$, the mean temperature of Earth is 186 K, or −87 °C. Thus, if snow and ice covered the whole surface of the planet even for a short period of time, the global mean temperature would be approximately 100 °C below the present-day value of 15 °C. This estimate shows that the snow cover might exert a drastic effect on the thermal regime.

A number of authors (Budyko, 1961; Rakipova, 1962; Donn and Shaw, 1966; Fletcher, 1966) have attempted to evaluate the effect of polar sea ice on the thermal regime of the Arctic. Using data on the energy balance of the central regions of the Arctic Ocean and approximate results from the semi-empirical theory of climate, it has been established that the polar ice reduces the mean air temperature in the central Arctic by several degrees in summer and by approximately 20 °C in winter. In this connection, the conclusion has been made that in our epoch an ice-free regime can exist in the Arctic Ocean, but this regime would have been very unstable and the ice cover could develop as a result of comparativey slight climatic fluctuations. Therefore, since even a relatively small area of permanent ice cover exerts a considerable effect on the thermal regime of the atmosphere, it should be properly taken account of in studies of climatic change.

As the studies of the general theory of climate cannot always meet the requirements which have to be satisfied by a realistic theory of climatic change, in a number of works attempts have been made to apply semi-empirical models in which large-scale atmospheric processes are schematized to make it possible to meet the necessary requirements. This line of research includes a semi-empirical model of the thermal regime of the atmosphere (Budyko, 1968) developed for examining climatic change and based on the following considerations.

In the absence of an atmosphere, the mean temperature of the Earth's surface will be determined by the conditions of radiation equilibrium, i.e. the longwave emission from the surface will be equal to the absorbed radiation. Let us write down this condition as

$$\delta\sigma T^4 = \tfrac{1}{4} S_0(1-\alpha), \tag{3.8}$$

where $\delta$ is a coefficient showing the difference in the properties of the emissive surface from those of a black body; $\sigma$ the Stefan constant; $T$ the temperature of the surface; $S_0$ the solar constant; and $\alpha$ the mean albedo of the Earth. Since $\delta = 0.98$ and $S_0 = 1368$ W m$^{-2}$, we find from (3.8) that, for $\alpha = 0.30$, the mean temperature of the Earth is 255 K, or −18 °C.

According to observational data, the mean air temperature near the Earth's surface is about 15 °C. Thus, with the same value of the albedo, the atmosphere increases the mean air temperature by approximately 33 °C. This increase in temperature is associated with the greenhouse effect, i.e. with the greater atmospheric transparency for shortwave radiation than for longwave radiation.

It should be emphasized that this estimate is conditional: without the atmosphere, the planetary albedo could not be equal to the assumed present value of 0.30. In contemporary conditions the mean albedo of the Earth's *surface* is 0.14. One might think that, before the atmosphere appeared, the albedo of the Earth's surface was smaller and probably differed little from the albedo of the Moon which is close to 0.07. With this value of the albedo, the mean temperature of the Earth's surface would have been equal to 3 °C.

In order to evaluate the effects of incoming solar radiation and albedo on the mean temperature near the Earth's surface under real conditions, it is necessary to consider how the temperature distribution depends on the longwave emission at the outer boundary of the atmosphere. This kind of dependency can be established using data on the longwave emission obtained either through observations or calculations.

Comparing these data with different meteorological elements, it has been established that the mean monthly values of outgoing radiation depend mainly on the air temperature near the Earth's surface and on the cloudiness.

This dependence has been expressed as an empirical formula:

$$I_s = a + bT - (a_1 + b_1 T)n, \tag{3.9}$$

(where $I_s$ is the outgoing emission in watts per square metre, $T$ the air temperature in degrees Celsius, $n$ the cloudiness measured in fractions of a unit), and the corresponding coefficients are $a = 223$, $b = 2.2$, $a_1 = 47.8$, $b_1 = 1.6$.

To verify formulas relating the outgoing radiation to meteorological factors, we should use the condition of equality between the outgoing radiation for the entire globe and the absorbed radiation:

$$Q_{sp}(1 - \alpha_{sp}) = I_{sp}, \tag{3.10}$$

where $Q_{sp}$, $\alpha_{sp}$ and $I_{sp}$ refer to the planet as a whole.

We note that, in accordance with our considerations concerning the effect of snow and ice cover on the thermal regime, it can be shown that Equations (3.9) and (3.10) have at least two solutions, one of which corresponds to contemporary climatic conditions and the other to the regime of a 'white Earth' totally covered by snow and ice.

Let us assume that, if the temperature of the surface air layer at all latitudes is less than zero, then the Earth will be completely covered with snow and ice. The global albedo of the Earth—atmosphere system in this case corresponds roughly to the albedo of the Antarctic, i.e. 0.6—0.7. It follows from (3.9) and (3.10) that the mean planetary temperature near the Earth's surface will vary from −47 °C (at lower albedo) to −70 °C (at higher albedo). At these values of the mean global temperature, complete glaciation of the Earth is inevitable. This corroborates the view expressed earlier that the existing climate-forming factors correspond, not only to present climatic conditions, but also to a climate of complete global glaciation.

From (3.9) and (3.10) we find the formula for the mean temperature of the Earth:

$$T_p = \frac{1}{b - b_1 n}\,[Q_{sp}(1 - \alpha_{sp}) - a + a_1 n]. \tag{3.11}$$

This equation allows us to draw some conclusions concerning the influence of climate-forming factors on the mean temperature near the Earth's surface.

A 1% change in the solar radiation alters the mean temperature, with cloudiness equal to 0.50 and the present value of the Earth's albedo, by approximately 1.5 °C. This estimate is twice as much as that of the same effect without the atmosphere. Thus, the radiative properties of the atmosphere considerably increase the effect of changes in radiation on the thermal regime of the Earth's surface.

A 0.01 change in the albedo modifies the mean temperature by 2.3 °C. It is quite evident that the thermal regime greatly depends on variations in albedo provided these variations are not produced by changes in cloudiness. As shown in earlier studies, the influence of cloudiness on the thermal regime associated with changes in the albedo is to a great extent compensated for by corresponding variations in the outgoing longwave radiation.

In individual latitudinal zones, the thermal regime is substantially dependent on horizontal heat redistribution in the atmosphere and hydrosphere. To take a quantitative account of the effects of horizontal heat redistribution on the atmospheric thermal regime we can use data on the components of the energy balance of the Earth—atmosphere system. The energy-balance equation of the system has the following form:

$$Q_s(1 - \alpha_s) - I_s = C + B_s, \tag{3.12}$$

where $C$ is equal to $F_s + L(E - r)$, i.e. to the sum of heat inflows caused by horizontal motions in the atmosphere and hydrosphere.

For mean annual conditions, the term $B_s$ (which denotes the accumulation or loss of heat over the given time period) is equal to zero, and $C$, the heat inflow, corresponds to the value of the radiation balance of the Earth—atmosphere system. Since the values of this balance can be found either from observational data or by calculations, it is evident that one can calculate the magnitudes of the horizontal heat redistribution at the same time.

It might be supposed that the values of $C$ are related in some way to the horizontal distribution of the mean tropospheric temperature. Taking into account the fact that tropospheric air temperature deviations from mean temperature vertical distributions are small compared to the horizontal variability of temperature, it may be concluded that the mean air temperature in the troposphere is closely connected with that at the surface of the Earth. Thus, there are grounds for assuming that a relationship exists between horizontal heat transfer and temperature distribution near the Earth's surface.

Let us examine this relationship for the mean annual conditions of the latitudinal zones of the Northern Hemisphere. For this purpose, we calculate the radiation balance using the formula

$$R_s = Q_s(1 - \alpha_s) - I_s. \tag{3.13}$$

Having thus obtained the values of the radiation balance of the Earth–atmosphere system, since these are equal of $C$ it is possible to calculate the meridional heat flux.

Since meridional heat transfer is effected in the form of heat transfer from warmer to colder regions, we can assume that $C$ depends on $T - T_p$, where $T$ is the mean temperature at a given latitude and $T_p$ is the mean planetary temperature of the lower air layer. To study this relationship, we have compared the values of $C$ to corresponding values of $T - T_p$; it has been established that there is a distinct relationship between them, which can be written as an empirical formula:

$$Q_s(1 - \alpha_s) - I_s = \beta(T - T_p), \tag{3.14}$$

where $\beta = 3.74$ W m$^{-2}$ K$^{-1}$.

The existence of such a relationship greatly simplifies the consideration of meridional heat redistribution in the model of the thermal regime. Using (3.9) and (3.14) we find that

$$T = \frac{Q_s(1 - \alpha_s) - a + a_1 n + \beta T_p}{\beta + b - b_1 n}. \tag{3.15}$$

With the help of this formula, we can calculate mean annual temperatures at different latitudes.

It follows also from the same equation that, with the present values of heat inflow to the outer boundary of the atmosphere, a stable climate with complete glaciation of our planet ('white Earth'), with very low temperatures at all latitudes, is quite possible. The stability of this regime is explained by the very high albedo of the surface covered with snow and ice. In such a case, climatic conditions, which are now typical of the Antarctic, could exist over the whole surface of the Earth.

The conclusion that a stable 'white Earth' regime is possible follows from any realistic theory of climate. To provide more accurate calculations of the thermal régime of the atmosphere it is necessary to study in more detail the parameters of the climatic model presented above.

In a previous study (Budyko, 1968) two values of the albedo were used in the calculation: 0.32 for an ice-free region and 0.62 for the Arctic polar ice region. The albedo of the polar ice zone was assumed to be constant if the area of the polar ice changed. While considering this hypothesis, it was established that the albedo of the Earth—atmosphere system is a function of the angle of incidence of solar rays: it decreases with increasing angle. This dependence, however, is relatively weak when albedo is high, which is typical of the ice-covered regions. At the same time with an increase in the area of the ice-covered region, the albedo tends to increase due to the change in climatic conditions in the ice zone. These factors were expected to more or less compensate for each other. Therefore, we might restrict ourselves to taking into account the average albedo for the ice-covered zone, irrespective of its extent (Budyko, 1969).

The actual values of the albedo in different latitudinal zones, obtained by Ellis and Vonder Haar (1976) from satellite observations, are shown in Fig. 3.7, where curves 1 and 2 represent albedos for the Northern and Southern Hemispheres, respectively. The $\Delta\varphi_N$ and $\Delta\varphi_s$ latitude belts correspond to the transitional zones from ice-free to completely ice-covered regions in each of the hemispheres.

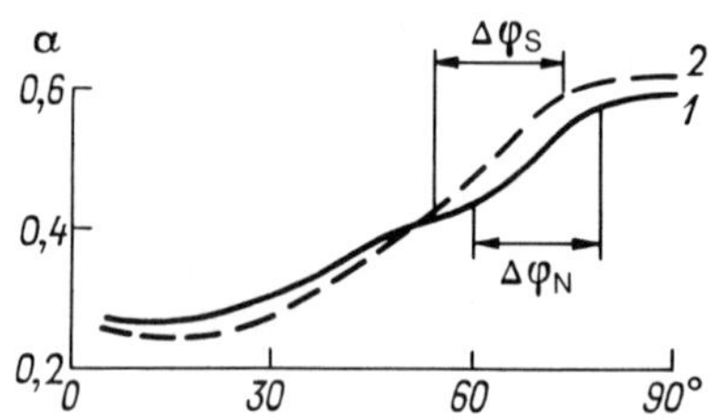


Fig. 3.7.

As seen in Fig. 3.7, the Antarctic ice-covered region, which far exceeds in area the Arctic ice-covered zone, has a higher average albedo value. This shows that, in this case, the influence of the altitude of the Sun on the albedo is more than compensated for by the effects of climatic change in the ice-covered zone.

The use of albedo data from Fig. 3.7 in calculating the modern temperature distribution presents no problem. The relationship between albedo values and ice position resulting from these data can be used for estimating albedo values for varying ice areas. In both hemispheres, although there is a considerable difference in polar ice area and in glaciation features, the albedo value for a zone covered with permanent ice is close to 0.6 (in the Southern Hemisphere the value is slightly greater and in the Northern it is somewhat smaller). In transitional zones where the area is only partially covered with ice, the albedo decreases as distance from the pole increases. These zones cover about 20° of latitude in each of the hemispheres.

It can be supposed that, with changes in ice area, the albedo in a completely ice-covered zone would be 0.6. The boundary of the zone is assumed to be 10° closer to the pole than the average position of the margin estimated from the area of ice.

In the transitional zones that cover 20° of latitude, the albedo can be evaluated by interpolation between the value for a completely ice-covered zone and that at the interior boundary of the transitional zone, with the assumption that the albedo of the ice-free zone does not change.

For substantiating the albedo parametrization with variable ice area, we should consider the dependence of the latitudinal mean albedo on cloudiness. Cess (1976) presented data showing a considerable increase in cloud albedo with increasing latitude due to a decrease in the zenith angle of the Sun. For this reason, the question arises: to what extent does an increase in albedo at high latitudes depend on the presence of ice cover and on the conditions of solar radiation reflection at the top of the clouds? In discussions of this problem, it should be kept in mind that, in high latitudes over vast ice-covered areas, there are stable regions with high pressure, low temperature and low air humidity where there is usually little cloudiness. The central Antarctic at present and the regions of Quaternary glaciations in the past are examples of such climatic conditions. In this connection, a decrease in the mean albedo over a completely ice-covered zone as the ice moves to lower latitudes seems to be highly improbable. It is conceivable that, for a 'white Earth' with very low temperatures at all latitudes, the atmosphere would be practically transparent to radiation. Therefore, the Earth's albedo would be close to the very high value of the albedo over clean snow cover (Budyko, 1974).

In computing the latitudinal albedo distribution, we should consider the position of the average boundary of permanent ice cover, which is dependent on thermal conditions. Since the temperature of the Earth's surface usually differs slightly from that of the lower layer of air, it can be supposed that, in the absence of an annual cycle of temperature, the boundary of permanent snow and ice cover would coincide on the continents with the 0 °C isotherm. This assumption may be compared with calculated temperatures at the boundary of permanent snow and ice cover in different northern latitudes. The estimates were obtained from data on the height of the snow line above sea level in mountains at these latitudes, with mean values of the vertical temperature gradient taken into account. The results of this calculation are shown in Fig. 3.8. As seen, this temperature actually approaches zero in the equatorial zone. In addition, in higher latitudes, where there is an appreciable annual temperature cycle, the snow and ice boundary corresponds to a negative mean annual air temperature that reaches −11 °C in high latitudes on the continents. Approximately the same mean annual temperature is observed at the boundary of permanent ice cover in high latitudes on the oceans.

Since continental climate conditions prevail over the ice surface, the mean annual air temperature at the ice-cover boundary in different latitudinal belts

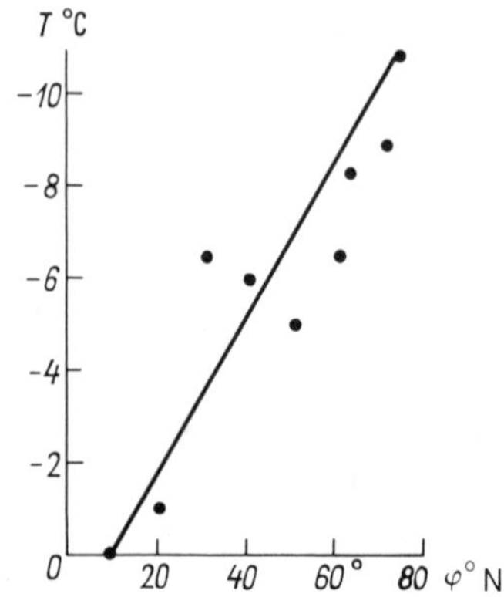


Fig. 3.8.

is assumed to correspond to a similar temperature at the snow boundary on the continents, as presented in Fig. 3.8.

The coefficients entering into the empirical Equation (3.9) for determining longwave radiation were initially obtained from a rough calculation of monthly means of outgoing emission in different geographical regions. Later, it became possible to find these coefficients by using satellite observations of outgoing emission (Cess, 1976). Similar computations were carried out by Beeva and Vinnikov (1981), who used more accurate data on cloud cover for different latitudinal zones. Taking their results into account, (3.9) can be written in the form

$$I = 258 + 1.69T - 76n \tag{3.16}$$

(where $I$ is in watts per square metre, $T$ in degrees Celsius and $n$ the cloudiness in fractions of a unit). The coefficients of this formula differ somewhat for the Northern and Southern Hemispheres, but the difference is not large. Mean values of the coefficients for the Earth as a whole are therefore used in formula (3.16).

The coefficient $\beta$ in (3.14) was calculated from data on the relationship between $(T - T_p)$, the difference between the temperatures of individual latitudinal belts and the Earth as a whole, and $C$, the value of meridional heat redistribution, equal (for average annual conditions) to the radiation balance of the Earth—atmosphere system as determined from satellite observations. The corresponding dependence is given in Fig. 3.9 with points depicting the values for every 10° of latitude, except for the Antarctic ice sheet, the surface of which is high above sea level. As seen from this figure, the hypothesis that values of $C$ are proportional to those of $T - T_p$ is well corroborated by empirical data, the coefficient $\beta$ being 3.69 W m$^{-2}$ K$^{-1}$. This value coincides almost exactly with that obtained earlier (Budyko, 1968).

By using values of the parameters given here, we can compute the distribution of mean air temperature at different latitudes. The results of this computation are plotted in Fig. 3.10 as curve 2.

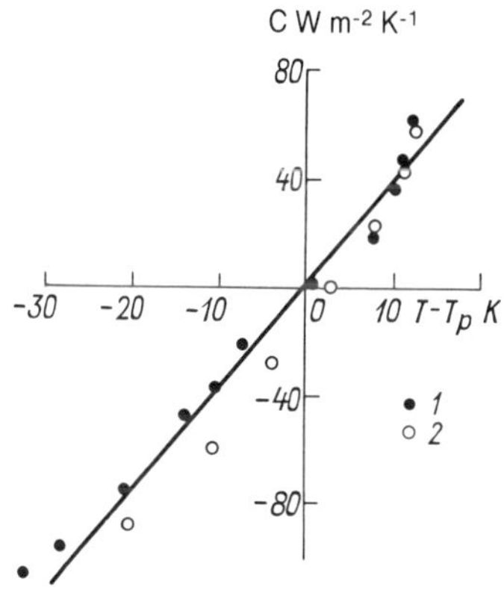


Fig. 3.9. 1, Northern Hemisphere; 2, Southern Hemisphere.

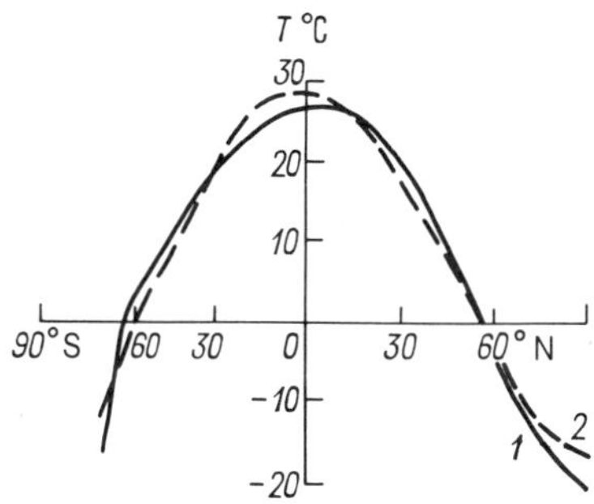


Fig. 3.10.

Comparing these results with the empirical data (curve 1), we can see that the average deviation of the measured and calculated values is 1.2 °C. This value is small compared to the range of variations in the mean latitudinal air temperatures, which indicates that the model used provides a satisfactory description of real climate conditions.

Figure 3.10 does not show the mean air temperature at the Earth's surface in central Antarctica. Here, observed temperatures are much lower than the calculated results. This difference is explained by the fact that the central Antarctic region is elevated high above sea level and there is no heat transfer by oceanic currents in this zone.

## 3.4. The Hydrological Cycle

### 3.4.1. THE WATER BALANCE

Water is the basic constituent of living organisms; all of them (except seeds and spores) contain more than 50% water, often more than 90%. There is no life, in whatever form, without water.

A considerable part of the biosphere's space is occupied by water. Water forms the hydrosphere. It is present in the atmosphere (as water vapour, water drops of mist, fog and clouds, droplets of water solutions of different aerosol substances, and snow crystals) and in the lithosphere, where, in addition to glaciers, soil and ground waters or deep springs, an enormous mass of water occurs in a bound state in the composition of different minerals. Still more water lies within the Earth's mantle from which it is released into the biosphere by volcanic eruptions and different hot springs.

Water exchange between the biosphere and the lithosphere, which was of great importance in the evolution of the hydrosphere, is much less intensive than the annual hydrological cycle in the biosphere itself. Still less intensive is the water exchange between the Earth and outer space, which takes place in the upper atmosphere.

The hydrological cycle in the biosphere includes an exchange of water (in its different states) between the hydrosphere, atmosphere, upper layers of the lithosphere and living organisms. This process can be described in terms of water balance equations for the biosphere's individual components.

The water balance equation for the land surface expresses the condition that the algebraic sum of all forms of gain and loss of water in solid, liquid and gaseous states received at a horizontal surface from ambient space over a specified time interval is equal to zero. This equation has the following form:

$$r = E + f_w + G, \tag{3.17}$$

where $r$ is the precipitation, $E$ the difference between evaporation and condensation at the Earth's surface (usually simply called evaporation), $f_w$ the surface run-off and $G$ the flow of moisture from the Earth's surface to deeper layers.

Equation (3.17) is more often used in a slightly modified form, which can be derived by considering the fact that the vertical flow of moisture $G$ is equal to the sum of ground water flow $f_p$ and the change in the moisture content of the upper layers of the lithosphere $b$. This equality corresponds to the equation of water balance of a vertical column that extends through the upper layers of the lithosphere down to the depths where moisture exchange practically does not take place.

Since the sum of surface run-off $f_w$ and ground water flow $f_p$ is equal to the total run-off $f$ we find that

$$r = E + f + b. \tag{3.18}$$

Equation (3.18) can also be used in the calculation of water balance of water bodies or their individual sectors. In this case, $f$ will describe the total horizontal redistribution of water over the period under consideration in the water body itself and underlying layers of the ground (if there is any noticeable redistribution of moisture). Similarly, for water bodies, the value of $b$ is equal to the total change in water quantity in the water body itself and underlying ground layers, if there are appreciable changes in moisture content. In many cases $b$ for water bodies corresponds to changes in water level. For an average annual period, $b$ is often rather small, and the water-balance equation acquires the following form

$$r = E + f. \tag{3.19}$$

For the water balance of the entire globe the horizontal redistribution of moisture is of no importance, and the water-balance equation will be simple:

$$r = E. \tag{3.20}$$

The same form of the water-balance equation holds true for the average annual period in the case of land areas without run-off, such as deserts.

Finally we shall give the equation of water balance for the atmosphere. By adding up all forms of moisture gain and loss within a vertical column passing through the atmosphere, we can easily obtain the equation

$$E = r + C_a + b_a, \tag{3.21}$$

where $C_a$ is the quantity of moisture that is gained or lost by the vertical column as a result of air currents and horizontal turbulent exchange, and $b_a$ is the change in the quantity of water within this column. Since the atmosphere can retain only relatively small quantities of water in any state, the value of $b_a$ is usually much smaller than the other components of the balance. Its average annual value is always close to zero.

Let us consider data on the components of the water balance of the continents. Information concerning precipitation on land is obtained through observations carried out by a network of meteorological stations. We should remember that the instruments used for measuring precipitation often record smaller amounts of precipitation than those that exist actually. This is particularly the case for the measurements of snow, which, due to the wind, is not completely caught in the rain gauges. The resulting systematic errors in measuring precipitation are different for different measuring instruments. When constructing maps of precipitation, attempts have been made to introduce corrections of the most substantial errors of this kind, yet the accepted adjustments proved to be insufficient. In recent works it has been established that in regions of heavy solid rainfall, the actual sums of precipitation might have been several tens of per cent greater than earlier adopted estimates. Measurements of liquid precipitation usually have smaller systematic errors, although these, too, may occasionally be appreciable.

Struzer and Sharova have constructed a world map of precipitation in which corrections have been made to exclude the influence of the principal systematic errors in observing precipitation. According to this map, the estimates of precipitation for the entire land are almost 10% higher than those of the earlier studies.

The average annual estimates of precipitation on different continents based on the precipitation map of Struzer and Sharova (*The Water Balance of the World*, 1974) are given in Table 3.3. The table does not include data on the precipitation over Antarctica and all islands, but they are taken into account when calculating the mean precipitation for the entire land.

TABLE 3.3

Precipitation and evaporation on the continents

| Continent | Precipitation (cm yr$^{-1}$) | Evaporation (cm yr$^{-1}$) | Run-off (cm yr$^{-1}$) |
|---|---|---|---|
| Europe | 77 | 49 | 28 |
| Asia | 63 | 37 | 26 |
| Africa | 72 | 58 | 14 |
| North America | 80 | 47 | 33 |
| South America | 160 | 94 | 66 |
| Australia | 45 | 41 | 4 |
| All land | 80 | 48.5 | 31.5 |

The estimates of the run-off that are presented in Table 3.3 have been obtained by generalizing the available data of observations in *The Water Balance of the World.*

Evaporation on each continent is equal to the difference between precipitation and run-off. Evaporation from the surface of the continents has been calculated in the same study by Zubenok using standard meteorological observations. This proved to be quite close to the difference between precipitation and run-off.

As seen in Table 3.3, the ratios of evaporation and precipitation for different continents differ greatly. In Australia the amount of evaporation approaches that of precipitation. On all other continents, except Africa, evaporation is less than two-thirds of the total precipitation.

Using the world map of precipitation of Struzer and Sharova, and the world map of evaporation from the oceans in *The Atlas of the World Water Balance* (1974) constructed by Strokina, we can obtain magnitudes of the water balance components of the oceans (Table 3.4). The magnitude of run-off in Table 3.4 multiplied by the area of oceans coincides with the product of the value of run-off in Table 3.3 and the area of land.

As already mentioned, the difference between evaporation from the surface of the World Ocean and precipitation is equal to the river run-off from the continents into the ocean. For individual oceans this difference is

TABLE 3.4

Water balance of oceans

| Ocean | Precipitation ($cm\ yr^{-1}$) | Evaporation ($cm\ yr^{-1}$) | Run-off ($cm\ yr^{-1}$) |
|---|---|---|---|
| Atlantic | 101 | 136 | 23 |
| Pacific | 146 | 151 | 8 |
| Indian | 132 | 142 | 8 |
| World Ocean | 127 | 140 | 13 |

the sum of the river run-off and the horizontal transfer of water from other oceans through circulation processes. It is difficult to determine the magnitude of this transfer by direct methods, because it is a small difference between two large values, namely the inflow and outflow of water, both of which are subject to significant errors. It is somewhat easier to evaluate the exchange of water between the oceans as a residual component of the water balance of each ocean, although in that case the accuracy of determining the corresponding estimates is also low.

Table 3.4 does not include data on the water balance of the Arctic Ocean (which are less accurate than those for other oceans), but they are taken into account in determining the components of the water balance for the World Ocean as a whole.

As the table shows, the sum of precipitation and run-off for the Atlantic Ocean is smaller than evaporation. It follows that the Atlantic Ocean receives water from other oceans, including the Arctic Ocean, where evaporation is noticeably lower than the sum of precipitation and river run-off. In the Indian Ocean the sum of precipitation and run-off is slightly smaller than evaporation. In the Pacific it exceeds the evaporation, the difference corresponding to the transfer of surplus water into other oceans.

Making use of these values of precipitation and evaporation for land and oceans, we find that, for the Earth as a whole, the total annual precipitation, which is equal to the total annual evaporation, is 113 cm. This value is larger than similar estimates obtained in earlier studies.

By using the world maps of precipitation and evaporation, it is possible to determine the values of the water balance components for various latitudinal zones. The results of the calculation show that, in different latitudinal belts, the income of water vapour to the atmosphere from evaporation may be either larger or smaller than its expenditure on precipitation. Water vapour in the atmosphere is primarily provided by high-pressure zones where evaporation greatly exceeds precipitation. That surplus of water vapour is expended in the zones adjacent to the equator and also in middle and high latitudes, where precipitation exceeds evaporation.

It is evident that the difference between precipitation and evaporation is also equal to the difference between the inflow and outflow of water vapour in the atmosphere resulting from horizontal air motions. The large values of this difference in many regions are, to some extent, an indicator of the importance of atmospheric water vapour transfer in the formation of precipitation. The dependence of precipitation on this water vapour transfer is examined in studies of the hydrological cycle in the atmosphere, and some of the results of these studies are considered further in this work.

It should be mentioned that the data on precipitation from *The World Water Balance* somewhat differ from the earlier data given at the beginning of the chapter.

The components of the water balance on the continents and oceans are not constant; they vary as a result of climatic fluctuations and other factors. Although annual variations in the water balance components are small compared to their absolute values, they are of great importance in the study of the evolution of the hydrosphere.

### 3.4.2. RELATIONSHIP BETWEEN ENERGY AND WATER BALANCES ON LAND

An important characteristic of the hydrologic regime on land is the normal run-off, i.e. the average annual amount of water flowing off from a unit of land surface in the form of different horizontal streams. Another important indicator of hydrological conditions is the run-off coefficient, i.e. the ratio of the normal run-off to the annual sum of precipitation.

Since the annual run-off depends to a great extent on the process of evaporation, which is also one of the basic processes in the transformation of solar energy at the Earth's surface, it is evident that the normal run-off and the run-off coefficient are definitely connected with the principal components of the energy balance. The examination of this relation should help to clarify the principles that determine specific features of the hydrologic regime of different geographical zones.

The relationship between the components of the energy and water balances has been established through the following considerations (Budyko, 1948). It is obvious that the average sums of evaporation from the surface of land, $E$, depend on the amount of precipitation, $r$, and the income of solar energy, evaporation increasing with an increase in precipitation and the radiation balance, $R$. When soil is very dry, almost all water received in the form of precipitation is retained by molecular forces on the soil particles and finally expended on evaporation. In such cases (for example, in deserts) the run-off coefficient $f/r$ approaches zero. Since the average dryness of the soil increases with an increase in the radiative heat income and a decrease in precipitation, we can deduce that

$$\frac{f}{r} \to 0 \quad \text{or} \quad \frac{E}{r} \to 1 \quad \text{at} \quad \frac{R}{Lr} \to \infty . \tag{3.22}$$

With a decrease in the ratio $R/Lr$ (we shall refer to this as the radiative dryness index) the value of $E/r$ will decline and a certain amount of run-off appears. When the sums of precipitation are sufficiently large and the radiative heat income is sufficiently small, a state of complete moistening of the upper layer of the soil will be reached. In this case, the maximum possible amount of heat energy from the available sources will be expended on evaporation. Its value can be estimated by considering the valve-like character of turbulent heat exchange between the underlying surface and the atmosphere.

Experimental studies have shown that the turbulent heat conductivity of the lower air layer depends substantially on the direction of the vertical turbulent heat flow. When the turbulent flow is directed from the earth to the atmosphere, a relatively higher intensity of turbulent mixing permits the flow to reach large magnitudes, which are comparable to other principal components of the radiation and energy balances. With reverse direction of the turbulent flow, the inverse temperature distribution considerably reduces the intensity of the exchange, and as a result, the turbulent heat flow turns out to be relatively small.

At middle latitudes in winter, the predominance of inversion conditions in the vertical temperature distribution results in a smaller turbulent heat exchange compared to its maximum summer values which are determined by the conditions of a superadiabatic distribution of temperature in the lower air layer. We notice a similar pattern in the daily cycle when comparing the values of turbulent heat flow at night and in the daytime.

Because of the valve-like effect, the annual sums of turbulent heat flow turn out to be positive, i.e. the average turbulent flow of heat is directed from the Earth's surface to the atmosphere in almost all climatic zones on land.

Since the annual sums of the turbulent heat flow cannot provide a significant income of energy to the surface, we must assume that the heat consumed by evaporation is compensated for only by the radiation balance. As a result, the upper limit of increasing $LE$ is equal to $R$. In other words, we may conclude that, under fully-moist conditions:

$$LE \to R \quad \text{at} \quad \frac{R}{Lr} \to 0. \tag{3.23}$$

Conditions (3.22) and (3.23) determine the form of the function $\Phi$ which relates $E/r$ to $R/Lr$

$$\frac{E}{r} = \Phi\left(\frac{R}{Lr}\right) \tag{3.24}$$

for $R/Lr \to 0$ and for $R/Lr \to \infty$.

A relationship among the water-balance components was established long ago by Schreiber (1904) and Oldekop (1911) on the basis of data on precipitation and run-off. This was expressed by Oldekop as

$$E = r(1 - e^{-E_0/r}) \tag{3.25}$$

(the Schreiber—Oldekop equation, where $E_0$ is the greatest possible value of evaproation under given conditions, i.e. potential evaporation) and

$$E = E_0 \tanh \frac{r}{E_0} \tag{3.26}$$

(the Oldekop equation, where tanh is the hyperbolic tangent). It is easy to prove that both formulas meet conditions (3.22) and (3.23), if following the foregoing considerations, we assume that $E_0 = R/L$.

One should should keep in mind, however, that potential evaporation at a given site is governed by the radiation balance corresponding to the level of moisture that is sufficient for vegetation to grow. The value of the radiation balance under conditions of sufficient moisture may differ from its values at the same site with a deficiency in moisture.

Let us point out the reasons for the dependence of the radiation balance of the active surface on the level of moisture. In most geographical regions with more or less humid climate, the albedo of the Earth's surface varies only slightly with variations in moisture. Since the differences between the temperatures of the Earth's surface and air are, on average, rather small under conditions of sufficient moisture, it should be possible to deduce the value of potential evaporation from the radiation balance determined for the actual state of the surface.

However, the situation is different in a dry climate when both albedo and surface temperature change if the surface is moistened, and the temperature approaches, at sufficient moisture levels, the temperature of air. It is apparent in this case that potential evaporation should be determined using the value of the radiation balance for the albedo of moistened surface and for the surface temperature equal to air temperature.

In order to test our considerations concerning the form of dependence of the ratio $E/r$ on $R/Lr$ for both small and large values of the latter parameter, different empirical data were used. As an example, let us present the results of a comparison of the values of radiative dryness index, $R/Lr$, with the values of the ratio $E/r$ obtained from data for the river basins of different continents (mountain areas excluded).

The dependence of $E/r$ on $R/Lr$ is presented in Fig. 3.11 by the line $OA$ for small values of $R/Lr$ in accordance with condition (3.23), and by the line $AB$ for large values of $R/Lr$ according to condition (3.22). The points representing values of the ratio $E/r$ are calculated from water-balance data by averaging $E/r$ over specific intervals of the parameter $R/Lr$. The experimental points show that there is, in fact, a smooth transition from line $OA$ to line $AB$ which are, as suggested earlier, limiting values for the relation between $E/r$ and $R/Lr$.

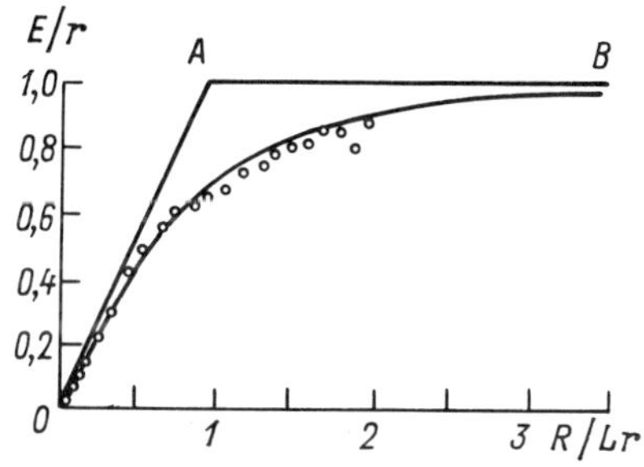


Fig. 3.11.

In order to give an analytical form for Equation (3.24), one can use formulas similar to (3.25) and (3.26)

$$E = r(1 - e^{-R/Lr}), \tag{3.27}$$

and

$$E = \frac{R}{L} \tanh \frac{Lr}{R}. \tag{3.28}$$

One may also employ the geometric mean of the right-hand terms of these equations

$$E = \sqrt{\frac{Rr}{L} \tanh \frac{Lr}{R} \left(1 - \cosh \frac{R}{Lr} + \sinh \frac{R}{Lr}\right)}, \tag{3.29}$$

(cosh and sinh are the hyperbolic cosine and sine). This is represented in Fig. 3.11 by a curve that agrees slightly better with the experimental data than either of the first two relationships. Equation (3.29) represents the dependence of the average annual evaporation on precipitation and radiation balance.

Since the normal run-off is equal to the difference between precipitation and evaporation, it is obvious that this formula allows us to determine run-off from data on precipitation and the radiation balance. Equations representing the run-off and run-off coefficient in terms of formula (3.29) will have the form:

$$f = r - \sqrt{\frac{Rr}{L} \tanh \frac{Lr}{R} \left(1 - \cosh \frac{R}{Lr} + \sinh \frac{R}{Lr}\right)}, \tag{3.30}$$

and

$$\frac{f}{r} = 1 - \sqrt{\frac{R}{Lr} \tanh \frac{Lr}{R} \left(1 - \cosh \frac{R}{Lr} + \sinh \frac{R}{Lr}\right)}. \tag{3.31}$$

Let us look into the physical meaning of the equation representing the relationship between the energy and water balances. As seen in Fig. 3.11, the relationship on which this equation is based is generally determined by two limiting conditions, one of which arises from the idea of the valve-like character of turbulent heat exchange in the surface air layer and the other corresponds to the obvious small value of the run-off coefficient in a dry climate. The selection of an interpolation function for the transition from the first condition to the second is not particularly important, since, for most of the range of variations in the parameters of energy—water balance equation, the appropriate relation deviates little from a boundary condition.

Thus, although the equation is a semi-empirical relation, the role of empirical data in its justification is minor in comparison with the general limiting considerations mentioned earlier. It should be also noted that the equation represents a supplementary relation that is independent of the energy and water balance equations.

Formulas (3.30) and (3.31) establishing the relationship between the components of the energy and water balances have been verified in a number of studies. Here, we mention the calculations of the run-off coefficient for European rivers (Budyko, 1951). For 29 rivers with drainage areas more than 10000 km$^2$ the mean error in determining the run-off coefficient was 0.04, the magnitudes of this coefficient ranged from 0.13 to 0.64. The relative error of the run-off coefficient for rivers with moderate values (more than 0.30) was 7%. This is close to the accuracy of measurements of both precipitation and run-off. It follows that, for large basins with moderate run-off coefficients, the difference between 'climatic' and actual run-off is insignificant, and in some cases could not be determined given the existing level of accuracy in measurements of precipitation and run-off.

A more comprehensive test of the equation relating the energy and water balances on land was carried out using data from 1200 continental regions, for which evaporation was found in terms of water balance as a difference between precipitation and run-off (Budyko and Zubenok, 1961). It was established that the mean discrepancy between the ratio of evaporation to precipitation calculated from the equation and that derived from the water balance was 10%.

It was noted in the same study that, in certain cases, there are systematic deviations from the mean dependence represented by the equation. The nature of these deviations depends on the annual march of potential evaporation and precipitation. If variations in monthly values of potential evaporation and precipitation run parallel, the mean annual values of the ratio of evaporation to precipitation increase to some extent. However, when an increase in potential evaporation in its annual course coincides with a decline in precipitation, the mean values of the ratio of evaporation to precipitation decrease. In most cases, these deviations are within the accuracy of calculation of evaporation by the energy—water balance equation.

This Equation (3.29) makes it possible to derive a general form of the dependence of run-off on the annual sums of precipitation and the radiation balance, which is presented in Fig. 3.12 as a graph. This regularity explains a number of empirical relations between run-off and precipitation that were established in earlier studies. The equation permits us to construct the dependence of run-off on precipitation for the values of radiation balance corresponding to certain local conditions. Such a dependence for the average conditions of the European plain is shown in Fig. 3.13 as curve *A*. For the purpose of comparison an empirical relationship found by Keller (1906), which associates the normal run-off with precipitation based on observational data on the rivers of Western Europe (line *B*), and an empirical curve of Sokolovsky (1936), which relates run-off to precipitation on the basis of observations of the rivers of Eastern Europe (curve *C*), are also shown. The good agreement between the results of Equation (3.29) and these findings confirms once more its universal character.

Using the energy—water balance equation, it is also possible to explain the considerable differences in the relations of run-off to precipitation that appear in data from different regions (Polyakov, 1946). A substantial variability of the radiation balance in middle latitudes causes the run-off of basins with a large radiation balance (i.e. in more southern regions) to be much smaller than that of basins with a smaller radiation balance (i.e. in more northern regions), if the sums of precipitation are the same. The rate of the

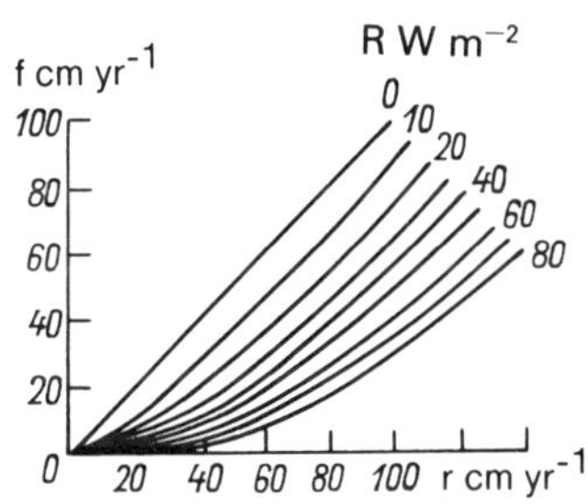


Fig. 3.12.

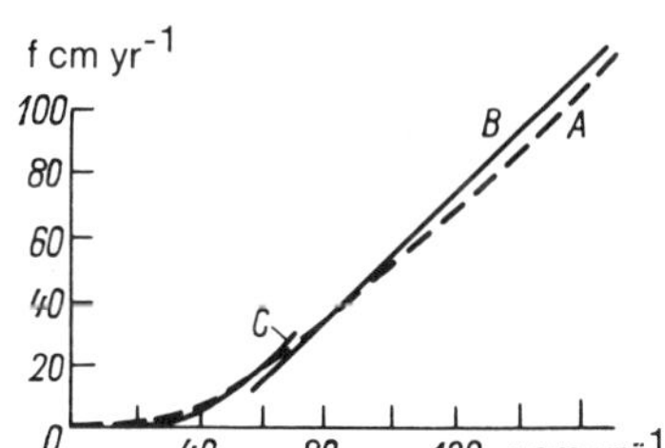


Fig. 3.13.

run-off change with increasing precipitation, d$f$/d$r$, varies at the same time. According to Equation (3.29), in northern basins it must be greater than in southern ones. This agrees well with empirical data.

Good agreement between the results for run-off calculated using the equation for the relationship between the energy and water balances and observational data confirms the decisive role of energy factors in the formation of annual sums of run-off. It is necessary to underline that this conclusion concerns the river basins of considerable area comparable with the scale of geographical zones. For small areas, river run-off may vary appreciably as a result of local factors.

### 3.4.3. THE WATER CYCLE IN THE ATMOSPHERE

Among the different effects that living things produce on modern climate, the process of transpiration of plants on the continents is of greatest importance. Since transpiration usually represents several tens of per cent of the total evaporation, variations in transpiration might influence to a certain extent the water cycle in the atmosphere, and consequently the amount of falling precipitation. In studying this influence it is expedient first to clarify the principles that govern the water cycle in the atmosphere.

The first studies of the atmospheric water cycle were carried out at the beginning of this century. Some of the authors, for instance, Brückner (1901), suggested that on land in every region precipitation is formed by a quantity of water vapour brought in from outside, which does not exceed the difference between precipitation and evaporation. Since the total amount of precipitation on land usually greatly exceeds this difference, it was often supposed that the water cycle is a multiple internal process. In other words, it was postulated that the water vapour brought from outside into a certain region of land repeatedly falls in this region as rainfall before being carried away further by atmospheric circulation. This concept of a multiple cycle for the water circulation implied that a comparatively slight variation in evaporation from the land surface might bring about a drastic change in the total precipitation.

The idea of a multiple water cycle was rejected as a result of the development of a quantitative theory of the water cycle (Budyko and Drozdov, 1950; Kashin and Pogosyan, 1950). We shall set forth here the derivation of the principal equations of the water cycle (Budyko and Drozdov, 1953).

Let us consider the transfer of water vapour in the atmosphere above a certain territory with an average linear scale $l$. The flux of water vapour brought by air currents to this territory is $wu$, where $w$ is the mean water content of the atmosphere at the windward side of the area under consideration and $u$ is the average velocity of the air flows carrying water vapour across the territory. Along the path of the air currents the water content of the atmosphere changes according to the difference between the consumption of water on precipitation and its income from evaporation.

It is evident that the flux of water vapour carried away from the given territory by air currents is equal to $wu-(r-E)l$, where $r$ is the total precipitation, $E$ is the total evaporation over a specified period of time, and $l$ is the linear scale of the territory.

The total flux of water vapour transported over a given territory is composed of two fluxes; namely, one of external (advective) water vapour formed by evaporation outside this territory and a flux of water vapour formed by local evaporation.

The first flux at the windward contour of the territory will be equal to $wu$ and at the leeward side (when the flux leaves the territory) it is $wu-r_a l$, where $r_a$ is the total precipitation formed from the external (advective) water vapour. The second flux is equal to zero at the windward side and $(E-r_M)l$ at the leeward side (when leaving the territory), where $r_M$ is the total precipitation produced from water vapour of local origin.

Thus, on the average, above the territory under consideration, there are two fluxes: the flux of external water vapour $wu-\frac{1}{2}r_a l$ and the flux of local water vapour $\frac{1}{2}(E-r_M)l$, producing an overall flux, $wu-\frac{1}{2}(r-E)l$. It should be kept in mind that $r_a+r_m=r$. Since the molecules of water vapour of local and external origin are fully mixed in the atmosphere by the process of turbulent exchange, it is evident that the ratio of total precipitation formed from local and external water vapour is equal to the ratio of the corresponding quantities of water vapour molecules in the atmosphere. In other words, it may be assumed that

$$\frac{r_a}{r_M}=\frac{wu-\frac{1}{2}r_a l}{\frac{1}{2}(E-r_M)l}. \tag{3.32}$$

From this, the following two equations can be derived:

$$r_a=r\frac{1}{1+(El/2wu)}, \tag{3.33}$$

$$r_M=r\frac{1}{1+(2wu/El)}. \tag{3.34}$$

From Equation (3.33) we can also determine the value of the water cycle coefficient $K$ which is equal to the ratio of total precipitation to the amount of precipitation of external (advective) origin:

$$K=\frac{r}{r_a}=1+\frac{El}{2wu}. \tag{3.35}$$

With the help of these formulas it is possible to analyse the dependence of the characteristics of the water cycle on the basic factors that influence it. For instance, it follows from (3.35) that the water cycle coefficient depends on the factors determining the balance of water vapour in the atmosphere and does not depend directly on the magnitude of the river run-off. We point out also that there is a dependence of the sums of the external precipitation, local precipitation and the water cycle coefficient on the scale of given territory, which follows from Equations (3.34), (3.33), (3.35). With an increase in scale $l$ the total local precipitation and the water cycle coefficient also increase. In this case, the dependence of the water cycle coefficient on the scale of the territory (for territories that are not too small) is not linear; with an increase in the area, the effect of curvature of the trajectory of the movement of air particles slightly lowers the mean velocity of water vapour transfer $u$.

To estimate the effect of water vapour transfer in the atmosphere on the formation of precipitation, we give here calculated data on the components of the water cycle for the European part of the U.S.S.R. (Table 3.5). The table shows that precipitation from local water vapour makes up only a very small fraction of the total precipitation. Both for the year as a whole and for individual months, the water cycle coefficient is not much greater than one, which shows that the concept of a multiple internal water cycle was erroneous.

Even over such a great area as the European part of the U.S.S.R. only a very small portion (about 12%) of the total precipitation falls a second time as a result of using water vapour from local evaporation. The main portion of the precipitation falling on a limited territory of land originates from water vapour brought from outside.

TABLE 3.5

Annual course of the components of the water cycle for the European part of the U.S.S.R.

| | I | II | III | IV | V | VI | VII | VIII | IX | X | XI | XII | Year |
|---|---|---|---|---|---|---|---|---|---|---|---|---|---|
| $E$ (cm month$^{-1}$) | 0.5 | 0.5 | 1.0 | 3.6 | 5.0 | 5.4 | 5.0 | 3.9 | 2.2 | 1.1 | 0.7 | 0.5 | 29.4 |
| $w$ (cm) | 0.4 | 0.4 | 0.6 | 0.9 | 1.5 | 2.0 | 2.3 | 2.2 | 1.6 | 1.2 | 0.8 | 0.5 | 1.2 |
| $u$ (m s$^{-1}$) | 7.7 | 7.8 | 7.8 | 7.2 | 6.6 | 6.2 | 5.8 | 6.3 | 6.9 | 7.5 | 7.7 | 7.6 | 7.1 |
| $K$ | 1.07 | 1.08 | 1.09 | 1.24 | 1.22 | 1.19 | 1.17 | 1.12 | 1.08 | 1.05 | 1.05 | 1.06 | 1.12 |
| $r$ (cm month$^{-1}$) | 2.7 | 2.3 | 2.4 | 2.8 | 3.8 | 5.5 | 6.3 | 5.9 | 5.1 | 4.9 | 3.8 | 3.2 | 48.7 |
| $r_a$ (cm month$^{-1}$) | 2.5 | 2.1 | 2.2 | 2.3 | 3.1 | 4.6 | 5.4 | 5.3 | 4.7 | 4.7 | 3.6 | 3.0 | 43.5 |
| $r_M$ (cm month$^{-1}$) | 0.2 | 0.2 | 0.2 | 0.5 | 0.7 | 0.9 | 0.9 | 0.6 | 0.4 | 0.2 | 0.2 | 0.2 | 5.2 |

Drozdov and his collaborators have studied the atmospheric water cycle above all the continents (*The World Water Balance*, 1974). Their results are given in Table 3.6. It can be seen that, even on the largest continent, Asia, most of the precipitation is formed from the water vapour carried to the continent from outside. Thus, the major source of moisture for the continents is water vapour from the oceans.

Because of the large scale of moisture transfer from the oceans in the atmosphere the contribution of local evaporation from the surface of the continents to the total amount of precipitation is relatively modest, particularly for areas whose size is less than 1 million km$^2$. Yet this does not mean that the influence of local evaporation on the total precipitation is limited to

TABLE 3.6

Atmospheric water cycle over the continents

| Continent | Precipitation ($km^3\ yr^{-1}$) | Advective precipitation ($km^3\ yr^{-1}$) | Local precipitation ($km^3\ yr^{-1}$) | Coefficient of water cycle |
|---|---|---|---|---|
| Europe | 7540 | 5310 | 2230 | 1.42 |
| Asia | 33 240 | 18 360 | 14 880 | 1.81 |
| Africa | 21 400 | 15 100 | 6300 | 1.42 |
| North America | 16 200 | 9800 | 6400 | 1.65 |
| South America | 28 400 | 16 900 | 11 500 | 1.68 |
| Australia | 3470 | 3040 | 430 | 1.14 |

variations in the components of the atmospheric water cycle. In addition to a direct effect of local evaporation on the water cycle, there is also an indirect effect provided by a relationship between precipitation and the relative humidity of the air.

In the studies of Drozdov and his coworkers (Drozdov and Grigorieva, 1963) a semi-empirical formula for determining the amount of precipitation was derived:

$$r = aw\varphi(h), \tag{3.36}$$

where $h$ is the mean relative humidity of the air within the layer 0—7 km, $w$ the moisture content of the atmosphere and $a$ is the coefficient of proportionality ($a$ is equal to 1 if an amount of precipitation of $0.1w$ falls per day with $h = 100\%$). The empirical function $\varphi(h)$ is of the same nature for many regions in middle latitudes. For instance, the following data (Table 3.7) are characteristics of this function for the European part of the U.S.S.R.

TABLE 3.7

| $h\%$ | 40 | 45 | 50 | 55 | 60 | 65 | 70 | 75 | 80 | 85 | 90 | 95 | 100 |
|---|---|---|---|---|---|---|---|---|---|---|---|---|---|
| $10^2\varphi(h)$ | 0.7 | 1.4 | 3.7 | 5.7 | 6.8 | 7.5 | 8.2 | 8.6 | 9.1 | 9.5 | 9.7 | 9.9 | 10.0 |

These data show that precipitation is insignificant with relative humidity below 40% and its amount increases rapidly as the relative humidity rises to 50—55%. The increase in precipitation is less rapid with further rise in the humidity, and with high air humidity its variations influence the amount of precipitation only slightly.

By using formula (3.36) it is possible to find the dependence of precipitation on local evaporation. In this case, the values of $w$ and $h$ are determined in accordance with the local evaporation effect on the total amount of water vapour that is being carried over a given territory. In this connection it becomes apparent that the effect of local evaporation on precipitation within a certain humidity range may be several times greater than its contribution to the total amount of water vapour in the atmosphere.

The foregoing theory of the atmospheric water cycle which describes the origin of water falling as precipitation takes account of only the direct influence of local evaporation on the amount of precipitation. As the studies of Drozdov have shown, a large role is often played by the indirect influence of local evaporation on precipitation. Without this influence the continental regions that are situated far from the ocean would have been transformed into deserts.

Formula (3.36) makes it possible, in principle, to estimate the dependence of precipitation on vegetation cover, since this determines the amount of local evaporation. However, the relation between the total evaporation and the nature of vegetation has not yet been studied sufficiently. For instance, the effect of forests on evaporation has been discussed for many years, but the relationships still remain unclear. Quite probably forests do not substantially affect the total evaporation as compared with other types of vegetation. But, when mossy swamps are drained and a forest vegetation develops in such areas, the total evaporation seems to increase, and a reverse effect may take place when lowland grassy swamps are drained.

It may be assumed that a change from one type to another type of vegetation may modify the total evaporation by approximately 10%. Using Equation (3.36) and the ratios given above, we find that the effect of variations in evaporation on precipitation will depend on the size of the region within which the evaporation has changed. If this has occurred within a geographical zone whose linear scale is of the order of 1000 km, the effect of fluctuations in evaporation on precipitation will be insignificant. The effect, however, might be quite substantial if evaporation changes over the entire continents or large parts of the continents.

## 3.5. Geographical Zones

### 3.5.1. FACTORS OF GEOGRAPHICAL ZONALITY

It was Dokuchaev who first drew attention to the existence of geographical zones, showing that, within wide territories (zones) on the continents, natural conditions retain many common features that change perceptibly from zone to zone. He noted that, because of the spherical shape of the Earth, "climate, plants and animals are distributed on the Earth's surface from north to south in a strict order . . . And since the distribution of soil-forming agents is governed by specific rules and is related to specific belts, their product, i.e. soils, must also be distributed throughout the globe in a certain zonal sequence that follows latitudinal circles with only small deviations" (Dokuchaev, 1948).

Dokuchaev identified the following main geographical zones and provided each with a brief description: the boreal, or tundra zone, the north forest zone, the forest-steppe zone, the steppe zone, the zone of arid steppes, the desert zone, the laterite zone. (The names of these zones somewhat differ in his different works.)

Among subsequent studies of geographical zones, we mention those of Grigoriev, who presented a series of works later published as collections (Grigoriev, 1966, 1970) where he considered the spatial structure of the geographical environment. Grigoriev distinguished four types of geographical environment associated with different latitudinal belts. The first of them is the equatorial belt where high air temperatures and abundant precipitation are observed throughout the year. The specific features of this belt are humid tropical forest, high run-off coefficient and intensive erosion.

The second belt is the tropical zone situated at latitudes where the altitude of the Sun differs greatly in summer and winter months, thus producing a distinct annual cycle of air temperature. In this belt, like in the previous one, the radiation balance of the Earth—atmosphere system is positive. The amount of precipitation in the torpical belt varies within a wide range, which promotes the formation of three different zones on this territory.

In the external tropical zone, the amount of total precipitation is small. We observe here, therefore, semi-deserts and deserts with scanty vegetation of xerophytes, insignificant run-off coefficient and intensive physical weathering.

In the zone adjoining the equatorial belt, natural conditions in the summer half-year are similar to those of the equatorial belt; but in the winter half-year they are closer to those of the external equatorial zone. In this connection, sharp changes in moisture conditions are characteristic of this zone throughout the year, and in a number of regions in the mid-summer season there is a second, less pronounced, dry period. Vegetation is diverse in this zone; it includes light tropical and savanna forests (deciduous and evergreen) as well as open savannas. The river run-off varies during the year in accordance with precipitation change. The process of weathering is of a seasonal nature, chemical weathering prevails in rainy seasons and physical erosion is more typical of dry seasons.

The third zone of the tropical belt, identified by Grigoriev as intermediate, is characterized by a prevalence of droughts throughout the year with a relatively short rainy season, during which a limited quantity of precipitation falls. In some respects, this zone is similar to the external tropical zone, although conditions here are more favourable for the development of plants.

The next latitudinal belt, according to Grigoriev, covers subtropical and middle latitudes. Here the annual magnitudes of the radiation balance of the Earth—atmosphere system are negative, but the air temperature in the warmest months does not fall below 5 °C. The differences in atmospheric circulation stipulate the amount of falling precipitation, which ranges from high values to values that are close to zero.

Different combinations of heat and moisture regimes in the third latitudinal belt correspond to different types of vegetation, such as tundra, forest-tundra, coniferous, mixed and hardwood forests, forest-steppe, steppe, semi-desert and desert. Depending on moisture conditions, the magnitude of the run-off coefficient ranges from zero to values that are close to unity. The process of weathering substantially varies throughout the year according to the annual course of temperature.

The fourth and the last latitudinal belt on the continents in the Northern Hemisphere, called the Arctic zone by Grigoriev, is located at high latitudes, where low temperatures prevail throughout the year. Within this zone scanty vegetation is found only in peripheral regions that encircle the area of permanent snow and ice — a desert of ice. The ice-free regions are marked by conditions of excessive moisture with a high run-off coefficient.

A number of authors have studied the patterns of the formation of geographical zones within different latitudinal belts.

Dokuchaev pointed out that the boundaries of geographical zones are determined to a great extent by climatic factors, and that they are greatly dependent on moisture conditions in particular. He estimated the ratios of precipitation to potential evaporation for the main geographical zones. The studies of Dokuchaev marked the beginning of numerous works devoted to examining the relation between the distribution of soil and vegetation and the ratios of precipitation to potential evaporation (or its characteristics).

Of the studies along this line, we mention that of Vysotsky (1905), which contains calculated values of the ratios of annual precipitation to potential evaporation (which is assumed equal to the evaporation from water surfaces). The values obtained were used by Vysotsky while analysing conditions for the formation of different types of soil and plant cover in different geographical zones.

Grigoriev (1946, 1954) suggested a different approach to the study of moisture conditions based on data relating to the radiation regime. He thought that moisture conditions are dependent on the total atmospheric precipitation and the radiation regime.

The following observation was made in a study by the author (Budyko, 1948). The equation of the energy and water balances on land for annual conditions may be written in the form:

$$\frac{R}{Lr} = \frac{E}{r} + \frac{P}{Lr}, \tag{3.37}$$

$$1 = \frac{E}{r} + \frac{f}{r} \tag{3.38}$$

(the terms of the energy balance equation are divided by $Lr$ and of the water balance by $r$). These formulas may be supplemented with the equation relating the energy and water balances:

$$\frac{E}{r} = \Phi\left(\frac{R}{Lr}\right). \tag{3.39}$$

These three equations establish a relationship among four relative values of components of the energy and water balances. It is therefore sufficient to know only one of them in order to determine the others.

Taking account of the special form of the energy—water balance equation it may be concluded that either $R/Lr$ or $P/Lr$ can be selected as the parameter that determines all relative values of the components of the energy and water balances. (Neither $E/r$ nor $f/r$ is a decisive factor under the conditions of a dry climate, when slight variations in $E/r$ or $f/r$ correspond to great changes in $R/Lr$ or $P/Lr$.)

Since the estimates of the turbulent heat flow are usually less accurate than those of the radiation balance, it is evident that $R/Lr$ should be taken as a principal parameter determining the relative values of the components of the energy and water balances. This parameter may be regarded either as the ratio of potential evaporation $R/L$ to precipitation $r$, or as the ratio of the radiation balance of the Earth's surface to heat consumed to evaporate the total annual amount of precipitation.

Thus, by using this parameter, we combine both the idea of Dokuchaev and Vysotsky about an indicator of moistening and that of Grigoriev concerning the importance of the ratio of the radiation balance and precipitation for moisture conditions.

While the relative values of the components of the energy and water balances are determined by only one parameter $R/Lr$, their absolute values require two parameters, for instance $R/Lr$ and $R$. In order to clarify the physical meaning of the radiative dryness index (i.e. $R/Lr$), we should consider the relation between the potential evaporation and the radiation balance of the Earth's surface.

The magnitude of potential evaporation from the surface of land under conditions of sufficient moisture may be found by the formula describing evaporation from a water surface or from a fully moistened surface:

$$E_0 = \rho D(q_s - q), \tag{3.40}$$

where $\rho$ is the air density, $D$ is the integral coefficient of diffusion, $q_s$ is the specific humidity of the air saturated with water vapour for a given temperature of the surface from which evaporation takes place, $q$ is the specific humidity of the air at the standard level of meteorological measurements. To obtain the temperature of the evaporative surface, which determines the values of $q_s$ it is possible to use the energy balance equation (see Chapter 4).

Figure 3.14 presents the results of a comparison of the values of $E_0$ based on formula (3.40) with the values of $R/L$ for 44 points in different climatic conditions. These data show that calculations of potential evaporation by formula (3.40) and by the mean annual values of the radiation balance yield similar results, for which the correlation coefficient is 0.98.

The relation between potential evaporation and the air humidity deficit (whose values are often used when calculating potential evaporation) is shown in Fig. 3.15. The average annual values of the air humidity deficit are

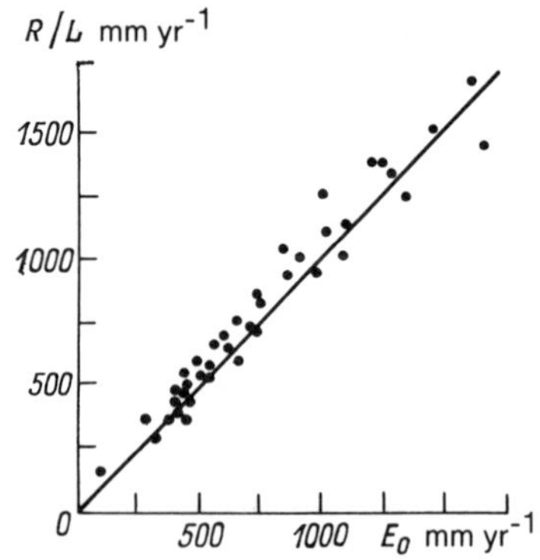


Fig. 3.14.

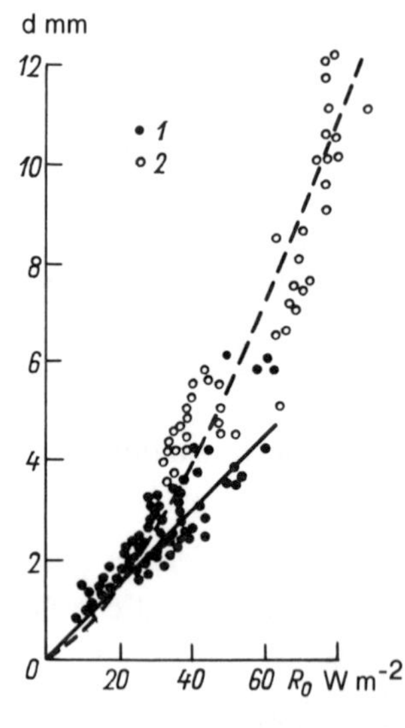


Fig. 3.15.

presented as a function of the radiation balance; they refer to the conditions at 187 points in different geographical zones in the U.S.S.R. Data pertaining to regions of a dry climate (semi-deserts and deserts) are designated by circles, while the rest of the data refer to steppe, forest-steppe, forest and tundra zones and are denoted by dots. As seen in Fig. 3.15, there is a rather close relationship between the radiation balance and the air humidity deficit in humid climatic conditions, which approaches a direct proportionality. This relationship is violated in the case of dry climates.

Thus, the radiation balance of the Earth—atmosphere system may be helpful in determining potential evaporation. The method for calculating potential evaporation from the air humidity deficit as well as a number of other empirical methods for estimating potential evaporation yield less satisfactory results.

### 3.5.2. A PERIODIC LAW OF GEOGRAPHICAL ZONALITY

In order to study the influence of climatic factors on geographical zonality, a world map of the radiative dryness index has been used (Budyko, 1977a). As seen in the map, the distribution of isolines of the radiative index of dryness on the continents is characterized by highly variable values of this index on each continent. The most humid conditions, corresponding to the smallest values of the index of dryness, are largely found in high latitudes, where potential evaporation is rather low. The largest values of the dryness index are found in semi-deserts and deserts.

Comparing the map of the radiative dryness index with the available geobotanical and soil maps, it is easy to see that the positions of isolines of the dryness index agree well with the locations of the main geographical zones. The smallest values of the index of dryness (up to $\frac{1}{2}$) correspond to tundra, values from $\frac{1}{2}$ to 1 to the forest zone, from 1 to 2 to the steppe zone, more than 2 to the semi-desert zone and more than 3 to the desert zone.

Leaving aside for the moment some specific conclusions, (which can be drawn while comparing the map of the dryness index with geobotanical and soil maps), we will underline the main point, namely that the radiative dryness index (which determines the relative values of the components of the energy and water balances) agrees well with the boundaries of the major geographical zones. It is evident that, at different latitudes within the same zones, one observes considerable differences in the conditions under which the natural processes develop. These differences are associated with the fact that the energy base of natural processes which can be described by the values of the radiation balance $R$ differ at various latitudes.

Thus, while it may be possible to use a single parameter, $R/Lr$, in describing general zonal conditions of natural processes, in order to define the absolute values of the intensity of natural processes, it is necessary to apply two parameters, namely $R/Lr$ and $R$. These two parameters determine the absolute values of the components of the energy and water balances.

The relation of geobotanical conditions with the parameters $R/Lr$ and $R$ may be represented as a graph with $R/Lr$ and $R$ plotted along its axes, and on which the main geobotanical zones are divided by dotted lines. A schematic form of such a graph is given in Fig. 3.16. The continuous line in the graph gives the boundary for observed values of $R$ and $R/Lr$ (except for mountainous areas). Within these boundaries, specific values of $R/Lr$, shown as vertical lines, separate the main geobotanical zones: tundra, forest, steppe, semi-desert and desert. Great differences in the values of the radiation balance for the forest zone and somewhat smaller differences for the steppe zone correspond to appreciable geobotanical changes within these zones, although the general features of the vegetation cover remain unchanged.

Since the zonality of soils is closely connected with that of the plant cover, the conclusions obtained concerning the relationship of botanical zones with certain values of the parameters $R$ and $R/Lr$ may be fully applied to soil zones as well. In this connection it is possible to find that, with an increase in $R/Lr$, the soil types change in the following sequence: (a) tundra soils; (b) podzolic soils, brown forest soils, yellow earths, red earths and lateritic soils (the variety of soils within this group corresponds to variations in $R$ over a wide range); (c) black earths and black savanna soils; (d) chestnut soils; (e) grey desert soils. The relation of soil zonality with climatic parameters $R/Lr$ and $R$ may be shown in a general form as a diagram, similar to the graph of botanical zones in Fig. 3.16.

The relation of the zonal hydrological regime on land with the parameters $R$ and $Lr$ may be established in both quantitative and qualitative forms. It

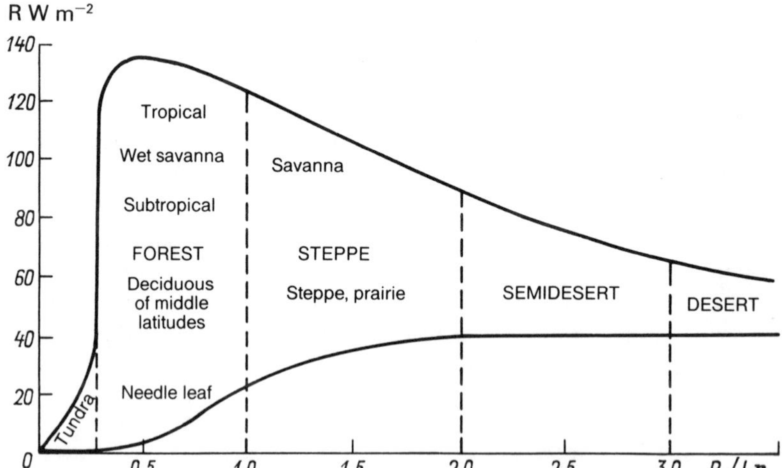


Fig. 3.16.

follows from the energy—water balance equation that to each gradation of the values of $R/Lr$ there corresponds a certain gradation of the values of the run-off coefficient. As a result, for tundra conditions were $R/Lr < \frac{1}{2}$ the run-off coefficient must be more than 0.6, for the forest zone with $\frac{1}{2} < R/Lr < 1$ the run-off coefficient must be from 0.3 to 0.6, in the steppe zone (where $1 < R/Lr < 2$) it ranges from 0.1 to 0.3, in the semi-desert and desert zones it is less than 0.1. The observational data confirm these results very well.

Thus, a consideration of the influence of energy factors allows us to explain in quantitative form zonal variations in the run-off coefficient. The absolute values of the total run-off are determined by two parameters, namely $R$ and $R/Lr$. Therefore, the distribution of total annual run-off values, shown in Fig. 3.17, is similar to that in Fig. 3.16. This graph presents the absolute values of the run-off (in cm/yr) in different geographical zones.

All this demonstrates that there is a close relation of geographical zones with two climatic parameters, namely the radiation balance of the Earth's surface and the radiative index of dryness. This relation was examined in the works of Grigoriev and the author (1956, 1962) in which it is called the periodic law of geographical zonality.

In accordance with the foregoing concept, within each latitudinal zone there exists a definite correspondence of the boundaries of geographical zones and the isolines of particular values of the radiative dryness index. In different latitudinal belts, the same values of the radiative index of dryness correspond to the geographical zones that are similar in terms of several essential features. Thus, when moving from one latitudinal belt to another values of the radiation balance change and one observes in a number of geographical zones that correspond to changes in the moisture conditions, features of similarity as well as differences, which are attributed to changes in energy conditions.

A general scheme of this pattern is presented in Table 3.8 which represents a further development of the graph given in Fig. 3.16. It should be mentioned that, in constructing the table of geographic zonality, the values of the radiation balance describing the energy regime in different latitudinal belts refer to the conditions of a fully moistened surface. The values for regions with a dry climate are higher than the observed values of the radiation balance.

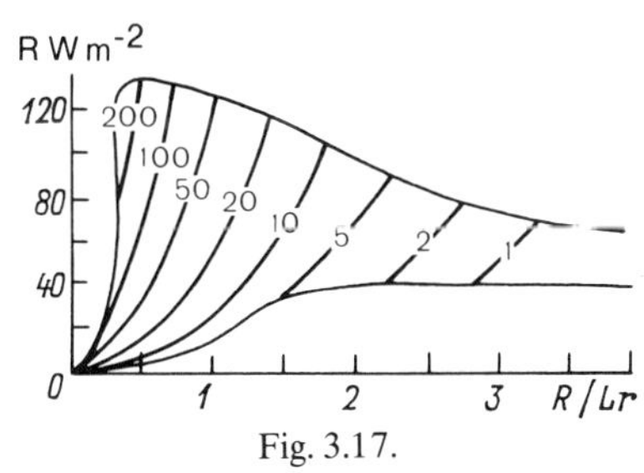


Fig. 3.17.

TABLE 3.8

Table of geographical zonality

| Energy basis — the radiation balance | Moisture conditions — the radiative index of dryness | | | | | | | | |
|---|---|---|---|---|---|---|---|---|---|
| | Less than 0 (highly excessive moisture) | From 0 to 1: Excessive moisture | | | | | From 1 to 2 (moderately insufficient moisture) | From 2 to 3 (insufficient moisture) | More than 3 (extremely insufficient moisture) |
| 1 | 2 | 3 | 4 | 5 | 6 | 7 | 8 | 9 | 10 |
| Less than 0 (high latitudes) | I Eternal snow | — | — | — | — | — | — | — | — |
| From 0 to 70 W $m^{-2}$ (Southern Arctic, Subarctic and middle latitudes) | — | IIa Arcitc desert | IIb Tundra (with small islands of sparse forest in the south) bow forest is swampy areas | IIc Northern and middle taiga | IId Southern taiga and mixed forest | IIe Deciduous forest and forest-steppe | III Steppe | IV Semi-desert of middle latitudes | V Desert of middle latitudes |
| From 70 to 100 W $m^{-2}$ (subtropical latitudes) | — | — | VIa Regions of subtropical gemigilei with substantial marshland | VIb Subtropical rain forest | | | VII Hard-leaved subtropical forest and shrub, deciduous forest | VIII Subtropical semi-desert | IX Subtropical desert |
| More than 100 W $m^{-2}$ (tropical latitudes) | — | — | Xa Regions of predominance of equatorial forest swamps | Xb Strongly superhumid equatorial forest | Xc Moderately superhumid equatorial forest | Xd Equatorial forest, passing to light tropical forest and forest-savanna | XI Dry savanna, deciduous forest | XII Desert-like savanna | XIII Tropical desert |

This table enables us to draw the following conclusion concerning the general principles regulating geographical zonality of lowland areas. Each column of the table, which shows different gradations of moisture conditions, corresponds to certain magnitudes of the run-off coefficient. At the same time, in each column, the annual normal run-off increases with an increase in the radiation balance, i.e. as the latitude decreases. (This does not apply to desert areas, where the run-off at low latitudes is close to zero.)

Similar types of plant cover correspond to each column of the table. Under conditions of excessive moisture, forest vegetation prevails at all latitudes, except in areas of highly excessive moisture (index of dryness less than $\frac{1}{2}$). In this case, forest vegetation is replaced by tundra at higher latitudes and by swamps and marshy forests at lower latitudes (since a dryness index below $\frac{1}{2}$ is observed only over limited areas in low latitudes, the regions of excessively humid conditions in low latitudes cannot be considered as individual geographical zones).

Each gradation of moisture conditions is characterized, not only by a definite type of vegetation cover, but also by the productivity of the vegetation cover. In the works of Grigoriev and the author mentioned earlier it was supposed that the productivity of the natural vegetation cover increases as moisture conditions approach optimum ones (with the same radiation balance), and with fixed moisture conditions the productivity grows as the radiation balance increases (the problem of the productivity of the vegetation cover in different geographical zones is examined more closely in Chapter 4).

Each column of the table corresponds to a definite sequence of changes in soil types, which possess substantially similar indications within the limits of the same column.

A special place in the table belongs to the area of eternal snow, whose characteristics are unique. This region is described by a negative radiation balance, a negative index of dryness, and practical absence of soils and vegetation. Similarly, the subzone of the Arctic deserts is characterized by a negligible value of the annual radiation balance and high humidity.

The general features of geographical zonality presented in the table can be explained by the following simple considerations.

Because of the spherical shape of the Earth, its surface is divided into several latitudinal belts that differ in terms of the radiation energy inflow to the Earth's surface. Within each of these belts (except the zone of eternal snow), different moisture conditions, ranging from excessive to highly insufficient, can be observed. It is easy to understand that geographical conditions with similar indices of moisture in different latitudinal belts possess a certain number of common characteristics (as well as a number of differences caused by variations in the inflow of radiation energy). If we successively compare regions in two latitudinal belts in which humidity (or dryness) increases, these common characteristics will be periodically repeated.

The table of geographical zonality establishes a relationship between geographical zones and the mean annual values of the elements of the climatic regime. It should be kept in mind that, as a rule, these values are the result of averaging variable values of the elements of climate that are characteristic of individual seasons of the year.

Taking into account the special features of the climatic regime in different seasons, we can identify the effect of climatic factors on geographical zonality more distinctly (Grigoriev and Budyko, 1962).

To specify climatic conditions of the seasons, we shall single out the following basic types of climatic regime:

(1) The Arctic climatic regime, characterized by the snow cover, negative air temperatures and a negative or near-zero radiation balance.
(2) The climatic regime of tundra, with mean monthly temperatures from zero to 10 °C and a small positive radiation balance.
(3) The climatic regime of the forest zones, with mean monthly temperatures exceeding 10 °C with a positive radiation balance and sufficient moisture (with evaporation no less than half the potential evaporation).
(4) The climatic regime of the arid zones (steppe and dry savanna), where with a positive radiation balance evaporation ranges from one-tenth to one-half of potential evaporation.
(5) The climatic regime of deserts, where, with a positive radiation balance, evaporation is less than one-tenth of potential evaporation.

If any of these types of climatic regime were observed throughout the whole year, the climatic conditions of each month would have corresponded to the geographical zonality. But in most geographical regions, several types of climatic regime follow each other throughout the year. For instance, four types of climatic regime replace each other during the year in the steppe zone of the European part of the U.S.S.R.: in winter, there is an Arctic regime, at the beginning of spring it is replaced by the tundra regime, which gives way to a climatic regime characteristic of the forest zone, after which in summertime, a climatic regime typical of the steppe zones is observed. In this case, only the last of the seasonal types of climatic regime corresponds to the given type of geographical zone.

A general description of annual changes in the climatic conditions at different latitudes of the Northern Hemisphere is presented in Fig. 3.18(a). This figure depicts the lowland areas in the zone from 20 to 40°E in Europe and Africa. Figure 3.18(b) shows the changes in climatic conditions for the eastern outskirts of Asia, and Fig. 3.18(c) for the eastern areas of North America and for South America. The data shown in these figures reflect substantial differences in the seasonal changes of the climatic regime at different longitudes. These differences are particularly pronounced in middle and low latitudes, where the type of climatic regime depends on moisture conditions.

In Europe and Africa, a regime with reduced humidity is observed within a wide latitudinal belt, which shifts northwards during the summer and southwards in wintertime. This corresponds to the most favourable moisture conditions in the subtropics in winter and spring, and in the tropical latitudes

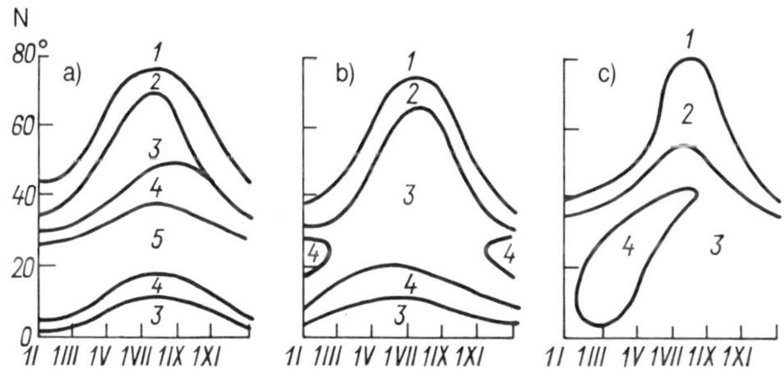


Fig. 3.18.

in summer. In eastern Asia and in North America, the structure of the moisture regime differs greatly from the first diagram, because of substantial differences in circulation processes.

Figure 3.18 shows that, in most regions at all longitudes, one finds either conditions of insufficient moisture (regimes 4, 5) or insufficient heat (regimes 1, 2) during the greatest part of the year. It is only in the narrow band near the equator that one observes throughout the entire year conditions corresponding to regime 3.

It is evident that, with regimes 2 and 4, and especially with regimes 1 and 5, the productivity of natural vegetation is reduced. In this case, we can see that under conditions of insufficient heat, the type of geographical zone is determined by the climatic regime of the period when the productivity of natural vegetation is the greatest, even if this period is relatively short. For example, the zonal landscape of tundra is determined by conditions of the warm seasons, which may last no longer than one-fifth to one-fourth of the year. In this situation, the climatic regime of the cold seasons, embracing the greater part of the year, does not determine the zonal character of the landscape.

This is, usually, also observed under conditions of insufficient moisture, when the more humid period appears to be decisive for the type of zone, even though this period might be shorter than the period of insufficient moisture. It should be noted, however, that in geographical zones of insufficient moisture, short periods with sufficient or surplus moisture may be observed. Such periods do not promote the development of typical forest vegetation, since this requires more or less prolonged soil moistening.

The indicated features that associate the type of geographical zone with seasonal variations in the climatic factors determining zonality, contribute to the concept of the influence of climatic factors on geographical zonality.

The formulation of the periodic law of geographic zonality set forth in 1956 is widely used in many contemporary studies. The possibility of applying this law to clarify the distribution of vegetation covers has been discussed in a number of papers (Armand, 1967; Bazilevich, Rodin and

Rozov, 1970; and others). It has also been shown that this law may be used in studying the zonality of geomorphological processes (Budyko *et al.*, 1970).

The significance of the periodic law for the study of the phenomenon of geographical zonality has been considered in the monograph by Kalesnik (1970).

CHAPTER 4

# PLANTS

## 4.1. Plants in the Biosphere

The people of ancient times divided living organisms into two groups: plants and animals. Today these groups are still considered the two kingdoms of living nature.

Plants and animals have some common features including a characteristic cell structure and a metabolism based on obtaining energy from the environment. The main distinctions between plants and animals are: (1) most plant cells have thick self-supporting cellulose walls (animal cells have thin delicate membranes and lack heavy supporting walls); (2) most plant growth continues throughout life (in animals growth usually slows or ceases at maturity); (3) most plants, particularly, land plants, are fixed in one place (animals generally move about freely, except for some inhabitants of the hydrosphere); (4) most plants are autotrophic, i.e. they make their own food from carbon dioxide, water and other simple chemicals in utilizing solar radiation energy during photosynthesis (all animals are heterotrophic, i.e. they feed on other animals or plants).

The role of autotrophic plants as the only source of organic matter and energy for the overwhelming majority of organisms in the biosphere is exceptionally important. The activity of these plants considerably influenced the changes in the atmospheric chemical composition, which in many respects determined the evolution of organisms. This question is discussed in detail in subsequent chapters.

As mentioned above, besides photosynthesis, there is another way of producing organic matter: chemosynthesis as carried out by some bactieria. These bacteria utilize the energy obtained in oxidizing certain mineral substances instead of solar radiation energy. Among the chemosynthetic organisms are included the nitrifying bacteria inhabiting the soil. These bacteria oxidize ammonia to nitrites and the latter to nitrates. The nitrifying bacteria produce soil nitrogen compounds available to plants. The nitrifying bacteria important for the nitrogen cycle are an example of the considerable effects that some chemosynthetic organisms have on large-scale processes in the biosphere. However, the total amount of organic matter generated in chemosynthesis is very small when compared with the global productivity of photosynthesis.

Photosynthesis, resulting in the production of organic matter, is carried out by land and water autotrophic plants utilizing a small portion (a fraction of a per cent) of incoming solar energy. The chemical energy and organic matter produced during the photosynthesis are expended partially on respiration and other life functions of a plant. The difference between the amounts of organic matter generated during the photosynthesis and that consumed during the lives of plants is the productivity of autotrophic plants.

The area where the autotrophic plants live is limited by the conditions necessary for photosynthesis. These conditions include the presence in the environment of carbon dioxide, water, some minerals (e.g. nitrogen and phosphorous compounds) and solar energy. Photosynthesis at a level required for plant productivity only occurs within a rather narrow range of temperatures. All these conditions are fulfilled for a restricted part of the biosphere, therefore the photosynthesizing plants exist within the boundaries of the 'living matter film', i.e. on the land surface and in the upper layers of water bodies. On a considerable part of the land surface there are no autotrophic plants: for example on the surface of ice sheets (because of a lack of fluid water and soil and because of low air temperature), in most arid desert regions (because of a lack of water) and on rocky surfaces (because of a lack of soil). In many oceanic regions, the productivity of autotrophic plants is limited because of a deficiency of the necessary mineral compounds for photosynthesis, resulting in insignificant photosynthetic productivity in these regions.

It should be mentioned that autotrophic plants comprise the basis of the 'living matter film' and in those regions of the Earth's surface where these plants do not grow or their activity is suppressed by unfavourable conditions, this film thins or even disappears.

The most ancient and primitive group of photosynthesizing organisms is the blue-green algae found in fresh water bodies, seas and oceans, in some hot springs, and on the soil surface. Some species of blue-green algae occur as unicellular organisms while other form multicellular colonies comprised of many unicellular organisms linked together. Nuclear matter in the cells of these algae is not bounded by a membrane (i.e. a distinct nucleus does not occur). Because of this the blue-green algae are referred to as the procaryotes.

Real algae, with a distinct nucleus, represent an extensive group including both unicellular and multicellular organisms. Though algae differ from higher plants by their reproductive mechanisms, they have common structural and physiological features.

Some brown algae, certain species of which can grow to 50 m long, are the largest algae. Green algae are widespread in fresh water bodies but are also found in the seas and oceans, soil and on trees. Red algae form an individual group because distinctive features of their chemistry enable them to carry out photosynthesis at comparatively great depths (to 100 m) in water bodies, where attenuated solar radiation consists mainly of the blue and violet wavelengths of the spectrum.

The land plants appeared much later than the hydrosphere plants. Therefore amongst the land plants, representatives of more evolutionary advanced groups dominate.

The most ancient group of land plants includes the mosses and related forms. Mosses bear great resemblance to blue-green algae, from which they probably originated. Most land plants are classed as vascular plants and include the ancient groups of club mosses and horsetails as well as the later appearing ferns, gymnosperms and flowering plants (or angiosperms). These groups differ from each other with respect to their reproductive mechanisms which are most complex in the seed plants (gymnosperms and angiosperms).

In seed plants, photosynthesis usually takes place in leaves covered with an epidermis which contains numerous microscopic openings (stomata) which can, in certain conditions, open and close. When stomata are open, atmospheric carbon dioxide penetrates inside the leaf where it is absorbed by chlorenchyma. Leaf cells contain the chloroplasts. Chlorophyll is able to absorb solar energy and to utilize this for synthesizing organic matter from carbon dioxide, water contained in plant tissues, and from mineral substances extracted from soil and dissolved in this water.

Seed plants, of which more than 200 000 species occur, form the majority of the biomass in the modern biosphere. Among the seed plants there are very small plants, but many trees are of considerable stature and in some cases can reach heights of more than 100 metres. The diverse group of seed plants are the characteristic representatives of typical vegetation covers in most geographical zones.

Heterotrophic plants devoid of chlorophyll are usually considered to belong to lower and, as a rule, more ancient groups. Among these, the fungi, which differ considerably from all other plants, are the most significant. Because of this, this group is sometimes placed in a special kingdom.

Microscopic bacteria with very small cells (of the order of 1 μm) play a significant part in the biosphere. Most bacteria are uniclellular organisms, some being composed of chains of cells connected in series. Unlike most plants, many bacteria can actively move in water.

Fungi, bacteria and other heterotrophic organisms breakdown organic matter primarily created by autotrophic plants. Their activity, together with the influence of abiotic factors, provides a rapid transformation of organic matter to mineral compounds, which is important in the cycling of carbon and other elements in the biosphere.

Plants can live only in a limited range of environmental conditions. For all plants without exception, the thermal and water regimes of the environment are of great importance. This is discussed in more detail in the second section of this chapter.

The main conditions necessary for the survival of photosynthesizing plants have been described earlier. It should be pointed out that some of the factors which enable plants to live in the modern biosphere are far from optimum, i.e. those which promote high productivity of photosynthesis. In particular, the carbon dioxide concentration in the atmosphere (close to 0.03% by volume) is considerably lower than that at which the highest productivity of photosynthesis for most autotrophic plants can occur. In addition, the amount of oxygen in the modern atmosphere (about 21% by volume) is much greater than that at which maximum photosynthetic productivity occurs.

Therefore, there is a supposition that autotrophic plants first appeared at a time when increased carbon dioxide and decreased oxygen concentrations occurred in the atmosphere, compared with those present today.

Photosynthesis is discussed in more detail in the third section of this chapter.

Since phototrophic plants are the basis of the 'energy pyramid' describing the energy transformations in the biosphere, the question of their productivity is of fundamental significance for studying global biospheric processes. This question is treated in the fourth section of this chapter.

Plants are the components of ecological systems which connect them with other biotic components of these systems as well as with abiotic factors of the environment. The question concerning these connections is addressed in Chapter 5.

The activity of autotrophic plants is a most essential factor for the evolution of the biosphere, in particular, due to the influence of this activity on the atmospheric chemical composition. Autotrophic plants absorb, in the course of photosynthesis, atmospheric carbon dioxide. Carbon dioxide in oxidizing organic matter does not fully return to the atmosphere because of the loss of a portion of this carbon dioxide in the formation of sediments. As a result, other things being equal, an increasing productivity of photosynthesis should cause a decrease in atmospheric carbon dioxide concentration.

Of great importance for evolution of the biosphere is the effect of photosynthesizing plants on atmospheric oxygen. Photosynthesizing plants utilize carbon from the atmosphere and give off oxygen. This creates a tendency towards increasing the concentration of atmospheric oxygen as the productivity of photosynthesis increases.

Changes in the chemical composition of the atmosphere due to the activity of autotrophic plants exert a pronounced effect on evolution of living organisms, climate genesis, erosion of exposed rocks and other global processes. These aspects of the activity of autotrophic plants are discussed in Chapters 7 and 8.

## 4.2. The Thermal and Water Regimes of Plants

The thermal conditions necessary for life are restricted to comparatively narrow limits. Most reactions in biological systems take place at temperatures of organisms from 10 to 50 °C. These limits are somewhat expanded for a number of organisms. In particular, photosynthesis of many plants can occur at lower temperatures, although in this case productivity is not great. It is known that some plants and bacteria adapt to thermal regime of hot springs with water temperature reaching 80 °C or even more. Other plants (cryophytes) can exist in snow and ice, i.e. at constantly below freezing temperatures. Nevertheless, optimum conditions for the existence of most living organisms fall within a narrow temperature range.

It should be pointed out that the temperature range of the environment in which organisms can exist is frequently wider than that favourable for their life activity. This is due to two causes. Firstly, the temperature of organisms often differs from the temperature of the environment in which organisms live. This difference is essential for plants, but is most significant for organisms able to regulate their body temperature as discussed below in detail. Secondly, organisms can, for short time periods, easily withstand the effects of higher or, more frequently lower temperatures than they normally experience. However, organisms can only survive for limited periods of time when exposed to high or low lethal temperatures, this time period varying widely for different organisms.

The rate of biological reactions is largely dependent on the temperature at which they occur. For quantitative estimation of such dependence, so-called temperature coefficients are employed which show how many times the rate of the reaction increases with a 10 °C temperature rise. The magnitude of these temperature coefficients for different biological processes varies, ranging from 1.2—1.4 to, in many cases, about 2. Usually, the rate at which reactions occur with rising temperature increases to a certain level after which the growth of the reaction rate ceases and can start to decrease.

Temperature exerts a considerable influence on almost all biological processes in plants. This occur through a number of different mechanisms. For instance, the solubility of the major plant metabolic compounds (e.g. carbon dioxide and oxygen) in the cell fluid noticeably decreases with increasing temperature. Changing the carbon dioxide concentration in the cell fluid alters the fluids acidity, which in turn influences a number of other biochemical processes in plants.

The thermal conditions considerably affect the activity of ferments and the major process of life activity of autotrophic plants — photosynthesis that determines their productivity.

In addition to having specific temperature requirements for growth, plants also require favourable moisture conditions. The majority of the plant's biomass consists of water which is constantly replenished. Without a sufficient amount of water, the process of photosynthesis and many other important physiological functions in autotrophic plants would be impossible.

The thermal and moisture regimes of plants inhabiting the hydrosphere do not require detailed discussion. In this case, plant temperature does not differ appreciably from the water temperature. Within the hydrosphere the water temperature can range from close to zero to approximately 30 °C. This range in temperatures enables plants to exist in all regions of the hydrosphere, although the distribution of individual stenothermal species is frequently restricted to rather narrow temperature ranges, which is reflected in the limited spatial distributions of these species. Of considerable importance to water plants is the chemical composition of the water and, in particular, the content of mineral compounds necessary for plant's growth and development. As mentioned earlier, the lack of such compounds (phosphorus, in particular) results in extreme improverishment of plant communities in many regions of the oceans. At the opposite extreme, the upper layers of a water body can be overgrown with plants if there is an excess of mineral substances necessary for their survival and growth.

The thermal and water regimes of land plants are much more complicated. For autotrophic plants, the dependence of photosynthesis on temperature is important, especially at temperatures close to freezing point. Under these conditions, photosynthesis is largely limited by enzyme reactions with reaction rate temperature coefficients of more than 2, therefore the photosynthesis rate rapidly increases with increasing temperature. At high temperatures the increase in the rate of photosynthesis with increasing temperature slows down. Optimum temperatures for photosynthesis are between 20 and 30 °C. A further increase in temperature leads to decreasing rates of photosynthesis.

It should be noted that in some plants, photosynthesis can also occur when temperatures are below freezing point, although the rate of photosynthesis under such conditions is very low. The dependence of plant respiration on temperature is more simple with the rate of respiration increasing with high temperature coefficients as temperatures rise.

Temperature has a pronounced effect on the growth and development of plants. At temperatures from 0 to 15 °C, the rate of growth rapidly increases with rising the temperature; at 15—30 °C the increase in the growth rate slows down and at temperatures higher than 30 °C it begins decreasing. The lowering of temperature at night frequently produces a favourable effect on the growth of plants. The mechanism responsible for this effect has been insufficiently studied. The only point clear is that growth is not limited by the decrease in plant respiration due to lowered temperatures during the period without photosynthesis.

Since ancient times, the dependence of plant growth on temperature has been taken into account in agriculture when estimating the heat resources of different regions. In the middle of the 19th century de Candolle proposed to sum the temperatures of individual days to estimate heat resources. The method of 'temperature sums' and its variations are still used in agrometeorology today.

The effects of temperature on the development of plants are rather complicated. The character of these effects varies at different stages of a plant's life. For instance, many plants need periods of noticeable temperature decrease or otherwise the ability to flower is lost or other functions of the plants are disturbed.

For all plants, there are the temperature limits within which they can live.

At temperatures below freezing, plants are injured by the formation of ice crystals in their tissues. The ice crystals destroy the walls of cells both directly and as a result of water movement in the tissues during the freezing process.

The size of a temperature decrease which different plants can withstand varies considerably for different plants. The size of this temperature decrease also depends on the rate at which the temperature is lowered (plants can more easily withstand a slow cooling than a rapid one) and on the degree of acclimatization of the plant (i.e. the number of coolings the plant has previously experienced).

An increase in temperature above a certain limit can also harm plants. For example this can cause the destruction of chlorophyll, leaf burn, and other damage that with a further rise of temperature can lead to decompositon of the plant. The decay of individual protein compounds in plant tissues is an important mechanism of damage from plant overheating.

Resistance to high-temperature damage is different for different plants. Succulent desert plants can endure the highest temperatures without damage.

Thermal factors thus exert a strong and controlling influence on the life activity of plants. The geographical distribution of plants, their seasonal changes, the species composition and productivity of vegetation cover all depend on thermal conditions.

Air temperature data, recorded at a certain height in a meteorological box, is frequently used in studying the influence of thermal factors on plant's growth and development. At the same time, it has long been known that the air temperature only approximately describes the thermal conditions that directly affect plants, due to the complicated interactions between heat exchange processes.

This opinion has been expressed in particular, by K. S. Veselovsky more than 100 years ago in his prominent work *On the Climate of Russia* (1857). He wrote: "Since physiological processes in plants and, in general, phenomena in organic life occur more or less under the direct action of solar rays, the study of climate in relation to thermal conditions would be incomplete if we restricted ourselves to the air temperature only. Therefore, in the chapter that treats heat we will speak, first, of air heat and then of heating power of solar rays or the temperature that the bodies on the Earth's surface gain when heated by the Sun." In developing this thought, Veselovsky pointed out: ". . . even by different causes that influence the air heating and by the conditions that determine the heating of the Earth's surface we can anticipate that . . . annual isotherms for air and for the Earth's surface will not correlate everywhere." To illustrate this proposition, Veselovsky gave several examples demonstrating sharp differences in the state of the vegetation cover at a constant air temperature but with different influxes of solar energy. It should be admitted that these views of Veselovsky were considerably ahead of the scientific thinking of his time and that the problems he outlined concerning the determination of the parameters of the thermal regime directly influencing plant life were not solved for a long time.

Experimental data on the actual temperature of plants can be obtained using special instruments; thermoelectric thermometers (with sensors looking like thin nets) or radiation thermometers, taking temperature readings by measuring the flux of longwave radiation. Some results of measurements made with such devices are plotted in Figs 4.1, 4.2 and 4.3 (Budyko, 1958). Figure 4.1 shows the average daily variation in the difference between the surface and air temperature from observations of a grass plot near Leningrad. Solid curves represent the observation results (1 — the temperature of plant surface taken by radiation thermometer and 2 — by thermoelectric thermometer).

The data in Fig. 4.1 show the magnitudes that differences between the surface and air temperature can reach in regions with excessive moisture. As seen from the graph, in moist climate conditions the average daytime differences between these temperatures may exceed 10 °C. From the data in Fig. 4.1 it follows that even under conditions of abundant moisture, the temperature of plants may differ from that of the air by a figure comparable with the geographic variation of temperature over distances of thousands of kilometres. This difference is, obviously the result of heating of the plant surface by the Sun's rays.

Figure 4.2 presents the observational data on the difference between the mean surface and air temperatures (solid curve) for a desert region and, in Fig. 4.3, for an oasis in central Asia. The data in Fig. 4.2 are interesting as they are characteristic of the limiting temperature difference that can exist between the plant and the atmosphere in desert conditions for plants that are close to the surface and which lose little heat in evaporation. Figure 4.3 demonstrates the difference between the temperature of cotton leaves and the air under conditions of abundant moisture from irrigation. As seen from Figs 4.2 and 4.3, the heat conditions of plants in deserts differ considerably,

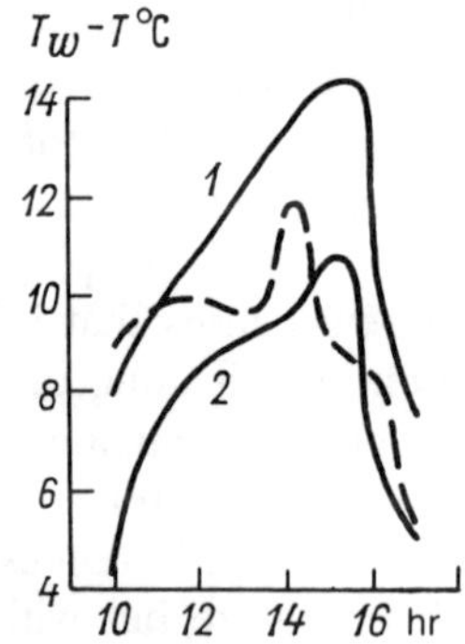


Fig. 4.1.

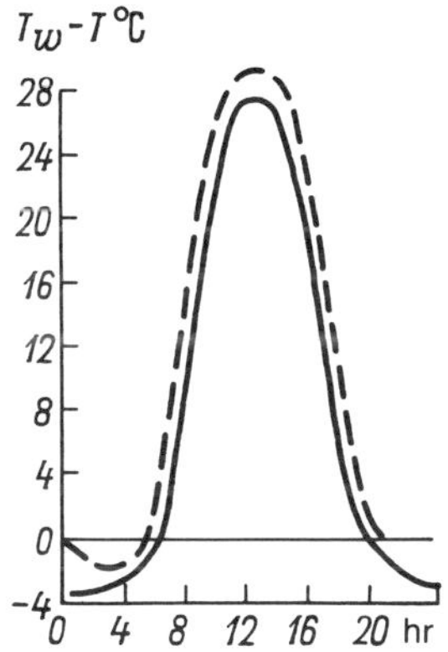


Fig. 4.2.

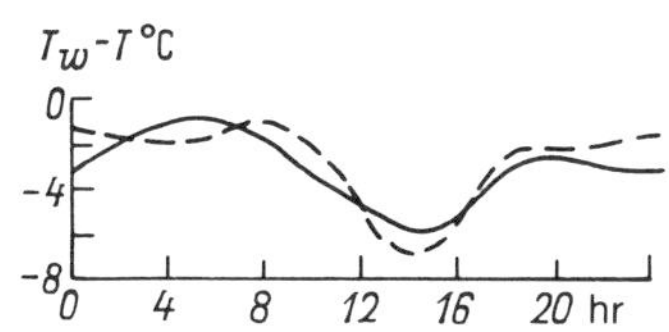


Fig. 4.3.

depending on the moisture conditions. Without irrigation, daytime plant temperatures may be considerably higher than air temperatures, but with irrigation providing a means to lose excess heat through evaporation, plant temperature are noticeably lower than air temperature.

Thus it becomes clear that the actual temperature of plant tissue often does not coincide with the air temperature. This conclusion is confirmed by observations carried out in many studies under a variety of climatic conditions. Since mass observation data on plant temperature obtained by special instruments which could be used instead of these on air temperature in studying the climatic effects on plants are not available to us, the question arises whether it is possible to obtain such information by calculations based on standard observation material from meteorological stations. This problem might be solved by determining the energy balance of the vegetation cover (Budyko, 1958).

Determination of plant temperature by the energy balance method is based on the following considerations. The equation of the energy balance of the land surface has the form

$$R = LE + P + A, \tag{4.1}$$

where $R$ is the radiation balance, $L$ the latent heat of vaporization, $E$ the evaporation, $P$ the turbulent heat flux, and $A$ the inflow of heat of soil. After transforming the components characterizing radiation balance and turbulent heat exchange, the equation can be presented in the form

$$R_0 - 4\delta\sigma T^3(T_w - T) = LE + \rho c_p D(T_w - T) + A, \tag{4.2}$$

where $R_0$ is the radiation balance of the Earth's surface obtained in determining the net longwave radiation from air temperature, $\delta$ the coefficient characterizing the difference between the properties of the radiating surface and a black body, $\sigma$ the Stefan constant, $T$ the air temperature, $T_w$ the temperature of the Earth's surface, $\rho$ the air density, $c_p$ the heat capacity of the air, and $D$ the integral coefficient of diffusion. It follows that

$$T_w - T = \frac{R_0 - LE - A}{\rho c_p D + 4\delta\sigma T^3}. \tag{4.3}$$

This formula can then be used to calculate the temperature of the Earth's surface ($T_w$) in cases where the air temperature ($T$) the energy balance components ($R_0$, $LE$, $A$) and the integral coefficient of diffusion ($D$) are known.

The value of $D$ is dependent on wind speed and various properties of the Earth's surface and can often be considered constant in approximate calculations. Its average value is about 1 cm s$^{-1}$ for daytime conditions, and about 0.6 cm s$^{-1}$ for mean daily conditions in summer.

In some cases, especially under conditions of abundant moisture, the temperature of the active surface can be derived from (4.2) on the assumption that

$$E = \rho D(q_s - q) \tag{4.4}$$

(where $q_s$ is the saturation specific humidity at the temperature of the active surace, and $q$ the specific humidity), taking into account the dependence of $q_s$ on $T_w$.

In a closed vegetation cover, the leaves of plants usually constitute a basic part of the 'Earth's surface', where heat and water exchanges with the atmosphere take place. In this case, the mean temperature of the plant leaves can be considered approximately equal to the temperature of the Earth's surface as found from (4.3).

The results of applying the energy balance method to calculate the Earth's surface temperature are presented in Figs 4.1—4.3, where they are shown by a broken line. It is apparent from the figures that the results obtained are in good agreement with the observed data. This clarifies the possibility of using the energy balance method for determination of the temperature of the vegetation cover.

In this connection, the problem has been posed of obtaining a great body of data on the Earth's surface temperature and constructing maps of its distribution. For this purpose, Yefimova has calculated the monthly means of the difference between the surface and air temperature in the European

U.S.S.R. She then constructed a series of maps of this difference and of the Earth's surface temperature for each month of the warm season. In her calculations, she used a mean value of the coefficient $D$ equal to 0.63 cm s$^{-1}$. One of these maps, showing the monthly means of the difference $T_w - T$ in July, is presented in Fig. 4.4. The distinctive features of the geographical distribution of the difference $T_w - T$ seen in this map are also typical of other months of the warm half-year.

As seen from the map, the difference $T_w - T$ increases from the central and western regions of the European U.S.S.R. towards the south and south-east. Such changes can be attributed to a reduction in heat loss through evaporation, owing to a decrease in soil moisture in these areas. At first glance, the increase in the temperature difference in northern regions of the European U.S.S.R. is somewhat unexpected. This increase is, apparently, connected with conditions associated with the transformation of air masses with comparatively low temperatures that come into the continent from the ocean in summer and are then gradually heated as they move towards the interior.

The energy balance method can also be used for investigating another problem relating to the temperature of vegetation cover. As is well known, the growth of various types of vegetation on more or less steep north and south facing slopes is dependent on the different thermal conditions. However, observations of air temperature made at the level of the meteorological box do not reveal any substantial difference between the thermal conditions of north and south facing slopes in regions with hilly relief. It is evident from

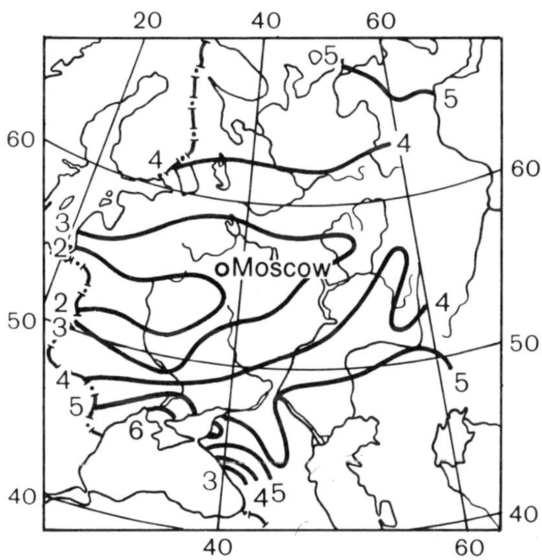


Fig. 4.4.

general considerations that the difference between the heating conditions on the two aspects can be attributed to the difference in the radiation regime and therefore should be most obvious at the Earth's surface. At higher levels, the turbulent mixing in the atmosphere smoothes out the temperature differences over specific elements of the Earth's surface. Thus, for a quantitative estimation of the thermal effects of aspect on vegetation, the data on surface temperature of slopes of different steepness and orientation are necessary.

These data can be derived from calculations based on the application of the afore-mentioned energy balance equations. In this case, the changes in the radiation balance values on slopes, resulting from variations in direct solar radiation and in net longwave radiation, should be taken into account. It is evident that the results of such calculations can be useful in explaining the microclimatic effects on vegetation cover under conditions of rugged relief.

The water regime of land plants mainly depends on the moisture content of the soil layer in which plant roots are located. Observational data on soil moisture are available for the U.S.S.R. However, these data refer only to agriculture fields, i.e. to a restricted part of the territory. The quantities of soil moisture for the entire continental surface can be determined by taking into account the soil moisture—evaporation relationship (Budyko, 1971).

At the beginning of the 20th century, it was known that the process of evaporation from the soil surface occurred in several different phases. In the first phase of evaporation, when a considerable amount of moisture is present in the soil, the evaporation rate depends little on soil moisture but rather, is determined, by external meteorological factors. This observation has been corroborated in the studies of Alpatiev (1950, 1954), who showed that for agricultural fields, the first phase of evaporation occurs over a comparatively wide range of soil moisture conditions. In this case, evaporation is equal to the value of potential evaporation, which is defined by meteorological conditions and to a limited extent depends on the properties of the evaporating surface.

The value of potential evaporation from land surface, if it is sufficiently moist, can be determined by a method analogous to that used for estimating evaporation from a water surface. In this case, to calculate potential evaporation we can use Equation (4.4) in the form

$$E_0 = \rho D(q_s - q), \tag{4.5}$$

where $\rho$ is the air density, $D$ the diffusion coefficient, $q_s$ the saturation air specific humidity at the temperature of the evaporating surface, $q$ the atmospheric specific humidity at the level of measurements in mass observations.

As can be seen from (4.5), to calculate potential evaporation it is necessary to know the atmospheric humidity, the integral coefficient of diffusion, and the temperature of the evaporating surface determining the $q_s$ value.

As noted above, to evaluate the temperature of the evaporating surface, the energy balance Equation (4.1) can be used. The value of the radiation balance can be presented as

$$R = R_0 - 4\delta\sigma T^3(T_w - T),$$

where $\delta$ is the coefficient characterizing the properties of the radiating surface, $\sigma$ the Stefan constant, $T$ the air temperature, $T_w$ the temperature of the active surface, and $R_0$ the radiation balance for a moist surface estimated in determining the net longwave radiation from the air temperature. For the sensible heat flux, in analogy to (4.4), we can write the formula $P = \rho c_p D(T_w - T)$, where $c_p$ is the specific air heat capacity at constant pressure. Taking these expressions into account, we obtain from (4.1)

$$R_0 - A = L\rho D(q_s - q) + (\rho c_p D + 4\delta\sigma T^3)(T_w - T). \qquad (4.6)$$

From (4.6), having data on air humidity $q$ and temperature $T$ and knowing the $D$ and $R_0 - A$ values, we can calculate the $T_w$ and $q_s$ values connected by a known physical relationship (the Magnus formula). This calculation will permit us then to derive potential evaporation $E_0$ from (4.5).

A similar method has been proposed in the studies by Penman (1948, etc.), who used it to determine evaporation from a water surface. Penman supposed that potential evaporation from a land surface is proportional to that from a water body under the same external conditions.

As the soil becomes drier and finally reaches its critical moisture content, evaporation passes into a second phase. In this phase the evaporation rate rapidly decreases with reduction in the soil moisture content. There are many experimental data which show that the relationship between the rate of evaporation from the soil and the soil moisture content can be considered as close to a linear one. On the basis of these considerations, we can construct a quantitative model, dependent on meteorological elements observed at a network of stations allowing the calculation of evaporation for individual periods (Budyko, 1950; Budyko and Zubenok, 1961).

Assume that with the moisture of the upper soil layer $w$ being equal to or more than a certain critical value $w_0$ evaporation $E$ equals potential evaporation $E_0$. When $w < w_0$ evaporation is reduced as the moisture decreases by the formula

$$E = E_0 \frac{w}{w_0}, \qquad (4.7)$$

the $w$ and $w_0$ values characterizing the amount of available water in the soil (that is, accessible to plants).

In determining the values of critical soil moisture $w_0$ it has been found that for the upper soil layer of 1 m depth, the value of $w_0$ is usually equal to a layer of available moisture of 10—20 cm, these values varying slightly and depending on geographical conditions and season. As a result of analysis of data on soil water balance, Zubenok (1968) obtained the following rounded values of the $w_0$ parameter for various climatic conditions in the U.S.S.R.

In the forest zone, the value of $w_0$ varies from 20 cm at a monthly mean air temperature below 10 °C in spring, to 15 cm in summer. In the forest-steppe, steppe, and semi-desert zones, $w_0$ in spring amounts to 17 cm. In summer, the values of $w_0$ are reduced to 12 cm in the forest-steppe zone and to 10 cm in the steppe and semi-desert zones. With an autumn decrease in monthly mean temperature below 3 °C in all zones, an increase of $w_0$ is typical. Under desert conditions, $w_0$ is about 15 cm, increasing to 30 cm in those summer months when the potential evaporation exceeds 20 cm month$^{-1}$. Lack of any noticeable annual cycle of critical humidity is characteristic of the permafrost zone, where the monthly means of $w_0$ are approximately equal to 20 cm.

Variations of the value of $w_0$ are associated with changes in the state of vegetation and to some extent show the activity of the root system of plants. The more developed the root system and the more efficiently it extracts water from soil, the smaller is the value of $w_0$.

Since the value of soil moisture $w$ in most regions cannot be obtained from observational data, in order to calculate evaporation in individual months, it is also possible to use the water balance equation. This equation has the following form

$$r = E + \Delta w + f, \tag{4.8}$$

where $r$ is precipitation, $\Delta w$ the change in water content of the active soil layer (in which water cycle mainly occurs), and $f$ the horizontal flow, equal to the sum of surface and ground run-off for the given time interval and region in question.

It should be borne in mind that the $f$ value does not necessarily coincide with the value of stream run-off in the given time period, since the formation of stream run-off is a rather long process. The horizontal flow included in (4.8) approaches the value of stream run-off only for the mean annual period.

Calculation of evaporation by (4.7) and (4.8) is noticeably simplified for climates with insufficient moisture as there is no horizontal flow. Under these conditions, the water balance equation has the form

$$r = E + \Delta w. \tag{4.9}$$

If the mean moisture of the active soil layer is designated by $w_1$ at the beginning of a month and by $w_2$ at the end of it, then (4.9) can be rewritten as

$$r = E + w_2 - w_1. \tag{4.10}$$

In this case,

$$E = E_0 \frac{w_1 + w_2}{2w_0} \quad \text{at} \quad 0 < \frac{w_1 + w_2}{2} < w_0 \tag{4.11}$$

and

$$E = E_0 \quad \text{at} \quad \frac{w_1 + w_2}{2} \geqslant w_0. \tag{4.12}$$

To calculate evaporation in each month, a method of successive approximations can be used. Choosing at random a value for soil moisture at the beginning of the first month, we can calculate, with the help of (4.10), the moisture at the end of this month $w_2$ and determine evaporation from (4.11) and (4.12). A similar calculation can then be carried out for the next month, assuming that the value $w_1$ is equal to the value $w_2$ for the first month, and so on. After summing up all the calculated monthly values of evaporation the result can be compared with the annual sum of precipitation. If the latter is more than the computed evaporation, it means that the $w_1$ value chosen for the first month was underestimated, and if it is less, then $w_1$ was overestimated. Changing $w_1$ in the appropriate direction and repeating the calculation, it is possible to find by the method of successive approximations a value of $w_1$ in the first month for which the annual sums of precipitation and evaporation will coincide. In this case, the obtained monthly values of evaporation represent the desired result.

Such method of calculation can be applied to areas with insufficient precipitation when in each month precipitation is much less than potential evaporation. Such conditions are observed in many desert, savanna, and dry steppe regions in low latitudes.

It must be noted that this simplified scheme can be used for determination of evaporation both with run-off, in case it appears chiefly as a result of spring snow melting, and if its annual mean is known from observational data. Calculation of evaporation and soil moisture, in this case, is carried out for the warm season with the mean temperature in every month above zero. When determining evaporation for the first month with temperatures above freezing, it is necessary to add to the total precipitation in this month the sum of the precipitation during the period with temperatures below zero and subtract from this sum the annual run-off. In determining the water balance equation by the method of successive approximations, the evaporation over the year should be compared with the difference between the annual sums of precipitation and run-off. This method can be used, for example, to calculate evaporation for many regions of the U.S.S.R. with a climate characterized by either insufficient or moderate moisture supplies.

The above models for calculation of monthly evaporation values do not solve the problem of determining evaporation in cases when the annual sums of precipitation are close to the values of potential evaporation and when run-off is formed in different seasons. To determine evaporation under different conditions of moisture, a more general model can be used which takes into account the water balance formula for every month as the full Equation (4.8).

Since the value of horizontal flow in every month, as a rule, cannot be found from observational data, it must be derived from other hydrometeorological elements. This task is complicated by the fact that horizontal flow is, to a considerable extent, associated with extreme values of the elements of the hydrometeorological regime and to a lesser degree, depends on the monthly means used in our calculations. However, considering certain relationships between mean and extreme values of the hydrological elements we can propose an approximate scheme for calculating run-off values, which is based on the following statements.

For each month the run-off coefficient $f/r$ varies from 0 to 1 (when calculating the run-off coefficient during the melting of snow, the sum of solid precipitation during the cold season should be added to the sum of precipitation for this period). It is evident that the run-off coefficient depends on the soil moisture. It might be considered that for a completely dry soil at $w/w_k = 0$ run-off is also equal to zero ($w_k$ designates the greatest supply of available moisture that can remain in the upper layers of soil with no connection with the ground water). With increasing moisture the run-off coefficient increases, reaching its maximum at $w/w_k = 1$. This relation can be presented in the form

$$\frac{f}{r} = \mu \frac{w}{w_k}, \tag{4.13}$$

where $\mu$ is the dimensionless coefficient of proportionality. It is obvious that the value of this coefficient will depend on the intensity of precipitation and will increase in regions experiencing frequent convective rains.

Equation (4.13) can be used for estimating run-off under conditions of insufficient moisture, when the monthly value of potential evaporation is greater than precipitation ($E_0 > r$). Under conditions of plentiful moisture, when precipitation is greater than potential evaporation, run-off is immediately influenced by the value of the difference $r - E_0$. With complete wetting of the soil, when $w/w_k = 1$, the value of run-off will approach the value of $r - E_0$ (i.e. in this case, the run-off coefficient will be close to the value $1 - E_0/r$). At $w/w_k < 1$ and $r > E_0$ the run-off coefficient will be less than the indicated value, its value being dependent both on the parameter $\xi = 1 - E_0/r$ and also on the intensity of the rains.

We present this dependence in the form

$$\frac{f}{r} = \frac{w}{w_k} \sqrt{\mu^2(1 - \xi^2) + \xi^2}\,. \tag{4.14}$$

Thus we obtain an equation for calculating the horizontal flow:

$$f = r\frac{w}{w_k}\sqrt{\mu^2\left[1-\left(1-\frac{E_0}{r}\right)^2\right]+\left(1-\frac{E_0}{r}\right)^2} \quad \text{at} \quad r > E_0, \qquad (4.15)$$

$$f = \mu r\frac{w}{w_k} \quad \text{at} \quad r < E_0. \qquad (4.16)$$

Note that these equations take into account two forms of run-off with different physical meanings. The first, having taken place at $r < E_0$, is a consequence of short falls of heavy precipitation that create a brief over-saturation of the soil and are, to a known degree, disposed of as run-off. The sum of this form of run-off is equal to $\mu rw/w_k$. In cases when $r > E_0$, second form of run-off also occurs. This appears as a result of surplus moisture being available throughout the month. Such run-off does not depend on the character of the rainfall, and with high soil moisture approximates to the difference between the precipitation and the potential evaporation.

It must be indicated that Equations (4.13)—(4.16) are interpolative and their accuracy is limited. Application of such formulas to calculation of evaporation is justified in those sufficiently frequent cases when changes in run-off have relatively little effect on the values of evaporation.

The accuracy of calculations of the annual variations in run-off by Equations (4.15) and (4.16) can be increased if there are reliable observational data on annual run-off normals in the given region. In this case, the annual mean of the $\mu$ coefficient can be calculated by comparison of the annual sum of calculated run-off with observational data. If there are no observational run-off data, the $\mu$ parameter can be derived from data on the annual sum of run-off obtained by climatological calculation methods, or from approximate values of this parameter for different geographical zones.

The calculations have shown that the coefficient $\mu$ is at a minimum in high and middle latitudes. In subtropical and tropical latitudes, it increases as a result of frequent rain showers. In calculations necessary for constructing world maps of evaporation, we used the mean value of $\mu = 0.2$ for regions north of latitude 45°N and south of latitude 45°S. In the belt between 45°N and 45°S the coefficient $\mu$ was assumed equal to 0.4—0.8, depending on moisture conditions.

Comparisons of monthly evaporation means calculated by models using water balance data are described in the paper by Budyko and Zubenok (1961).

In Zubenok (1956, 1968, and others), these models have been used for calculating evaporation in individual months of different years. The results obtained were compared with those in evaporation calculated by data from gradient observations of the temperature and humidity of the lower air layer at stations in various regions of the U.S.S.R. This comparison showed that independent methods for determining evaporation provide sufficiently close results which are in particularly good agreement in regions with sufficient moisture.

The application of these methods has resulted in the calculation of monthly and annual evaporation values for the surface of continents. These values have been used in constructing maps of evaporation which together with the maps of potential evaporation have been published in the *Atlas of the World Water Balance* (1974). By using the data from these maps and Equation (4.7) it is possible to find the characteristic values of soil moisture for different geographical regions.

Another method for determining the spatial distribution of soil moisture is based on using models describing the general circulation of the atmosphere. Manabe and Wetherald (1980) and Manabe and Stouffer (1980) have constructed such maps. The results of these studies are similar to those calculated by the method described in this Section.

In conclusion, the data obtained on the distribution of soil moisture show a decrease from geographical zones with excessive moisture towards these with increasing aridity (steppes, semi-deserts and deserts).

## 4.3. Photosynthesis

In the biosphere, organic matter is largely created by phototrophic plants during the process of photosynthesis. In order to build organic matter, photosynthesizing land plants utilize carbon dioxide, water and a small amount (compared to the total mass of synthesized matter) of mineral substances from the soil. As well as receiving carbon dioxide from the air, plants may also receive some carbon dioxide from the soil, but it is the first of these sources that is important. A portion of the energy of incoming shortwave solar radiation is also used in photosynthesis.

The basic components of the overall photosynthesis reaction can be expressed in a simplified way in the equation

$$6CO_2 + 6H_2O = C_6H_{12}O_6 + 6O_2.$$

In numerous experimental investigations, it has been found that the efficiency of photosynthesizing vegetation (that is, the ratio of energy spent in synthesizing biomass to the total amount of incoming solar energy) is very small and usually does not exceed 0.1—1%. Under favourable conditions, the coefficient can increase up to a value of several per cent (Nichiporovich, 1973).

At the same time it should be indicated that according to the extensive experimental data, vegetation makes use of water resources rather uneconomically. The productivity of transpiration (the ratio of an increase in plant dry weight to water expenditure in transpiration for a given time interval) usually ranges from 1/200 to 1/1000 (most often about 1/300). The view has long been expressed that such copious transpiration does not correspond to physiological needs of the plant and is, to a considerable extent, a waste of water (Maksimov, 1926, 1944, and others).

These two fundamental factors indicate that in natural conditions, vegetation assimilates only a small part of the available energy and water resources. It is evident that for investigating plant productivity, the factors that strongly restrict the utilization of natural resources in the synthesis of biomass should be identified. For this purpose, we turn to an analysis of the physical mechanism of carbon dioxide-assimilation and transpiration (Budyko, 1949, 1956).

The assimilation organ of a photosynthesizing plant — a leaf — represents a case made of more or less thick tissue pierced by numerous small openings, stomata. The case encloses a very large surface of cells containing chloroplasts. This surface is exposed to the atmosphere through intercellular spaces and stomata.

It is essential for photosynthesis that the inner surfaces of the leaf should be maintained in a wet state, since carbon dioxide can be assimilated only in dissolved form. For this reason, the relative humidity of the air in the intercellular spaces is high and usually exceeds the relative humidity of the atmosphere by a large amount. Diffusion of carbon dioxide into a leaf through the open stomata is inevitably accompanied with diffusion of water vapour in the opposite direction (i.e. plant transpiration).

If a vegetation cover is sufficiently closed, then for the layer containing the plants we can write the following energy balance equation

$$R = LE + P + lA, \tag{4.17}$$

where $R$ is the radiation balance, $E$ the evaporation (transpiration), $L$ the latent heat of vaporization, $P$ the sensible heat flux from the Earth's surface to the atmosphere, $A$ the assimilation, and $l$ the heat expenditure for assimilation of a unit mass of carbon dioxide.

Assuming that for a sufficiently dense vegetation cover evaporation is mainly determined by the process of transpiration, we will point out that the three components entering the right-hand part of Equation (4.17), namely, evaporation ($E$), sensible heat flux ($P$) and assimilation ($A$), depend on diffusive processes with similar mechanisms.

In plant transpiration, water vapour diffuses from the moist walls of the inner leaf surface to its outer surface through the intercellular spaces and stomata, the concentration of water vapour varying from $q_s$ (the mean concentration in the air at the internal surface of the leaf) to $q_0$ (the mean concentration at the outer leaf surface). The evaporation rate will be in this case

$$E = \rho D'(q_s - q_0), \tag{4.18}$$

where $\rho$ is the air density, and $D'$ is the diffusion coefficient for water vapour on the path from the inner to the outer leaf surface. It is obvious that the value $D'$ depends on the leaf morphology (the number and size of stomata, the thickness of the leaf, the density of the cuticle and so on). The second stage of water vapour diffusion is associated with the change in water vapour concentration from $q_0$ (on the leaf surface) to $q$, the concentration above the vegetation layer. For this, we can write the equation

$$E = \rho D''(q_0 - q), \tag{4.19}$$

where $D''$ is the coefficient of external diffusion, depending mainly on the intensity of turbulent exchange. Excluding $q_0$ from Equation (4.18) and (4.19), we obtain

$$E = \frac{\rho(q_s - q)}{\dfrac{1}{D'} + \dfrac{1}{D''}}. \tag{4.20}$$

The diffusion of carbon dioxide from the air above the vegetation cover to the absorbing surfaces inside the leaf can also be divided into two stages. Firstly, the diffusion rate (equal to the rate of assimilation) equals

$$A = \rho D'_c(c_0 - c_1), \tag{4.21}$$

where $D'_c$ is the effective coefficient of diffusion through the intercellular channels and stomata for carbon dioxide on the path from the outer leaf surface to the surface of the parenchyma, $c_0$ is the mean air concentration of carbon dioxide at the outer leaf surface, and $c_1$ the mean concentration of carbon dioxide in the air at the surface of the parenchymal cells. Secondly, the assimilation rate equals:

$$A = \rho D''_c(c - c_0), \tag{4.22}$$

where $c$ is the concentration of carbon dioxide outside the vegetation, and $D''_c$ is a quantity analogous to $D''$. From Equation (4.21) and (4.22) it then follows that

$$A = \frac{\rho(c - c_1)}{\dfrac{1}{D'_c} + \dfrac{1}{D''_c}}. \tag{4.23}$$

The ratio between the rate of assimilation and the rate of evaporation from (4.20) and (4.23) thus equals

$$\frac{A}{E} = a\,\frac{c - c_1}{q_s - q}, \tag{4.24}$$

in which it should be noted that the coefficient of proportionality $a$ as defined by the formula

$$a = \frac{D'_c D''_c(D' + D'')}{D'D''(D'_c + D''_c)}$$

can change only within very narrow limits. Since the process of diffusion of water vapour and carbon dioxide between the outer leaf surface and the free-flowing air is determined by turbulent mixing, then it is evident that coefficients $D''$ and $D''_c$ have the same values. It is natural to assume that the diffusion of water vapour and carbon dioxide between the inner absorbing and the outer surface of a leaf is of molecular nature. In this case, the ratio $D'_c/D'$ must be equal to the ratio of the coefficient of molecular diffusion of carbon dioxide in air to the coefficient of molecular diffusion of water vapour in air, that is, approximately 0.64.

Taking this into account, it can be concluded that, depending on the relationship between the values of $D'$ and $D''$, the coefficient $a$ can vary only from 0.64 to 1, approaching unity when external diffusion plays the leading role (a thin leaf with mostly open stomata), and reducing to 0.64 when internal diffusion dominantes (a thick leaf with poor ventilation). If the rate of evaporation from the leaf under average conditions is assumed to amount to approximately 50% of that from an equal wet surface, then it is easy to calculate that in this case the coefficient $a$ approximates to 0.8.

Considering that the turbulent heat flux from the vegetation cover to the atmosphere equals

$$P = \rho D'' c_p (T_w - T) \tag{4.25}$$

(where $c_p$ is the heat capacity of the air, $T$ the air temperature, and $T_w$ the temperature of the leaf surface), we derive from (4.17), (4.20), (4.23) and (4.25) a formula for the assimilation rate:

$$A = \frac{R(c - c_1)}{\dfrac{L}{a}(q_s - q) + l(c - c_1) + b(T_w - T)}, \tag{4.26}$$

where

$$b = \frac{c_p}{a}\left(1 + \frac{D''}{D'}\right).$$

Equation (4.26) shows that the assimilation rate depends on the difference between the concentrations of carbon dioxide above the vegetation cover and on the internal surface of the leaf $(c - c_1)$ and increases with an increase in this difference. Since the concentration of carbon dioxide above the vegetation cover is comparatively stable, it is evident that the assimilation rate will increase with a reduction of $c_1$ that is, with an intensification of carbon dioxide absorption by the internal leaf absorbing surfaces.

In the case when 'physiological absorption' occurs at a significant rate and the total rate of assimilation is limited not by physiological processes, but by the diffusion flux of carbon dioxide, the value $c_1$ must be much less than $c$ and the formula for assimilation takes the form

$$A = \frac{Rc}{\dfrac{L}{a}(q_s - q) + lc + b(T_w - T)}. \tag{4.27}$$

By using Equation (4.27), we can determine the quantity of solar energy required by a plant for the most effective assimilation of carbon dioxide from the air. Taking into account that $C$ on the average, is equal to $0.46 \times 10^{-3}$, $L = 2430\ \mathrm{J\,g^{-1}}$, $a = 0.8$, $l = 10\,300\ \mathrm{J\,g^{-1}}$, and $b = 2.5\ \mathrm{J\,g^{-1}\,K^{-1}}$, we find that

$$lA = \frac{1.2R}{770(q_s - q) + 1.2 + 0.62(T_w - T)}. \tag{4.28}$$

It can be considered that under optimal conditions of photosynthesis, when $c_1 \ll c$, the internal leaf surface is sufficiently moistened, and as a result, the value $q_s$ should approach the concentration of saturating water vapour at the temperature of the leaf.

The mean daytime difference between the temperatures of leaves and that of the free air under summer conditions at middle latitudes is of the order of several degrees. Assuming that this difference is 5 °C and that the mean daytime relative air humidity in summer is close to 50% and the mean air temperature is 20 °C, we obtain the values $q_s = 2.0 \times 10^{-2}$ (at a leaf temperature of 25 °C) and $q = 0.7 \times 10^{-2}$. Substituting these values into Equation (4.28), we find that $lA = 0.08R$, i.e. the energy required for assimilation under mean climatic conditions for middle latitudes can reach 8% of the radiation balance.

From available data, we can find that for most of the European U.S.S.R., the radiation balance in summer amounts to 55—60% of the total incoming shortwave radiation during this season. Taking this estimate into account, we can calculate that under average summer conditions at middle latitudes, the natural vegetation can assimilate approximately 5% of the incoming solar radiation, provided that carbon dioxide is absorbed in an effective manner. This theoretical estimate is in good agreement with empirical data on the 'coefficient of useful efficiency' of photosynthesis under the most favourable conditions (Nichiporovich, 1973).

However, the available empirical data show that such comparatively high coefficients of solar energy utilization are observed only in specific cases, while the mean ratio of energy expenditure in photosynthesis by the natural vegetation cover to the incident solar energy is usually of the order of 0.5% (see below).

On this basis, we should infer that under mean conditions, the relationship $c_1 \ll c$ is not fulfilled, and, on average, the difference $c - c_1$ makes up about 10% of the value of $c$. In other words, a comparison of the theoretical diffusion model with empirical data permits us to determine that plants usually use only a small part (of the order of 10%) of potential diffusion income of carbon dioxide.

It can be mentioned that the conclusion drawn agrees with the results of a number of experiments by Lyubimenko and other authors, who, on the basis of physiological investigations, have found that leaves do not fully utilize carbon dioxide from potential diffusion income. The studies carried out resulted in the conclusion that "under natural conditions, production of dry matter is limited not so much by the small content of carbon dioxide in the atmosphere as by an insufficiently fast rate of work of the enzymatic apparatus which controls the outflow of assimilators and their assimilation" (Lyubimenko, 1935).

The model of the physical mechanism of assimilation and transpiration described above allows an explanation for the uneconomical expenditure of water during plant growth and development (i.e. why the observed values of transpiration productivity are small).

If we substitute the mean values obtained for $q_s$ and $q$ under daytime summer conditions into Equation (4.24) and assume, according to the estimate derived $c - c_1 = 0.1c$, then the ratio $A/E$ turns out to be approximately equal to $1/360$. This means that when the plant is assimilating 1 g of carbon dioxide, it loses through transpiration, on average about 360 g of water.

It should be noted that, although the ratio $A/E$ does not exactly coincide with transpiration productivity,[1] these values may be assumed not to differ by very much. Thus, the estimate obtained explains theoretically the observed values of transpiration productivity.

A further important point which can be made from Equation (4.24) is that transpiration productivity is substantially dependent on the humidity deficit. This dependence has also been noticed by other investigators, who have observed an appreciable increase in transpiration productivity as the air humidity deficit is decreased. This observation has been made many times, both under greenhouse conditions and in natural conditions when moist climates replace dry climates.

The basic conclusion from the above may be formulated in the following manner. Plants usually utilize a very small fraction of the natural energy and water resources available. This fraction is small even in comparison with the small 'coefficient of useful activity' that could be reached at the greatest possible diffusion of carbon dioxide from the air. The calculation results given in this section show that, in the case of complete utilization of atmospheric carbon dioxide, vegetation can assimilate no less than 5% of the incoming solar energy and that, under these conditions, the transpiration productivity must equal not several thousandths (as it usually does), but several hundredths. Since, under natural conditions, such high indices of photosynthetic efficiency are not usually observed, we must conclude that

[1] The productivity of transpiration is connected with the increase in dry organic matter which differs somewhat from the value of assimilation. This arises because of the utilization by the plant of water and mineral substances from the soil and because of loss of a fraction of the dry matter during the development of plant through respiration and in loss of some tissues.

plant productivity is substantially limited by factors whose nature will be discussed below.

The formulas that have been mentioned earlier relate to a single leaf or a group of leaves which are growing under similar environmental conditions. In determining the total photosynthesis within a layer of vegetation it is necessary to take into account differences between the meteorological factors affecting photosynthesis at various levels in the vegetation.

This question was first addressed in the studies of Monsi and Saeki (1953) and Davidson and Philip (1958), who calculated the total amount of photosynthesis with regard to changes in the amount of radiation at different heights in the vegetation. However, they did not consider the relationship between photosynthesis and variations of carbon dioxide concentration. Subsequently, a more general theory of photosynthesis in the vegetation has been proposed which includes consideration of the influence of both radiation factors and changes in carbon dioxide concentration at various levels within the vegetation cover, on photosynthesis (Budyko, 1964 and others). The results obtained in these studies are set forth below.

Studies of photosynthesis at various levels in the vegetation must consider physical features of meteorological processes that take place in the air layer where the plants are found, i.e. within the vegetation itself. In studying physical processes within the vegetation, values of meteorological elements at various levels should be averaged horizontally, making it possible to exclude the effects of individual plants on the meteorological regime.

By applying such averaging, it can be found that contrary to the Earth's boundary air layer, the mean vertical fluxes of shortwave and longwave radiation, heat, water vapour and momentum essentially change with height within the vegetation. In this case, the quantity of shortwave radiation declines as it approaches the Earth's surface (owing to the absorption and scattering of radiation by the surfaces of plants). The total radiation flux (radiation balance) also decreases closer to the Earth's surface because of the screening effects of plants. The fluxes of water vapour and momentum change in a similar way, with the flux of water vapour increasing with height within the vegetation due to the influence of plant transpiration, while the momentum flux decreases when moving downwards from the top of the vegetation because of the inhibiting effects of vegetation on air transfer. This causes the attenuation of turbulent exchange within the vegetation as compared with higher air layers, with the coefficient of turbulent exchange within the vegetation decreasing when approaching the Earth's surface.

An illustration of the relationship between the height and the coefficient of turbulent exchange $k$, as well as the photosynthetically active radiation $Q$, is presented in Fig. 4.5. This is based on data obtained by Yefimova (1968) for a field under wheat 1.2 m high. It should be noted that the photosynthetically active radiation is a part of the total shortwave radiation flux which is utilised for photosynthesis. Its value is approximately one-half of that of shortwave radiation.

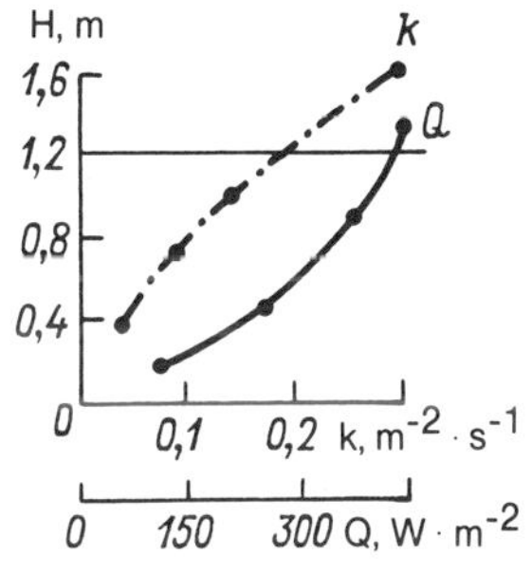


Fig. 4.5.

In developing a theory of photosynthesis in the vegetation, it is necessary to determine the quantitative dependence of photosynthesis for a single plant leaf upon the external factors influencing it. Many laboratory studies (e.g. Rabinowitch, 1951) have shown that for small values of photosynthetically active radiation $Q$ photosynthesis is proportional to the amount of radiation and is only slightly dependent on other external factors, i.e. that for

$$Q \to 0, \qquad A = \beta Q. \tag{4.29}$$

This condition corresponds to the so-called ascending part of the photosynthetic light curve.

With large amounts of radiation, photosynthesis depends only slightly on incoming radiation (the plateau of the light curve) but depends substantially on the concentration of carbon dioxide, temperature and other external factors. Experimental studies have shown that under conditions of 'light saturation' photosynthesis is proportional to the concentration of carbon dioxide for the range of concentrations observed in natural conditions. In view of that conclusion we will assume that for

$$Q \to \infty, \qquad A = \tau c, \tag{4.30}$$

where the coefficient of proportionality $\tau$ is a function of temperature and other external factors, but does not depend on radiation and carbon dioxide concentration.

Obviously the values $\beta$ and $\tau$ can be determined from experimentally calculated light curves of photosynthesis, with $\beta$ being a tangent of the angle of inclination of the initial sector of the light curve, while $\tau$ is associated with the height of the plateau of the light curve for a given carbon dioxide concentration. It should be remembered that the coefficient $\tau$ describes carbon dioxide-diffusion when photosynthesis is occurring under conditions of light saturation. These conditions are met by a formula (Budyko, 1964) describing the light curve of photosynthesis:

$$A = \frac{\beta Q}{1 + (\beta Q/\tau c)}. \tag{4.31}$$

This represents a generalization of the hyperbolic formula proposed by Tamiya in 1951 (Monsi and Saeki, 1953).

The vertical flux of carbon dioxide within the vegetation cover at the level $z$ is

$$A_z = \rho k \frac{\mathrm{d}c}{\mathrm{d}z}. \tag{4.32}$$

The change in the flux within the layer d$z$, resulting from the assimilation of carbon dioxide by leaves will, according to Equation (4.32), be as follows

$$\mathrm{d}A_1 = \frac{\beta Q s\,\mathrm{d}z}{1 + (\beta Q/\tau c)}, \tag{4.33}$$

where $s$ is the surface of leaves per unit volume. The change in the photosynthetically active radiation with height within the vegetation is obtained from the equation

$$\mathrm{d}Q = \varphi(Q) s\,\mathrm{d}z, \tag{4.34}$$

where the function $\varphi(Q)$ describes the relationship between the absorption of radiation within a layer of thickness d$z$ and the amount of photosynthetically active radiation at the top of this layer. Since at the top of the vegetation ($z = H$) the radiation flux is known ($Q = Q_H$), we find from Equation (4.34) that

$$\int_Q^{Q_H} \frac{\mathrm{d}Q}{\varphi(Q)} = \int_z^H s\,\mathrm{d}z. \tag{4.35}$$

Solving Equation (4.35) for $Q$ we find the following relationship

$$Q = \psi\left(Q_H, \int_z^H s\,\mathrm{d}z\right), \tag{4.36}$$

which describes the dependence of photosynthetically active radiation on height. If we consider that

$$\varphi(Q) = \nu Q \tag{4.37}$$

($\nu$ is the proportionality coefficient) this relationship will have the form

$$Q = Q_H \exp\left(-\nu \int_z^H s\,\mathrm{d}z\right). \tag{4.38}$$

By using Equation (4.38), we derive from Equation (4.34):

$$\mathrm{d}A_1 = \frac{\beta Q_H \exp(-\nu \int_z^H s\,\mathrm{d}z)}{1 + \frac{\beta Q_H}{\tau c} \exp(-\nu \int_z^H s\,\mathrm{d}z)} \cdot s\,\mathrm{d}z. \tag{4.39}$$

The change in the flux $A_2$ produced by plant respiration may be expressed in the form

$$dA_2 = -\varepsilon s\, dz, \tag{4.40}$$

where $\varepsilon$ is the coefficient of proportionality. The total change in the vertical flux of carbon dioxide is determined by relationships:

$$dA_z = dA_1 + dA_2. \tag{4.41}$$

From these formulas we obtain the non-linear differential equation

$$\frac{d}{dz}\left(\rho k \frac{dc}{dz}\right) = -\frac{\beta s Q_H \exp(-\nu \int_z^H s\, dz)}{1 + \dfrac{\beta Q_H}{\tau c}\exp(-\nu \int_z^H s\, dz)} - \varepsilon s. \tag{4.42}$$

In order to solve Equation (4.42), the following boundary conditions should be used

$$\left(\rho k \frac{dc}{dz}\right)\Bigg|_{z=0} = -A_0, \tag{4.43}$$

and

$$\left(\rho k \frac{dc}{dz}\right)\Bigg|_{z=H} = \rho D_H (c_\infty - c_H), \tag{4.44}$$

where $A_0$ is the carbon dioxide flux from the soil, $c_\infty$ the concentration of carbon dioxide in the air at height $z_1$, where its dependence on the properties of vegetation is small, $c_H$ the carbon dioxide concentration at level $H$, and $D_H$ the integral coefficient of turbulent diffusion within the air layer between levels $H$ and $z_1$.

The first of these conditions expresses the equality between the influx of carbon dioxide from the soil and the value of the turbulent flux of carbon dioxide in the lower air layer. The second condition corresponds to the equality between the turbulent flux of carbon dioxide at the top of the vegetation cover and the turbulent flux incoming from the atmosphere into the vegetation.

From the balance condition for carbon dioxide in the space between leaves for the entire vegetation cover we find that

$$A = A_H + A_0 + \int_0^H \varepsilon s\, dz, \tag{4.45}$$

where $A$ is the total value of assimilation, and $A_H$ the vertical carbon dioxide flux at level $H$.

The solution to Equation (4.42) yields the value of $A_H$ and then, by formula (4.45), we find the magnitude of total assimilation. In this case the dependence of $k$, $s$ and $\varepsilon$ on height within the vegetation cover should be taken into account. This can be accomplished by using the available empirical data. In the general case, Equation (4.42) can only be integrated numerically.

Application of the theory described above makes it possible to study the dependence of photosynthesis within the vegetation on the value of index $S$ (which is equal to the ratio of the surface of leaves to the area occupied by the vegetation cover) and on various external factors.

As well as this theory of photosynthesis within the vegetation, other methods for constructing numerical models of this process have been proposed (Ross, 1964; Monteith, 1965; De Wit, 1965; Uchijima, 1966; Duncan *et al.*, 1967; Tooming, 1974; and others). These are all based on the general idea of a transition from photosynthesis within a single leaf to photosynthesis in a horizontal homogeneous layer, but having different physical conditions at different heights. The differences between these methods relate to different degrees of detail in using the numerical models; some of the models are more general while others are more specific.

It is now possible to consider some results produced by application of the theory presented here for studying the properties of photosynthesis within a layer of vegetation. These results relate to the special case in which the values of $\tau$, $s$ and $\varepsilon$ do not depend on height and the vertical profile of the turbulent diffusion coefficient in the intercellular space is linear, i.e. $k = k_0 + k_1 z$, where $k_0 \ll k_1 H$.

Figure 4.6 shows the estimated dependence of assimilation rate $A$ on the flux of photosynthetically active radiation at the top of the vegetation cover ($Q_H$) for various values of the leaf-area index $S = \int_0^H s\,dz$ and of the ratio $\tau/\rho$. This relationship has been calculated for the following values of other parameters: $A_0 = \varepsilon = 0$; $D_H = 0.6$ cm s$^{-1}$, $k_1 = 3$ cm s$^{-1}$, $\nu = 1$, $\beta = 1.9 \times 10^{-5}$ g J$^{-1}$, $k_1 H/k_0 = 200$, $\rho = 1.29 \times 10^{-3}$ g cm$^{-3}$, $c_\infty = 0.46 \times 10^{-3}$ (this represents 0.03% volume).

The lines in Fig. 4.6 represent light curves. However, unlike the light curves developed for individual leaves, the light curves presented in Fig. 4.6

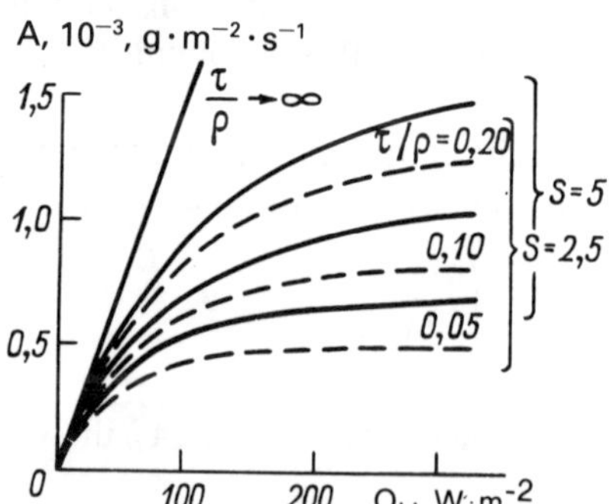


Fig. 4.6.

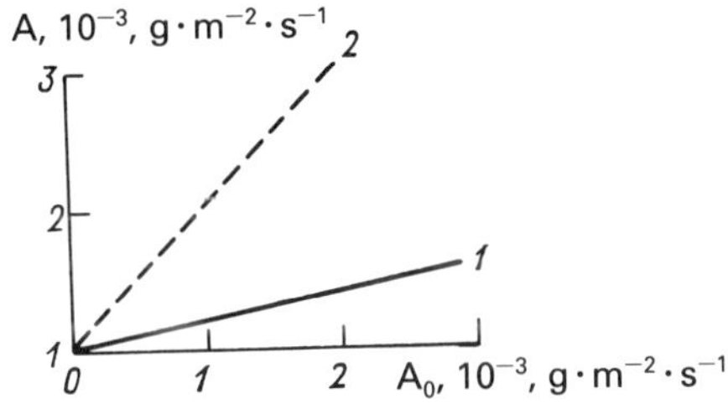


Fig. 4.7.

were obtained theoretically for the entire vegetation cover. These curves show a strong dependence of total assimilation on the radiation flux at low levels of radiation. In this case assimilation is not very sensitive to other factors. On the other hand, for large values of radiation, total assimilation is almost independent of radiation, and is determined by other factors, in particular by temperature which influences the coefficient $\tau$.

Figure 4.7 depicts the effects of carbon dioxide flux from the soil ($A_0$) on the total assimilation. Line 1 describes the actual relationship between $A$ and $A_0$, while line 2 corresponds to the ideal case when there is complete assimilation of carbon dioxide released from the soil. Comparison of lines 1 and 2 shows that $A$ increases as $A_0$ increases, but at a much lower rate than would occur if all additional carbon dioxide coming from the soil were expended on assimilation. It is easy to understand this, if one considers that the carbon dioxide flux from below first encounters leaves located in the shade and not able to be engaged in intensive photosynthesis because of insufficient radiation. The total increase in concentration produced by the influx of carbon dioxide from below is largely compensated for by a reduced contribution of carbon dioxide from above. As a result, the influence of the carbon dioxide flux from the soil on assimilation is usually small in comparison with the carbon dioxide flux from the atmosphere. This conclusion is in good agreement with that of Montieth, Szeicz and Yabuki (Montieth *et al.*, 1964) made on the basis of experimental studies of the diffusion of carbon dioxide in the vegetation.

Let us now turn to a consideration of the climatic factors that determine the productivity of the vegetation $\Pi$. We assume productivity to equal the difference between total assimilation and expenditure of organic matter on respiration, neglecting losses of matter by atrophy of individual plant organs as the plant develops.

Since the productivity of the vegetation cover depends substantially on the value of the leaf-area index, and declines for both small and large values of this index (see below), we will estimate the value of its productivity for an optimal value of the leaf index that corresponds to the highest values of productivity.

The equations mentioned above show that in addition to parameters characterizing the properties of the vegetation cover itself, the magnitude Π is also influenced by an appreciable number of meteorological parameters. It therefore seems appropriate to identify the more important ones. Under conditions of sufficient moisture two factors, the level of photosynthetically active radiation and temperature have a pronounced influence on the productivity of vegetation. The value of the radiation flux $Q_H$ is included in the equation. As for temperature $T$, it influences the parameters $\tau$ and $\varepsilon$, which are employed in the calculation procedure. Unfortunately, in spite of numerous studies in this area, we do not yet possess sufficiently abundant and reliable quantitative data concerning the influence of temperature on parameters $\tau$ and $\varepsilon$ for different types of plants. For this reason we have used the functions $\tau(T)$ and $\varepsilon(T)$ in a generalized form which is characteristic of certain grain crops.

Based on the indicated relationships and using the equations described previously we can calculate the dependence of total productivity Π on the flux of photosynthetically active radiation at the top of the vegetation cover and on temperature. This calculation shows that increases in the flux $Q_H$ produce in all cases an increase in productivity Π. At the same time, as temperature $T$ increases, the productivity first rises and then, after attaining a certain maximum value which is a function of $Q_H$, begins to decline. This is a consequence both of the similar relationship between $\tau$ and $T$, and of intensified respiration with increasing temperature.

As can be seen from these results, the effects of radiation and temperature on the productivity of vegetation are complex. Furthermore, in reality radiation is always a factor whose value is at a 'minimum'. This can be attributed to a constant insufficiency of radiation for lower, partly shaded, leaves of the vegetation. Temperature produces another influence on productivity; an increase in temperature above a certain limit substantially lowers productivity.

Let us pause to consider the causes for the reduction in the productivity values when compared with the estimates obtained in the calculation. The effects of climatic, biologic and soil factors contribute to these causes. The values for potential productivity estimated by the method described above relate to optimal moisture conditions, which in fact occur infrequently during the entire vegetation period, even in the most favourable climatic regions.

The values of potential productivity refer to plants whose structure is optimal over the entire vegetation period from the point of view of photosynthesis, and whose photosynthesizing organs display a constant level of activity throughout this period. In reality such conditions are not found. In addition, the potential productivity of plants corresponds to conditions with sufficient mineral nutrition, conditions that in practice seldom occur, even in cases when adequate forms of agricultural technology are employed. All these factors account for a substantial reduction in the productivity of both agricultural plants and natural vegetation in comparison with their potential productivity as estimated by the method described above. Therefore potential productivity can be used as conventional indicator of the organic matter production of autotrophic plants for various geographical regions depending on their climatic resources.

In reality, the moisture regime of the soil plays a particularly important role in influencing plant productivity. The results of an empirical study of the relationship between vegetation productivity and precipitation are cited below. Particular attention should be given to the possibilities of applying the numerical modelling methods for heat and water exchange processes in vegetation, to studying the effects of the water cycle on productivity.

Studies of this type include works of Menzhulin, who proposed that in conditions of insufficient soil water supply, the leaf surface area of plants declines, as does the stomatal conductivity of leaf surfaces. On the basis of a semi-empirical model describing these phenomena and taking the transpiration/total evaporation relationship into account, Menzhulin estimated the evaporation effects (and also the associated soil moisture effects) on the productivity of vegetation (Menzhulin, 1976).

As was noted earlier, the variety of vegetation types in different regions of the globe can largely be attributed to the influence of climatic conditions on the development of the plants. The numerous studies concerned with climatic effects on plant growth have largely been undertaken using empirical methods, which restrict the possibilities of identifying the causal mechanism through which the climatic factors affect the vegetation cover. However, it is possible to envisage studies of the dependence of properties of the vegetation on climatic conditions which are based on the general physical features describing the influence of atmospheric factors on the vital processes of plants. This approach would make it possible to arrive at theoretically based estimates of average values for such vegetation cover characteristics as the volume of biomass per unit surface area for geographical zones in different climatic conditions.

In this connection let us consider the balance of organic matter in the vegetation. In analysing this we shall rely on the following relationship:

$$M = \mu S + m, \tag{4.46}$$

where $M$ is the mass of organic matter contained in the vegetation cover per unit surface occupied by the latter, $S$ the leaf index, $\mu$ the mass of leaves per unit of leaf surface, and $m$ the mass of organic matter contained in all plant organs other than leaves per unit surface.

The productivity of vegetation cover per unit surface area is found by the formula

$$\Pi = A(S) - B(M), \tag{4.47}$$

where $A(S)$ is the photosynthesis within the vegetation dependent on the leaf index value $S$, and $B(M)$ is the expenditure of organic matter on respiration and through other forms of organic matter loss which depends on the mass of matter $M$.

It is reasonable to assume that the expenditure of organic matter on leaf respiration and on their atrophy is proportional to the mass of leaves, i.e. is equal to $v\mu S$, where $v$ is the coefficient of proportionality. In this case

$$\Pi = A(S) - v\mu S - B(m). \tag{4.48}$$

In order to determine the function $A(S)$ it is necessary to apply the afore-mentioned theory of photosynthesis. The nature of this relationship requires that $\partial A/\partial S$ is positive, declining monotonically as $S$ increases, and approaches zero for large values of the leaf index. In this connection, if $(\partial A/\partial S)|_{S=0} > v\mu$, then there exists such a value of the leaf index $S_0$ at which productivity of the vegetation cover reaches a maximum. The value $S_0$ can be determined from the conditions

$$\left.\frac{\partial A}{\partial S}\right|_{S=S_0} = v\mu. \tag{4.49}$$

It seems probable that, as a result of the prolonged evolution of autotrophic plants directed at increasing the effectiveness of their photosynthetic activity, in natural vegetation covers possessing sufficient moisture for transpiration the leaf index values will correspond more or less to the optimal value $S_0$. In order to verify this proposition by the afore-mentioned method, the value $S_0$ for various climatic conditions has been determined from Equation (4.49). This value turned out to be rather stable, slowly decreasing as latitude increases and approximately equal to 5 for leaf-bearing plants.

Such an estimate is in satisfactory accord with values of the leaf index obtained from observations of numerous natural vegetation types in regions of sufficient moisture (*Fundamentals of Forest Biogeocenology*, 1964), and this confirms the assumption made earlier.

The development of perennial plants continues until they attain a natural maturity, i.e. a state in which the intake of organic matter in the vegetation cover is compensated for by expenditures on respiration and atrophy, both of individual plant organs and of individual plants. In such cases the following condition is fulfilled:

$$A(S) - v\mu S - B(m) = 0. \tag{4.50}$$

Assuming that $S = S_0$ let us find the limiting value of organic mass $M_0$ that may be accumulated in a vegetation cover that has reached a stationary state, from the equation

$$A(S_0) = B(M_0). \tag{4.51}$$

This equation may be employed for theoretical estimation of the quantity of biomass contained in typical vegetation covers in different natural zones of sufficient moisture.

Following the idea proposed by Davidson and Philip (1958), it is possible to construct an equation that will describe changes in the biomass of vegetation covers over time. This will enable estimation of the rate of accumulation of biomass as the vegetation cover approaches a state of natural maturity. The form of this equation is similar to (4.48):

$$\frac{\mathrm{d}M}{\mathrm{d}t} = A(S) - v\mu S - B(m), \tag{4.52}$$

where $\mathrm{d}M/\mathrm{d}t$ is the rate of change in biomass over time.

Equations (4.51) and (4.52) have been employed for estimating the amount of biomass in the vegetation of zones with sufficient moisture, both in stationary states and when they change with time. Such calculations are made difficult by the insufficient comprehensiveness of available data on the intensity of the processes of biomass loss in the course of a plants life, necessary for determining $v$ and $B$.

Nevertheless the use of values for these parameters available in literature makes it possible to calculate by Equation (4.51) the values of biomass reserves in forests at different latitudinal zones. These results are in satisfactory agreement with empirical data obtained by Lavrenko, Andreyev and Leontiev (1955). It should be noted that such calculation reveals the specific climatic factors that influence the changes in biomass reserves as vegetation passes from one type to another.

In particular the formulas used show that a reduction in forest biomass reserves with increasing latitude is mostly dependent on a decrease in the duration of the vegetation period, while the effects of a reduction in the air temperature and changes in radiation during the vegetation period are less significant.

Estimations based on Equation (4.52) enable the general patterns of biomass change during the course of development of perennial vegetation to be determined. These show increased rates of biomass accumulation for the first years of life of the vegetation, which are eventually replaced by a gradual decline in the rate of accumulation. When the rate of accumulation becomes low the plant cover approaches a state of natural maturity.

These estimations have shown that Equation (4.52) determines the relationships between forest biomass reserves and the age of the forests, which agrees well with observational data.

## 4.4. The Productivity of Vegetation Cover

As mentioned in the above section, the effectiveness of natural vegetation in assimilating the energy of solar radiation varies considerably depending on natural conditions. The coefficient of useful assimilation of solar energy during photosynthesis attains a maximum under optimal conditions of heat and water supply and declines when these are reduced. This coefficient also depends on the physical and chemical properties of soil.

Quantitative calculations of the efficiency with which vegetation utilizes solar energy requires information on the incoming solar radiation and on the productivity of the vegetation. In studying the radiation regime at the Earth's surface, maps of both total radiation and photosynthetically active radiation have been constructed for various territories and for the world as a whole (Yefimova, 1965, 1969, 1977, etc.).

In determining mean values of the productivity of natural vegetation covers it is possible to employ data from various studies, including empirical determinations of productivity in individual geographical zones. In the studies of Lieth (1964), Rodin and Bazilevich (1965), Bazilevich and Rodin (1967), and of other authors, the first maps of plant productivity on a continental scale have been compiled. While recognizing the great importance of these studies for investigating the geographical patterns of vegetation productivity, it should be pointed out that because of a lack of basic data these maps were largely simplified. In particular, the maps constructed by Bazilevich and Rodin present productivity data in terms of average values of productivity for different geobotanical zones rather than plotting the isolines. The application of such maps for studying the spatial distribution of the coefficient of solar energy assimilation by plants is difficult because productivity may change, not only from zone to zone but also within individual zones.

Therefore, to estimate the assimilation of solar energy by vegetation cover, a map of the productivity of natural vegetation was compiled for the U.S.S.R. taking into account the relation between productivity and climatic conditions (Budyko and Yefimova, 1968). The observation that in conditions of sufficient moisture, plant productivity is a function of the radiation balance of the Earth's surface, is taken into account in compiling this map. The ratio of radiation balance to the amount of precipitation was used as an index of moisture conditions.

A corresponding graph was plotted in which empirical values of the productivity of natural vegetation over a year (yearly increments in organic matter both above and below ground) were compared with values of radiation balance and precipitation. This graph is shown in Fig. 4.8 with values for the productivity of dry organic matter (grams per square metre) over a year plotted along the vertical axis and with the ratio of annual values of radiation balance $R$ to heat amount $Lr$ required for a full evaporation of total annual precipitation (the radiation index of dryness) plotted along the horizontal axis. The curves on this diagram correspond to different values of the radiation balance (W m$^{-2}$).

While all the available data on vegetation productivity were used in constructing this graph, few observations were available for lower latitudes (i.e. for large $R$), and this reduces the accuracy of the graph under these conditions. However, for middle latitudes the relationship demonstrated in Fig. 4.8 can be considered more reliable. The system of isolines in the graph agrees well with the law of geographical zonality described in Chapter 3.

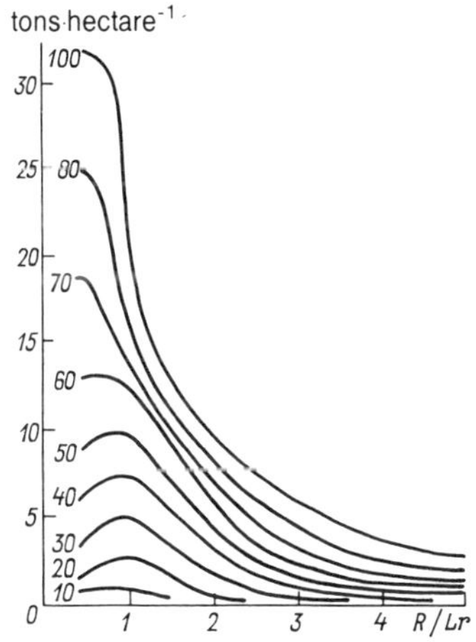


Fig. 4.8.

As seen from the graph, changes in the productivity depending on the dryness index at various levels of radiation balance are largely similar. Analogous relations between productivity and climatic factors were found in the studies of Bazilevich and Rodin (1969) and of Drozdov (1969). Some of the specific features of these relations call for additional studies. This includes investigation of the dependence of maximal productivity values on the radiation index of dryness for given magnitudes of the radiation balance. Figure 4.8 shows that it is not easy to determine the position of maxima precisely for any given radiation balance. In the study of Armand (1967) it was proposed that the maxima take place in climatic conditions of high and middle latitudes, whereas in the tropics the productivity increases without limit with decreasing values for the dryness index. Bazilevich *et al.* (1970) came to the conclusion that such maxima were observed at all latitudes, and, with increasing latitude the corresponding value of the radiation index of dryness rises.

In compiling a map of productivity based on Fig. 4.8 and on data concerning radiation balance and precipitation, productivity has been calculated for many localities in the territory of the U.S.S.R. and used in the construction of a map (Fig. 4.9). This map shows that values of the annual increment of biomass in the natural vegetation cover vary from less than 2 tonnes per hectare on islands and the Siberian coast of the Arctic Ocean to values of more than 20 tonnes per hectare on the Black Sea coast of the Caucasus. Furthermore, in zones with sufficient moisture it can be found that the productivity increases from north to south corresponding to the increased solar radiation income and longer vegetative period. A reduction in the productivity of vegetation cover is observed in regions of insufficient moisture, with the productivity in the deserts of Central Asia being decreased to 2 tonnes per hectare per year.

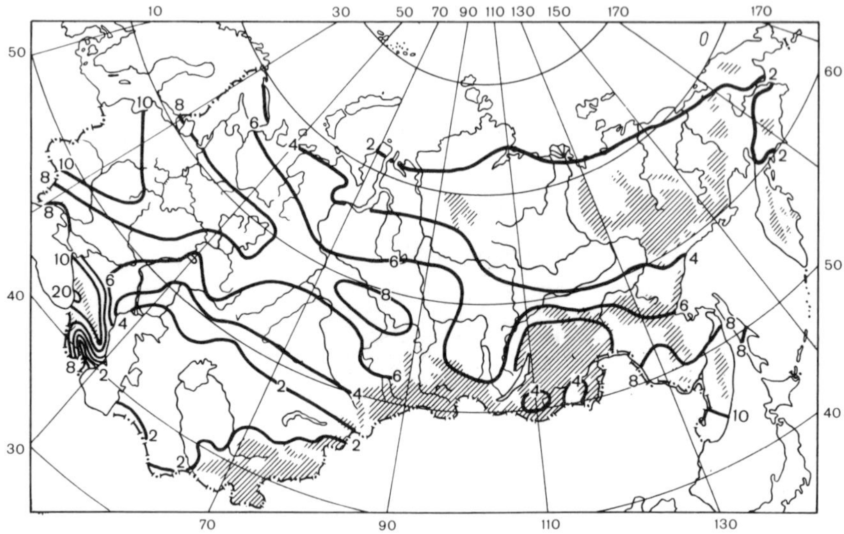


Fig. 4.9.

Comparison of the values for annual productivity of natural vegetation shown in Fig. 4.9 with those obtained by Bazilevich and Rodin, generally shows satisfactory agreement among various geobotanic zones. Some differences between these maps occur in zones of dry steppes, deserts and partly in regions of Eastern Siberia and can be attributed to an insufficient volume of empirical data for these territories. By using the map of vegetation productivity and the map of income of photosynthetically active radiation over the vegetation period (Yefimova, 1965) it is possible to calculate the share of incoming radiation energy that is assimilated in the organic matter formed during the year. An approximate value of the caloricity of dry organic matter of plants equal to 16 500 J $g^{-1}$ has been taken into account in these calculations.

Figure 4.10 represents a map of the distribution of energy contained in the annual increase in vegetation organic matter as a percentage of the incoming energy of photosynthetically active radiation during the vegetative period. As shown in this map, up to 2% of incoming photosynthetically active radiation is absorbed in the annual increase of organic matter. The largest coefficient of assimilation of solar radiation energy (1.8—2%) is found in the region of the maximal annual increment, namely on the Black Sea coast of the Caucasus. A relatively high value of energy utilization (1.0—1.2%) is found over almost the entire European part of the U.S.S.R., except for arid regions in the south and southeast. In the Asian part of the U.S.S.R., in the central part of Western Siberia, in the south of the Khabarovsk Territory and in the

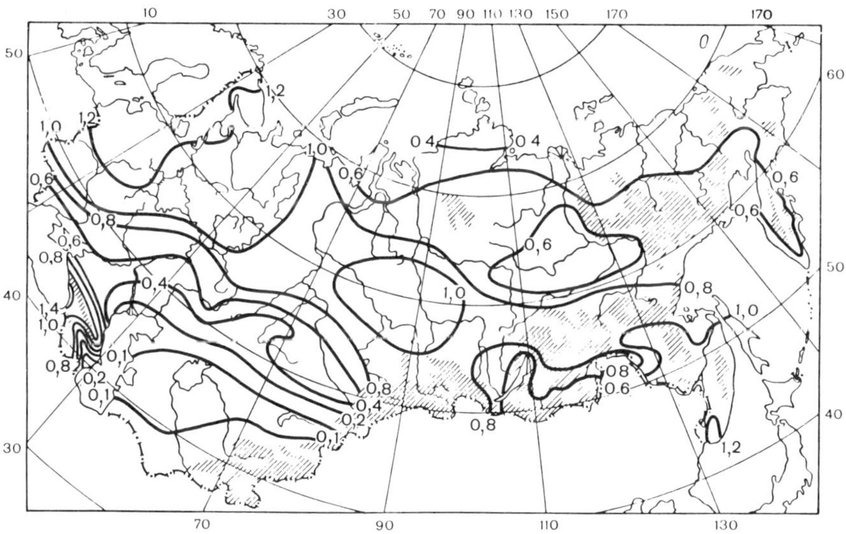

Fig. 4.10.

Primoriye Territory approximately 1% of the energy is employed. Over the remaining parts of the country the annual increment in biomass contains from 0.4 to 0.8% of the incoming photosynthetically active radiation. In the deserts and semi-deserts of Central Asia the assimilation of energy in the annual increase in biomass drops to 0.1—0.2%. Let us note that in terms of their content, these coefficients of utilization of solar energy are not the same as those mentioned at the beginning of the chapter which refer to the total incoming radiation rather than to the photosynthetically active radiation.

Smaller values for the coefficient of utilization of solar energy by plants in the eastern regions of the U.S.S.R. in comparison with western regions is attributable to less favourable moisture conditions and to a reduction in the vegetative period in many eastern regions.

The average value of the coefficient of assimilation of photosynthetically active solar radiation by natural vegetation cover is equal to approximately 0.8%. The magnitude of the corresponding coefficient for total solar radiation will be one-half of that value, i.e. roughly 0.4%.

In later studies, Yefimova (1976) made use of a more precise relationship between the productivity of the natural plant cover and meteorological factors in estimating the productivity and the coefficient of assimilation of photosynthetically active radiation for each continent and constructed corresponding maps. The results obtained are summarized in Table 4.1. As seen from this table, the average productivity per unit area for the five continents does not differ very much. It is evident that over large areas of each of these continents its magnitude is greatly limited by insufficient heat or moisture.

The South America continent represents an exception in this regard. Over the large parts of its territory, climatic conditions are very favourable to the life of plants. In Australia and Africa, the coefficients of assimilation of photosynthetically active radiation are low. This is due to insufficient moisture over large areas of these continents, which inhibits the utilization of solar radiation energy by the vegetation. The most favourable moisture conditions for plant life exist in Europe and South America; in Europe, located at higher latitudes and exposed to less solar radiation, its utilization is relatively greater.

TABLE 4.1

Productivity of natural vegetation cover

| Continents | Productivity | | Coefficient of assimilating the photosynthetically active radiation energy (as a percentage of total over the vegetation period) |
|---|---|---|---|
| | ($t \times 10^9$) | ($g\ m^{-2}$) | |
| Europe | 8.9 | 850 | 1.26 |
| Asia | 38.3 | 980 | 0.88 |
| Africa | 31.0 | 1030 | 0.59 |
| North America | 18.1 | 820 | 0.94 |
| South America | 37.2 | 2090 | 1.13 |
| Australia (including islands of Oceania) | 7.6 | 860 | 0.44 |
| Land as a whole | 141.1 | 950 | 0.86 |

The vegetation cover on the continents is characterized by a comparatively small value for the ratio of annual productivity to total plant biomass; this value being usually less than unity and frequently not exceeding several per cent. The vegetation in water bodies, for which the ratio of annual productivity of plants to their mass can be much more than unity, is a different situation.

There are less data on the geographical distribution of productivity of autotrophic plants in water bodies compared with those on the land. Observation data show that in seas and oceans the greatest volume of organic matter is produced by plankton located in a surface layer several dozens of metres deep. At greater depths the amount of solar radiation present is usually insufficient to support the process of photosynthesis. Productivity is substantially less in the open ocean than it is in shelf zones. Maximal values of the productivity can be attained in shallow water bodies with large quantities of the mineral compounds essential for plant growth.

According to data from Koblenz—Mishke and Sorokin (1962), the total value of productivity for the oceans, known only very approximately, is equal to about 45—55 billion tonnes per year. This corresponds to a productivity per unit area of ocean surface of approximately 150 g m$^{-2}$. The last figure is much less than the average productivity for a unit area of the land surface. Estimates of productivity of the oceans in other modern studies lie in most cases in the range from 50 to 200% of the value given above (ref. to Gershanovich and Muromtsev, 1982 and other reviews).

Taking this estimate into account we find that the annual value for the productivity for the globe is about 200 billion tonnes, i.e. about 400 g m$^{-2}$ of dry organic matter. This corresponds to the mean annual expenditure of energy equal to approximately 0.20 watt m$^{-2}$, i.e. about 0.1% of the amount of solar radiation incoming to the Earth's surface.

This value is too small to permit the process of photosynthesis to directly influence the major components of the energy balance of the Earth. In this connection it is usually believed that photosynthesis is not a substantial factor influencing climatic conditions. It can be noted that such influence is also restricted by a comparatively rapid return into the atmosphere of energy assimilated in the course of photosynthesis as a result of mineralizing organic matter that transforms this energy into heat.

There are, however, cases in which such an influence manifests itself in a rather noticeable form. This occurs in the case of fires which affect substantial areas of vegetation. In such situations, the energy of photosynthesis accumulated over long time periods is rapidly transformed into heat so that the volume of additional influx of heat per unit area is much larger than the energy of solar radiation. As a result, the meteorological regime over areas where large forest fires occur, changes sharply, and such changes may extend to a considerable height in the atmosphere.

Except for such special cases, however, it should be noted that the plant cover exerts its greatest influence on climate, not as a result of the expenditure of solar energy on photosynthesis, but in connection with the influence of transpiration on the balance of heat and moisture. Thus, data on the coefficients of the utilization of solar energy by vegetation cover are important mostly as characteristics of climatic conditions governing the development of plants and as indicators of the effectiveness of the photosynthetic activity of different types of vegetation.

Let us note that the afore-mentioned data on the productivity describe the quantity of organic matter that may be utilized by heterotrophic organisms. Primary productivity (i.e. the volume of assimilation) is always substantially greater than the total productivity considered above. This arises because a part of the organic matter is expended on plant respiration and other physiological functions of plants. In many cases, total productivity comprises approximately one-half of primary productivity.

CHAPTER 5

# ANIMALS

## 5.1. Animals in the Biosphere

The modern fauna of the Earth is characterized by an exceptional variety of animals. This variety is expressed first in the body sizes of animals from different species. The representatives of the contemporary fauna can be microscopically small or several tens of metres long.

The structure of animal bodies is also remarkably diverse: from the comparatively primitive structure of protozoa to the complex structure of higher animals, in particular the vertebrate. The age of modern species and genera of animals differ considerably: some are ancient and have changed little for tens or even hundreds of millions of years, others first appeared relatively recently.

Although modern plants, as already mentioned, are also rather diverse, the diversity of animals exceeds that of plants. Another distinct feature of animals when compared with plants is that progressive groups of animals possess more complex organization than the most advanced plants.

The zone in which animals exist in the biosphere, is far wider than the zone of plant life. This difference arises because the zone which animals are able to exploit includes a wide variety of environmental conditions and is not limited by the need to obtain a sufficient amount of solar energy, as is the case for photosynthetic plants.

An indispensable condition for the life of animals is a sufficient amount of food. Among other factors that limit the life zone of animals is the gaseous composition of the environment. Aerobic animals usually can only live in the presence of an appreciable amount of oxygen which they require for their respiration, while for anaerobic organisms oxygen is often poisonous.

The vital activity of animals substantially depends on their thermal regime; the nature of this dependence will be treated in the next section of this chapter.

In contrast to the majority of plants which are unable to move in space, there are three modes of the existence of motile animals in a certain locality of the biosphere, i.e. a permanent, temporary and episodic dwelling. Animals, as well as plants, can live permanently only in the hydrosphere or lithosphere (often at its surface). Although flying animals spend a part of their life in flight, they also pass a certain time at the Earth's surface, in the upper soil layers, or in water bodies.

The vast majority of animals occur in the surface 'film of living matter', i.e. in a comparatively thin envelope embracing the surface of land, soil, upper layers of water bodies and the surface air layer. Beyond this film, a relatively small number of animal species exist permanently in the deep layers of the hydrosphere.

According to habitat and movement, the population of water bodies is usually divided into three major groups: plankton, benthos and nekton. Plankton consists of minute organisms that have no organs for active movements and float with the motion of the waters. Some planktonic representatives possess a limited ability to move actively and can advance vertically for hundreds of metres. This activity is usually associated with daily cycle of radiation. Planktonic organisms include both animals (zooplankton) and plants (phytoplankton). Organisms that live at the bottom of water bodies belong to benthos group. These may also be of either animal or plant origin (zoobenthos and phytobenthos, respectively). The nekton group consists of actively swimming marine animals, including fishes, aquatic mammals and a number of invertebrates.

It is typical of many animals to travel considerable distances in the course of their life-cycle. This movement is often associated with changes in their environment. At the same time, a number of animals as well as plants do not move for the majority of their life. Among these animals are included representatives of relatively progressive groups, for instance the Tunicata (usually included into the Chordata phylum) as well as primitive organisms like sponges. During the larval stage of their life-cycle the sea-dwelling tunicates resemble lanceletes belonging to the chordates. Later, the majority of the tunicates loses a number of organs as a result of a complex morphophysiological metamorphose and are transformed into creatures attached to the sea bottom with a much more primitive structure than their larvae.

The ability of various species of animals to actively move, differs greatly. Some swimming and flying animals move regularly over enormous distances, sometimes exceeding 10000 km. For instance, a great many whales annually cover long distances to reach sites with abundant food to feed on. Similarly, great distances are covered by many fishes, in particular eels and salmon amongst others. These fish are basically sea-dwelling but lay roe in inland water bodies. Also well known is the seasonal migration of birds who often travel distances of thousands of kilometres.

A lot of relatively small animals are also able to passively traverse considerable distances, among which are included parasites carried by the animals they infest. Another form of passive movement by animals is floating with the sea and particularly air currents. Strong winds can carry protozoa (especially in the cyst form, when they protect themselves by a resistant cover against unfavourable conditions) for thousands of kilometres. Violent storms and tornadoes may lift into the air and carry for considerable distances larger animals, such as fishes and frogs.

The ability of animals to move is of great importance for settling new geographical regions.

It should be mentioned that it is possible for animals to spend short periods of time in regions where conditions are different from those where they have their permanent residences in the biosphere. For example, animals can be deprived of water or food or they may find themselves subjected to thermal conditions under which they could not survive for long time periods. The areas of the biosphere where animals are occasionally found but could not survive for long can only be referred to as a part of the biosphere with certain reservations.

The great diversity of modern fauna is the result of the availability of numerous ecological niches that differ from each other, both in the past when modern species of animals emerged and in the present epoch. Although many animals have wider life zones than plants, these zones are nevertheless limited, which is reflected in the small sizes of areas occupied by certain animal species.

Widespread distributions are typical of only a minority of animals (cosmopolites). However, a wide distribution of cosmopolitan animals such as some species of rats, marine birds and the insects associated with man, is better explained by their ability to adjust to a rather narrow band of conditions found in numerous ecological systems over vast areas of the continents or oceans rather than their adaptation to varied environmental conditions.

If we look at the biosphere as a three-dimensional system, it becomes more apparent how animals are limited in their ability to make even episodic movements to every part of a geographical region. For example, no animal species can penetrate into deep layers of the ocean (or relatively thinner soil layers on the continents) as well as into the upper layers of the troposphere.

Turning to the life zones of animals, we shall now consider in more detail the conditions that are typical of these zones.

For much of geological time, the life zone of the majority of the animal phyla has been the hydrosphere. For animals to exist in the hydrosphere, besides the principal conditions necessary for life as listed above, several conditions, including those associated with the chemical composition of water, need to be satisfied. From a general view point limitation of life in the hydrosphere by the chemical composition of the water is not of great importance as many plants and animals can adapt to a wide range of salt concentrations in natural waters. The exception is the very high salt concentrations that occur in limited number of natural water bodies (for instance, in the Dead Sea, the name of which itself exhibits the unfavourable conditions of this water body for biological processes). However, if we consider specific species rather than aquatic organisms in general, it became apparent that the salinity regime is, as a rule, very important. For example, the overwhelming majority of fish are either sea water or fresh water organisms. Certain species of fish can inhabit fresh waters at times, and at other time waters with comparatively low salinity (as compared to the ocean). The number of fishes, whose life cycle occurs partly in the ocean and partly in fresh water bodies, is rather small.

Another condition essential for animal survival in the hydrosphere relates to the depth at which they live. This determines the magnitude of pressure under which an animal lives and which increases with depth reaching a thousand atmospheres in the deepest oceanic trenches. The oceans are inhabited to the greatest depths, but each of the marine animal species can exist only in a certain, sometimes, rather limited range of depths. Therefore, both an extreme submergence of the surface water animals and bringing to the surface of deep-sea animals can lead to their death. A third condition influencing the survival of marine animals is illumination, although this plays a less important role than the foregoing factors. Deep-sea animals live in darkness, which they sometimes illuminate by a light source they can produce. The activity of animals inhabiting the surface water continues in the darkness produced by an ice layer at the surface of water bodies and also in high latitudes in winter when light is weak or absent altogether.

Besides the principal conditions mentioned earlier, the distribution of land animals depends on a number of additional factors. A large number of land animals need temporary or permanent shelters, which may be absent in some geographical regions. Animals dwelling in the ground are dependent on the nature of soil, which can be suitable or not so suitable for living in. Of some importance for land animals are conditions of illumination, although a great many animals can quite easily adapt to weak light (night animals) or even to a complete darkness (for example, the soil organisms).

Of general importance for all organisms is the absence in their environment of high (in terms of their endurance) concentrations of poisonous substances. The importance of this condition is considered in Chapter 10.

Along with the major features of the abiotic environment, the most important factor for the existence of living organisms is their mutual relations at different ranks in ecological systems. The most important among these relations are the trophic interrelations, which closely connect all animals with plants, the herbivores directly and carnivores through one or several intermediate links of the trophic chain. Besides simple trophic relations, there are more complex forms of mutual relations of animals with other organisms, including symbiosis (the coexistence of dissimilar organisms that can be beneficial for all participants or for some of them), commensalism (a relation in which one animal obtains food and the other consumes a part of this food) and widespread parasitism (when an organism lives in or on another organism, the host, which it usually injures by consuming its tissues or digested food). Parasitism occurs in many groups of animals particularly in protozoa, worms and the arthropods.

The interrelations among animals of the same species are diverse and may be either very close or relatively distant. A number of animals live in colonies (e.g. corals). Particularly close interrelations occur among social insects. For example, the representatives of many bee or ant species die very quickly when isolated from their communities. At the same time there are various animals which contact other representatives of their species only in the reproduction period. A number of primitive animals lacking sexual reproduction have no relations with other individuals of their species.

The role of animals in the biosphere is determined, first of all, by their participation in the breakdown of organic matter of both autotrophic and heterotrophic organisms. Although this process occurs partly as a result of the activity of plants, bacteria and abiotic factors, without the activity of animals, the rate of the organic carbon cycle and of the other constituents of living matter would be much lower. At the same time a large number of animals are important in the accumulation of a number of mineral and organic substances, which, following the death of these animals, result in the formation of sedimentary rocks. An example of such a process is the formation of chalk, limestone and other calcareous rock deposits, which consist largely of the remains of shells from extinct aquatic animals. Another way in which animals participate in the formation of sediments is through the destruction and erosion of the materials forming the upper layers of the lithosphere. Some animals in the course of their lives substantially affect soils. For instance, Darwin found that earthworms exert a profound effect on the structure of the soil they live in. This effect is undoubtedly essential for the growth of plants, and consequently has a bearing, directly or indirectly, on all organisms in the relevant regions.

Animals occupy an important place in ecological systems, producing an influence, via trophic links or otherwise, on different organisms belonging to these systems. These effects are often of a bilateral character. For example, the distribution of herbivorous animals subsisting on plants is limited by food resources, i.e. the number of these plants. Similar interrelations occur between the number of animals pursued by carnivorous animals and the number of carnivores themselves.

These simple trophic interrelations are only some of the various ways in which animals influence other organisms in ecological systems. Of a more complex nature are, for instance, the mutual relations between animals and plants whereby the pollen or seeds of plants are transported by animals. Such relations are particularly pronounced between insects and flowering plants. There are a lot of other examples of symbiotic relations among animals and plants belonging to the same ecological system.

The effect of changes in the biosphere on the evolution of both animals and plants is discussed in Chapter 8.

Of exceptional importance in the evolution of the biosphere was the appearance in the animal world of an intellectual creature (humans) who became the decisive factor in changing the biosphere. The role of man's activities in the biosphere's evolution is considered in Chapters 9 and 10.

## 5.2. The Heat Regime of Animals and Man

The zones which animals can inhabit are confined to a certain range of thermal conditions. For many animals (e.g. the stenotherms) this range is very small and does not exceed a few degrees. Thermal conditions are a vital factor in the geographical distribution of animals.

It should be emphasized that animals do not experience the effects of the environment temperature directly, rather they are affected by the temperature at the surface of their bodies. This temperature usually differs little from the environmental temperature in exothermal animals, i.e. animals who due to the low level of metabolism, mainly obtain heat from the environment. In endothermal animals considerable internal heat generation maintains a relatively high body temperature especially for their internal organs.

Besides the exothermal—endothermal division, animals can also be described with regard to the stability of their thermal regime. In this case, animals are divided into poikilotherms, whose body temperature varies over a wide range, and homeotherms, who maintain their body temperature at an approximately stable level. There is also an intermediate group of heterothermal animals, who can only control their body temperature to a limited extent or for limited intervals of time.

The body temperature of poikilothermal animals does not entirely coincide with the temperature of the external medium. This arises because of the effect of metabolic processes, which are associated with the secretion of heat. However, the heat production of these animals is, as a rule, insignificant, and consequently the temperature difference is usually small. There are some exceptions, which are due to periods of high activity by the animals. Thus, for example, the body temperature of many insects sharply increases during flight.

There are also other factors whose effects may contribute to a change in the body temperature of poikilotherms compared to the temperature of the external medium. For instance, the body temperature of these animals can fall as a result of heat loss caused by evaporation from the animals' body or at the expense of outgoing longwave radiation to the environment. The absorption of solar radiation can lead to a rise of the body temperature. In large poikilothermal animals (for instance, in some reptiles) fluctuations of the body temperature may noticeably lag behind changes in the temperature of the external medium because of the effect of thermal inertia of the animal's body.

The homeothermal animals include two higher classes: birds and mammals, which in the course of evolution have developed an effective mechanism to maintain a steady body temperature irrespective of external conditions. This mechanism is based on the activity of the heat regulation centre, located in one of the sections of the brain called the hypothalamus. When the external factors of the thermal regime change, this centre ensures an appropriate change in the systems of physical and chemical heat regulation within the body, which enables the animal to maintain a constant body temperature. The basic means of heat regulation is a high variability of metabolism, which sharply increases with decreasing temperature. Along with this means of heat regulation, many animals possess other means such as changes in evaporation from the body surface and condition of the hairy coat. Different homeothermal animals have different body temperatures. In birds it is usually 40—42 °C and in mammals it more often ranges from 36 to 39 °C. The stability of body temperature of homeothermal animals is relative. Their body temperature often varies slightly through day and night, with a change in muscular load, according to diet, and because of some other factors. Under environmental conditions that are abnormal for homeothermal animals, their body temperature can fluctuate appreciably.

The vital activity of both poikilotherms and homeotherms can be carried out within a certain range of body temperatures limited by so-called lethal temperatures. These lethal temperatures can differ significantly for the same species of animals depending on the acclimatization of the animal, i.e. on the conditions of higher or lower temperatures it has experienced previously. The lethal temperatures of poikilothermal animals are very diverse. Some animals die with a comparatively small fluctuations of their body temperature, others can endure temperature changes of tens of degrees without harm. In homeothermal animals, the difference between the upper and lower lethal temperatures lies most often within the range of 15—25 °C. For man, with a normal body temperature of about 37 °C, the lower lethal body temperature is 24—25 °C and the upper 43—44 °C.

The biological functions of both poikilothermal and homeothermal animals change considerably with fluctuations in their body temperatures. The effect of temperature is very great on the vital functions of poikilothermal animals, whose activity is possible only within a specific and sometimes rather narrow range of temperatures. To this effect, the behaviour of poikilothermal animals is often determined by the necessity of maintaining the optimum body temperature. Thus, for example, with a deficiency in heat, many insects, reptiles and other animals move into the sunlit sites and take a position that ensures the greatest radiative heating of their bodies. When the Earth's surface is overheated by the Sun, these animals abandon these sites and either burrow into holes or climb into branches of plants where the temperature is lower than that of the Earth's surface. It should be emphasized that the thermal conditions greatly influence the reproductive processes of poikilothermal animals, which in many respects determines their ecology in a particular period.

Although the ecology of homeothermal animals depends to a lesser extent on the thermal regime, under unfavourable conditions, however, these animals also use microclimatic features of the landscape to adjust their body temperature to the optimum. The need for such action is especially great for small animals, whose body thermal inertia is insignificant. They accordingly more easily cooled by low environmental temperatures because of the large value of the ratio between their body surface, through which the outflow of heat takes place, and their body mass, which determines the heat production from metabolism (Max Rubner's law).

The substantial influence of thermal factors on the vital activity of animals highlights the importance of environmental thermal conditions in the geographical distribution of animals and in their ecology.

The heat regime of animals, like that of plants, is closely connected with their energy balance. Since the basic physical principles of the thermal regime of animals and man coincide, we shall first consider the problem of the heat regime of the human body.

In experimental investigations (Büttner, 1951; Burton and Edholm, 1955), it has been established that the mean temperature of the surface of the human body varies in a relatively wide range (about 20 °C), depending on the relationship of the components of the energy balance. This body surface temperature usually differs from that of the internal organs (or from the temperature of the body as a whole), where the temperature is much more stable. Different values of mean surface temperature correspond to different conditions of heat perception: heat, cold or an intermediate state that is usually called the zone of comfort. The most favourable state of heat sensation is usually observed at a mean temperature of the body surface close to 33 °C. With a surface temperature below 29 °C, conditions of cold are felt, and with temperature above 34—35 °C, excessive heat.

It should be noted that in spite of numerous investigations in this field, the early methods applied make it impossible to answer the question about what heat sensation man experiences in a certain region under the influence of the whole complex of meteorological conditions, physiological factors and a given type of clothing. To solve this problem, a method for calculating the energy balance of the human body (Budyko, 1958, Budyko and Tsitsenko, 1960; and others) has been applied.

Under conditions of a stationary regime, when the temperature of the human body changes little with time, the equation for the energy balance of the human body has the form

$$R + M = LE + P, \tag{5.1}$$

where $R$ is the radiation balance of the surface of the body; $LE$ the consumption of heat on evaporation; $P$ the turbulent flow of sensible heat from the body surface to the atmosphere; $M$ the heat production, which is determined on the average by the quantity and calorific value of food assimilated during a given period. Equation (5.1) does not include components that are small compared to the values of the principal components of the balance. Of these it is worth mentioning heat loss due to respiration, which can reach appreciable values at low air temperatures.

In order to consider the thermally insulating effect of clothing on the heat balance, we must make up balance equations for the external surface of the body of a clothed man and for the external surface of his clothing. The first of these equations can be presented in the form

$$M = LE' + P', \tag{5.2}$$

where evaporation $E'$ and turbulent heat flux $P'$ are assumed to be equal to

$$E' = \rho D'(q_s - q_x)a \tag{5.3}$$

and

$$P' = \rho c_p D'(T_s - T_x) \tag{5.4}$$

respectively, where $D'$ is the coefficient characterizing the heat insulating properties of the clothing; $q_s$ the specific humidity of the air saturated with water vapour at the temperature of the body surface; $q_x$ and $T_x$ the mean specific humidity and temperature of the air at the level of the outer surface of the clothing; $T_s$ the mean temperature of the body surface; $a$ the coefficient equal to the ratio of the rate of evaporation from the body surface to the rate of evaporation from a wet surface under the same conditions.

Let us express the energy balance equation for the outer surface of the clothing in the following form

$$R_0 + \rho c_p D'(T_s - T_x) = (\rho c_p D + 4\delta\sigma T^3)(T_x - T). \tag{5.5}$$

In this equation $R_0$ designates the radiation balance of the surface of the clothing, calculated on the assumption that its temperature is equal to that of the environment. This value is added to the heat inflow from the body to the outer surface of the clothing, equal to $\rho c_p D'(T_s - T_x)$. The right-hand side of the equation represents the heat loss from the surface of the clothing as a sum of the turbulent sensible heat flow $\rho c_p D(T_x - T)$ and the radiative flux $4\delta\sigma T^3(T_x - T)$, with $D$ denoting the integral coefficient of turbulent diffusion. Equation (5.5) does not include the value of evaporation, because water vapour is not generated on the surface of the clothing.

Equations (5.2)–(5.5) should be supplemented with a formula which implies that the flux of water vapour passing through the surface of the clothing does not change in value. This can be expressed in the following form:

$$\rho D'(q_s - q_x)a = \rho D(q_x - q)a, \tag{5.6}$$

where $q$ is the specific humidity of the surrounding air. From Equations (5.2)–(5.6) we obtain the relationship

$$R_0 + M\left[1 + \frac{\rho c_p D + 4\delta\sigma T^3}{\rho c_p D'}\right] = L\rho D(q_s - q)a\left[1 + \frac{4\delta\sigma T^3}{\rho c_p(D + D')}\right] + \\ + (\rho c_p D + 4\delta\sigma T^3)(T_s - T). \tag{5.7}$$

Equation (5.7) allows us to calculate $T_s$, that is the mean temperature of the surface of the body, in terms of the meteorological parameters $R_0$, $T$, $q$, $D$, the physiological characteristics $M$, $a$, and the parameter $D'$ representing the properties of the clothing. In this calculation we should take into account the known dependence of $q_s$ on $T_s$ expressed by the formula of Magnus. It should be noted that the coefficient $D$ depends on the wind speed, increasing with increasing speed.

Having made such a calculation, we can evaluate the heat perception of a clothed man under different climatic conditions and also determine the comparative effect of different factors on heat perception.

Equation (5.7) is simplified if it is applied to calculate the heat regime of a man without clothing or whose clothing does not have a heat-insulating effect. In this case, we assume that $1/D' = 0$, which corresponds to clothing with an infinitely great heat conductivity. Under these conditions we obtain the formula

$$R_0 + M = L\rho D(q_s - q)a + (\rho c_p D + 4\delta\sigma T^3)(T_s - T). \tag{5.8}$$

An equation similar to equation (5.8) was found in the work of Büttner (1938), who employed it for evaluating the effect of individual factors on the heat perception of man.

The application of Equations (5.7) and (5.8) to calculate the thermal regime of the human body under different natural conditions is associated with the need to determine the radiation balance $R_0$. Thanks to the availability of extensive data on the radiation regime of the Earth's surface, this problem can be solved approximately as follows.

Let us assume that a human body in a vertical position has a form roughly corresponding to that of a relatively thin vertical cylinder. On the basis of simple geometrical considerations we can propose that on a unit area of the lateral surface of the vertical cylinder there falls a quantity of direct solar radiation equal to $(S \text{ cotan } h)/\pi$, where $S$ is the total direct radiation incident on a unit area of a horizontal surface and $h$ is the angular altitude of the Sun. The amount of scattered radiation from the sky incident on a unit area of the cylinder's surface is approximately half the amount of scattered radiation $Q'$ incident on the element of a horizontal surface. This value is supplemented with the income of radiation reflected from the Earth's surface which is about $\frac{1}{2}(S + Q')\alpha$, where $\alpha$ is the albedo of the Earth's surface.

Net longwave radiation at the unit of the cylinder's surface is equal to half the net longwave radiation at a horizontal surface $I_0$. To this is added the algebraic sum of two terms expressing the effect on net longwave radiation of the difference between the temperature of the cylinder's surface and the air and of the difference between the temperature of the cylinder's and Earth's surfaces. The first of these is equal to $\delta\sigma(T_s^4 - T^4)$ and the second, $\frac{1}{2}\delta\sigma(T_s^4 - T_w^4)$, where $T_w$ is the temperature of the Earth's surface. Their sum approximately equals $4\delta\sigma T^3(T_s - T) - 2\delta\sigma T^3(T_w - T)$. The first of these components has already been included in the energy balance equation.

Taking this into account, we obtain the following formula:

$$R_0 = \left[ S\frac{\cot \text{an}\, h}{\pi} + \frac{1}{2}Q' + \frac{1}{2}(S + Q')\alpha \right](1 - \alpha_0) - \frac{1}{2}I_0 +$$

$$+ 2\delta\sigma T^3(T_w - T), \tag{5.9}$$

where $\alpha_0$ is the reflectivity (albedo) of the surface of the human body.

In deriving Equation (5.9), among other simplifying assumptions, we have adopted a hypothesis about the isotropy of the fluxes of scattered radiation and longwave emission incident on the cylinder's surface.

In order to find the values of $S$, $Q'$ and $I_0$ in Equation (5.9) one can use either observational data or calculation methods. The magnitude of $T_w - T$ can be calculated by the method suggested in Chapter 4.

By substituting the value $R_0$ into Equations (5.7) and (5.8) we obtain the following:

$$\left[ S\frac{\cot \text{an}\, h}{\pi} + \frac{1}{2}Q' + \frac{1}{2}(S + Q')\alpha \right](1 - \alpha_0) - \frac{I_0}{2} + 2\delta\sigma T^3(T_w - T) +$$

$$+ M = L\rho D(q_s - q)a + (\rho c_p D + 4\delta\sigma T^3)(T_s - T); \tag{5.10}$$

and

$$\left[ S\frac{\cot \text{an}\, h}{\pi} + \frac{1}{2}Q' + \frac{1}{2}(S + Q')\alpha \right](1 - \alpha_0) - \frac{I_0}{2} +$$

$$+ 2\delta\sigma T^3(T_w - T) + M\left[ 1 + \frac{\rho c_p D + 4\delta\sigma T^3}{\rho c_p D'} \right] =$$

$$= L\rho D(q_s - q)a\left[ 1 + \frac{4\delta\sigma T^3}{\rho c_p (D + D')} \right] + (\rho c_p D + 4\delta\sigma T^3)(T_s - T). \tag{5.11}$$

To solve Equations (5.10) and (5.11), it is necessary to determine the physiological parameters they contain, namely, heat production $M$ and the characteristic of the evaporation conditions $a$. According to the results of a number of investigations, the heat production of a man in a state of rest and in conditions of a thermal comfort is more or less stable and approximately equal to 50 W m$^{-2}$.

The data from the work of Winslow *et al.* (1937a) permit us to clarify the relationship between the heat production and the mean temperature of the body surface. Figure 5.1 shows the dependence of heat production $M$ on the mean temperature of the body surface $T_s$ based on the experimental data presented by Winslow, Herrington and Gagge. It can be seen in the graph

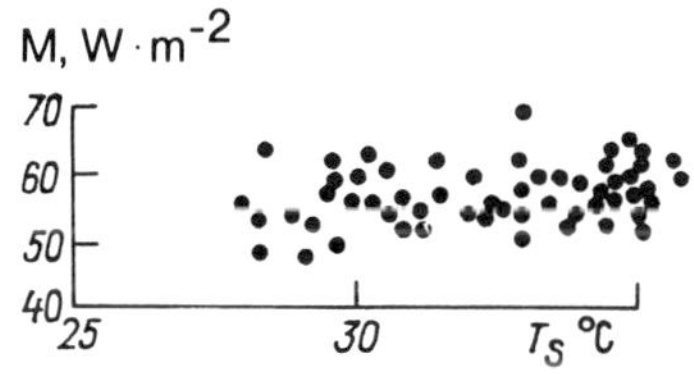


Fig. 5.1.

that heat production of a man not engaged in physical labour depends relatively little on the temperature of the body surface in the interval corresponding to the zone of comfort. The mean value of heat production appears to be about 55 W m$^{-2}$, which corresponds to the estimate given earlier.

The problem of determining conditions of evaporation from the body surface can be solved with the use of the material from another paper by Winslow *et al.* (1937b). On the basis of experimental data from their research, a graph, presented in Fig. 5.2, has been constructed. This shows the dependence of the coefficient $a$ on the temperature of the body surface. It should be noted that, unlike heat production, the parameter $a$ changes appreciably, even in a comparatively narrow range of body surface temperature. These variations correspond to differences in the functioning of the sweat secretion system, which is an important regulator of man's thermal regime. At body temperatures above 35 °C there is a discontinuous change in the dependence of $a$ on $T_s$, presumably associated with a disruption of the stable thermal state of the human body under conditions of overheating.

Comparison of Figs 5.1 and 5.2 shows that in the body surface temperature range of 29—35 °C, a change in the rate of evaporation from the body surface proves to be a more substantial factor in human heat regulation than a change in heat production.

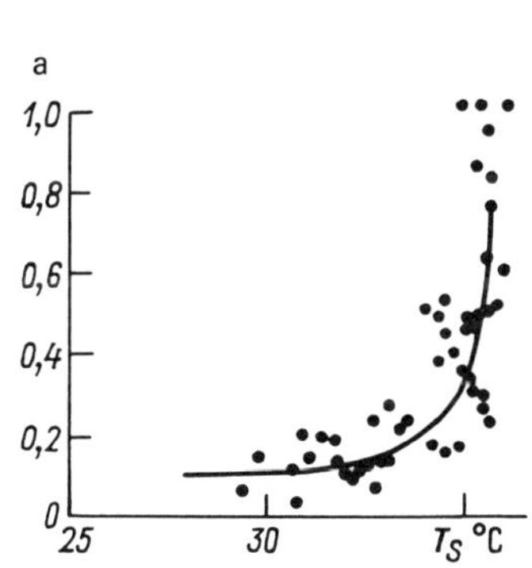


Fig. 5.2.

It is evident that in solving Equations (5.10) and (5.11) one must consider the dependence of the coefficient $a$ on the temperature of the body surface. This can be carried out with the help of the data in Table 5.1 (based on Fig. 5.1).

TABLE 5.1

Dependence of the coefficient $a$ on the temperature of the body surface $T_s$

| $T_s$ (°C) | 29 | 30 | 31 | 32 | 33 | 34 | 35 |
|---|---|---|---|---|---|---|---|
| $a$ | 0.10 | 0.11 | 0.11 | 0.12 | 0.13 | 0.18 | 0.31 |

Using Equations (5.10) and (5.11) it is possible to calculate the mean temperature of the surface of the human body, which defines its thermal state. In some investigations the results of such calculations have been compared to data from direct measurements of body temperature (Yefimova and Tsitsenko, 1963; Kandror *et al.*, 1966).

Experimental investigations carried out in different climatic zones of the U.S.S.R. have shown that with positive air temperatures there is usually good agreement between the measured and calculated values of the mean body temperature. At air temperatures below freezing point the correlation between the measured and calculated values decreases, which is evidently associated with the difficulty of achieving a stationary state of the human body under such conditions. However, this latter conclusion does not preclude the possibility of using calculated body temperatures at low air temperatures as a general characteristics of the degree of climate severity.

Equations (5.10) and (5.11) can be used in calculating the thermal state of the human body under specific climatic conditions. As an example, we give here the calculated results for the European territory of the Soviet Union for mean July conditions at 1.00 p.m. (Budyko and Tsitsenko, 1960). The coefficient $D'$ used in this calculation can evidently vary within a wide range depending on the type of clothing.

Applying Equation (5.11) it is possible to find which values of this coefficient correspond to clothing that provides the thermal comfort under common indoor conditions or under mean conditions of a summer day in a continental climate at middle latitudes. This calculation shows that the appropriate values of $D'$ are of the order of several tenths of one centimetre per second. In the study of the thermal state of the human body the coefficient $D'$ was assumed to be equal to 0.4 cm s$^{-1}$. To determine the coefficient $a$ the data from Table 5.1 were used and the heat production $M$ was considered equal to 55 W m$^{-2}$.

As a result of the calculations, maps of the mean temperature of the body surface of a man protected and unprotected by clothing at 1.00 p.m. in different months of the warm season have been constructed. Figure 5.3 shows the relevant maps for July and May. Isolines of the mean body temperatures equal to 29 and 34.5 °C are presented. As mentioned earlier, it can be assumed that at temperatures below 29 °C, conditions of heat deficiency are observed and at temperatures above 34.5 °C, those of heat excess. The data relating to a man in light clothes are shown by solid lines, those relating to a man without clothes, by a broken line.

It can be seen from Fig. 5.3(a) that in July the zone of a favourable thermal regime for a man in light clothes extends over the greater part of the European territory of the U.S.S.R. Conditions of heat deficiency are observed only in the extreme north, and an area of excessively warm conditions occurs in the northern Caucasus and in Lower Povolzhiye. In the absence of the heat-insulating effect of clothing, the heat-perception regime changes. North of latitude 60°N, there is an area of insufficiently warm conditions, but there is no area of excessively warm conditions in the territory under consideration.

A slightly different picture is observed in May (Fig. 5.3(b)). In this case conditions of heat excess are not observed at all. Heat conditions favourable for a man in light clothes are pronounced in the central and southern regions, while for a man unprotected by clothing, favourable conditions are found only in the extreme southeast.

Such calculations can be used for determining the thermal efficiency of different kinds of clothing under specific climatic conditions and also for evaluating the possibility of making use of such practices as sunbathing, airbathing and so on. Calculations based on Equations (5.10) and (5.11) also make it possible to estimate the comparative effect of different factors on the

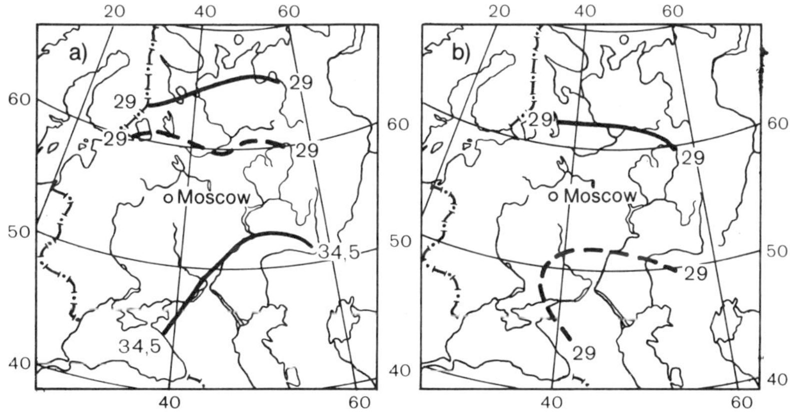


a Fig. 5.3. b

thermal state of man under actual climatic conditions. By these means it is possible to establish that, on a summer day, the main source of heat for man is not his own heat production but the income of shortwave radiation. For mean conditions of the European part of the U.S.S.R. in this case, internal heat production makes up 25—30% and the income of radiative heat 70—75% of the total income of energy.

The relative value of different forms of heat loss changes considerably depending on external conditions. With high temperatures, heat conversion for evaporation becomes a principal form of heat loss, which is evidently associated with the activity of the sweat-secretion system. In the far southeast of the European U.S.S.R. the amount of heat used for evaporation reaches 75—80% of the total heat loss. At lower temperatures, the relative value of heat loss for evaporation decreases sharply, reaching 20—25% and even lower values. In this case the turbulent flow of sensible heat, amounting to 50—60% becomes the principal means of heat loss. Heat loss in the form of net longwave radiation is smaller in all cases, varying in the range from 5 to 20% of the total heat loss.

If by using Equations (5.10) and (5.11) we evaluate the effect of different meteorological factors on the heat state of man, it turns out that under summer climatic conditions in the European part of the U.S.S.R., variations in air temperature, solar radiation and wind velocity are of vital importance, while fluctuations in air humidity exert a comparatively small effect on the thermal regime of the human body. The thermal regime of man is also essentially dependent on the level of metabolism. This effect can be considered in the afore-mentioned formulas, by using data on the secretion of metabolic heat with different levels of human activity taken from available tables (Oke, 1978).

The procedure set forth above for calculating the characteristics of the thermal state of man has been used in different bioclimatic investigations, among which the works of Liopo (1966, 1968) should be mentioned. These studies present a nomogram that allows us to estimate the characteristics of the heat regime of man as a function of meteorological conditions and physiological factors. In the papers of Liopo and Tsitsenko (1963), Aizenshtat (1965) and other authors, data have been obtained in relation to both the mean conditions of the thermal state of man in different regions of the U.S.S.R. and temporal variability of these conditions.

The procedures discussed here can also be applied, with slight modifications, to study the heat regime of both poikilothermal and homeothermal animals. In such studies one should first carry out a calculation of the radiation regime of the body of the animal taking into account its form. It is evident that in a similar way to the human body, this form can be assumed to comply with a simple geometrical model in rough calculations. The albedo of the surface of the animal's body, required for the determination of absorbed radiation, needs to be found experimentally.

Conditions of evaporation from the surface of the animal's body must then be taken into account. For many animals (for example, reptiles) evaporation is insignificant, which allows us to consider the coefficient $a$ in the above formulas as equal to zero. In the cases when the value of heat used for evaporation from the animal's body is comparable with other components of the energy balance, the coefficient $a$ should be found from empirical data.

If the animal under consideration has a hairy coat, the effect of this coat on the heat regime can be evaluated in a similar way to the effect of clothing on the heat regime of man. For this purpose, on the basis of experimental data, one must evaluate the magnitude of the coefficient $D'$, which in this case characterizes the heat conductivity of the hairy coat. Without this coat, the value $1/D'$ is equal to zero.

The value of the heat production itself for poikilothermal organisms can often be considered close to zero, and for homeothermal animals it should be determined from experimental data in relation to the level of the animal's activity.

Our considerations show that calculation of the energy balance and an evaluation of the thermal state of the body of an animal is, on the whole, no more complicated than a similar calculation for man. It is necessary, however, to mention that these calculations are sometimes impeded by the absence of the necessary experimental data.

It follows from calculated data of the thermal regime of poikilothermal animals (Budyko, 1959) that their body temperature can be 10—20 °C (and more) greater than the air temperature if they are heated by the Sun's rays, particularly when evaporation from their bodies is small. At night the body temperature of poikilothermal animals which are out of their shelters, can be lower than the air temperature by approximately 5—7 °C.

The body temperature of homeothermal animals, which greatly depends on the heat regulation, is less variable. Nevertheless, the influence of meteorological conditions on their thermal regime is substantial. An approximate idea of the quantitative regularities of this influence can be obtained from the above data on the thermal state of the human body.

Among the studies of the thermal state of homeothermal animals, the work of Yaroshevsky (1968) is noteworthy. Examining the heat balance of sheep, Yaroshevsky established that the amount of heat used in evaporation from the sheep's body increases noticeably as the body temperature of the sheep increases. Heat consumption for respiration plays an important role in the heat exchange of sheep, which in young animals can make up more than a half of the total heat loss. The thermal regime of sheep depends greatly on the state of the sheep's fleece, particularly, its length, thickness and degree of soiling. By moving about and changing their body position, sheep can alter their thermal state appreciably. For instance, when the air temperature is low, sheep cluster into tight groups or lie down on the ground, which leads to a decrease in the heat loss from the animal's body. Yaroshevsky emphasized that the thermal regime is of great importance in the vital activities of sheep and in many cases it considerably restricts their productivity.

We believe that the conclusions resulting from Yaroshevsky's experiments on sheep, are in many respects true for other mammals, the regularities of the thermal regime of which have not yet been investigated.

The methods applied for studying the thermal regime of the animals living today can also be employed in the calculation of the thermal state of animals that existed in the geological past. An example of such calculation is presented in Chapter 8.

## 5.3. Ecological Systems

### 5.3.1. THE ENERGY REGIME OF ANIMALS

The organic matter produced by autotrophic plants is almost completely consumed by heterotrophic organisms. Some of these heterotrophic organisms subsist on the living organic matter, while others, largely microorganisms, feed on the dying parts of the autotrophic plants. Only a relatively small portion of the total production of organic matter is transformed into mineral substances without the activity of living organisms. An example of this process is forest and steppe fires, in the course of which organic matter is converted into carbon dioxide, water vapour and certain mineral compounds. A very small portion of the total production of the biomass is buried in the upper layers of the lithosphere and at the bottom of water bodies in the form of coal, peat and other organic substances. Thus, almost the entire total production of autotrophic plants enters different food chains.

As already mentioned in Chapter 1, the transmission of energy from organisms to organisms within trophic chains is associated with great loss in energy. The ratio of the biomass of organisms to the amount of organic matter consumed by them is sometimes called the coefficient of ecological efficiency. This coefficient, as a rule, does not exceed 10—20%. Small coefficients of ecological efficiency give rise to rapidly reducing numbers of organisms located in consecutive links of the trophic chain. The number of organisms at a certain link of the trophic chain (in other words, at a certain trophic level) depends on their size.

The intensity of the processes of energy exchange (metabolism) per unit weight of a living organism is usually the higher the smaller this organism is in size. One of the reasons for this is that the processes of exchange are substantially dependent on the rate of diffusion of gases through the surface of organisms, which increases per unit of their biomass as the size diminishes. For this reason, the intensity of metabolism in microorganisms is for the same unit of weight many times greater than the metabolism of macroorganisms. Metabolism also depends on the nature of physiological processes in the tissues of organisms. For instance, the metabolism of wood is usually insignificant compared with the metabolism of leaves and thin roots of trees. Metabolism in mammals is generally tens of times higher than in reptiles of the same weight. This difference shows that the energy demand of mammals is much greater than that of reptiles.

The ratio of energy obtained in the course of nutrition and metabolism determines to a great extent the biomass of different organisms.

Great difficulties occur in trying to estimate the total biomass for the entire Earth. According to Kovda (1969), the biomass on the continents is close to $3 \times 10^{12}$ t of dry organic matter. The data of Kovda shows that more than 95% of this is plant biomass and less than 5% is animal biomass. A more detailed estimate of the biomass of organisms on the Earth is given in the work of Bazilevich *et al.* (1971). These authors consider that the mass of green plants on the continents is equal to $2.4 \times 10^{12}$ t and the mass of animals and microorganisms is $0.02 \times 10^{12}$ t. For the oceans the mass of plants makes up $0.2 \times 10^9$ t and the mass of animals, $3 \times 10^9$ t. According to Dobrodeev and Suyetova (1976), the mass of marine animals in the ocean is about $7 \times 10^9$ t.

It follows from these data that the global value of biomass fully depends on the magnitude of the biomass of green plants on the continents. It is noteworthy that although life in the ocean is much more ancient, the overall mass of organisms in the oceans is hundreds of times less than the mass of organisms on the continents.

Considering that the annual productivity of plants on the continents makes up $140 \times 10^9$ t, we find that the duration of one cycle of organic matter of the continents is about 20 years. It is evident that this average figure for land refers mostly to forests, which constitute the principal part of the biomass of plants. In other continental geographical zones the duration of the organic matter cycle is much shorter. This cycle is even shorter in the oceans where for phytoplankton it is presumably only a few days.

The reasons why forests form an overwhelming proportion of the total biomass of living organisms is because trees belong to the first link of the trophic chain. It is also due to the large size of each individual tree. Together with the specific properties of wood, this considerably decreases the rate of metabolism per unit biomass. Even though phytoplankton in the oceans yield a relatively high total productivity, the small size of the individual planktonic organisms increases their metabolism per unit weight so substantially that the total mass of the plankton appears negligible when compared to that of forests.

Let us assume that the total biomass of animals is $10^{10}$—$10^{11}$ t. Considering that animals assimilate about 10% of the total productivity of plants, we find that the average duration of one cycle of organic matter of animals is approximately one to several years. The actual length of life of one generation of animals varies over a very wide range. This estimate of the duration of one cycle of animal organic matter is for groups which constitute the main portion of the total animal biomass. Unfortunately, the distribution of biomass among different groups of animals is not yet fully understood.

The largest component of animal biomass is the invertebrates, among which soil organisms are of great importance. The biomass of large animals per unit area is comparatively small. Thus, it follows from the summary of Huxley (1962) that in African savannas the biomass of large wild animals may reach 15—25 t km$^{-2}$, in middle latitude forests it may be 1 t km$^{-2}$, in tundra 0.8 t km$^{-2}$ and in semi-desert 0.35 t km$^{-2}$.

Unlike the case of animals and plants, a much more precise estimate can be obtained of the biomass of human beings and of their energy consumption in the course of feeding. For the current population of mankind (exceeding four billion people) the total biomass is about $0.2 \times 10^9$ t. Taking into account the average daily energy consumption by one man, it is not difficult to find the value for the total consumption of energy by people. This value roughly corresponds to the current productivity of agriculture, even though it does not provide a sufficient amount of food for human populations in all regions of the world.

Man occupies one of the last positions in trophic chains, consuming both the primary production of autotrophic plants and the biomass produced by many herbivorous and carnivorous animals. Because of this, man's position in the energy pyramid lies at the upper levels, which are much narrower than the base of the pyramid (describing the productivity of autotrophic plants). From the data given above, it is easy to find that at present man consumes a small portion of the primary production of the organic world. However, in the near future this portion is expected to increase considerably due to the rapid growth of the human population.

The dimensions and shape of the biological energy pyramid on our planet constantly vary in accordance with the development of agricultural activities required to provide food to the world's growing population. Consuming a growing amount of the primary biological production, man also uses a large and ever-increasing quantity of technical energy, which is a new source of heat for our planet. Since the combustion of carbon-based fuels leads to the formation of great amounts of carbon dioxide, human activities have already changed the chemical composition of the atmosphere and this change is becoming more and more pronounced. As a result, it is very likely that global climate will change considerably in the near future, and this will exert a profound influence on all the components of the biosphere. This problem is discussed in Chapters 9 and 10.

### 5.3.2. THE PROPERTIES OF ECOLOGICAL SYSTEMS

The global ecological system (i.e. the biosphere) comprises the planetary energy system of living organisms. The source of energy for this system is mainly the photosynthetic activity of autotrophic plants, whose biomass is used by heterotrophic organisms. A large number of the organisms consuming the energy assimilated by autotrophic plants are themselves eaten by other organisms, thus transporting energy by way of the trophic chain. Homogeneous trophic links unite the populations of animals and plants into elementary energy systems that are part of the global energy system. Each of the elementary energy systems is described by a complex of mutual relations among the organisms it contains and between these organisms and the abiotic environment. This system is at the same time an elementary ecological system.

In order to study the influence of large-scale changes in the environment on living organisms, it is necessary to consider the stability of ecological systems containing these organisms. It should be borne in mind that ecological systems, in the same way as individual species of organisms, have been subjected to a long evolutionary process, in the course of which the less stable systems have vanished and only those systems remained whose stability was sufficiently high in relation to fluctuations in the external factors.

The concept of stability of ecological systems is often used in modern studies, but different authors define it in different ways (e.g. Svirezhev and Logofet, 1978; and others). Following the earlier approach (Budyko, 1974), let us consider the stability of an ecological system by analysing it using a numerical model. For this purpose, let us examine a highly simplified model of an ecological system, incorporating only four groups of organisms:

(a) autotrophic plants;
(b) herbivorous animals;
(c) carnivorous animals;
(d) parasites, consuming the biomass of the first three groups of organisms.

It is possible to make up balance equations for living organic matter, which are very similar to the energy balance equations. Let us designate the biomass of a certain group of organisms per unit area occupied as $B$, their productivity (i.e. the accumulation of the biomass per unit time minus the expenditure on the processes of vital activity of the organisms) as $\Pi$ and the consumption of the biomass by parasites as $D$. The equations of the balance of biomass have the following forms:

for autotrophic plants

$$\frac{dB_a}{dt} = \Pi_a - b_b B_b - D_a, \tag{5.12}$$

for herbivores

$$\frac{dB_b}{dt} = \Pi_b - b_c B_c - D_b, \tag{5.13}$$

for carnivores

$$\frac{dB_c}{dt} = \Pi_c - D_c, \tag{5.14}$$

where $b_b$ and $b_c$ are the coefficients describing the consumption of the biomass of organisms that are being eaten by herbivorous animals and carnivores per unit mass of these groups of animals. The derivatives on the left-hand side of Equations (5.12)—(5.14) determine the rate of change in the biomass with time.

Assuming in the first approximation that the coefficients $b_b$ and $b_c$ are constant, we note that Equations (5.12)—(5.14) include nine variables. Therefore, in order to solve them it is necessary to make use of six additional ratios. These ratios may be inferred from empirical data on various ecological systems.

For a wide range of systems it is possible to apply simplified relations having the form:

$$\Pi_a = \Pi_a(B_a, B_b), \tag{5.15}$$

$$\Pi_b = \Pi_b(B_b, B_a, B_c), \tag{5.16}$$

$$\Pi_c = \Pi_c(B_c, B_b), \tag{5.17}$$

$$D_a = D_a(B_a), \tag{5.18}$$

$$D_b = D_b(B_b), \tag{5.19}$$

$$D_c = D_c(B_c). \tag{5.20}$$

The nature of these relations has been examined in many empirical studies.

The productivity of the dense cover of autotrophic plants $\Pi_a$ increases with the growth of its biomass $B_a$ first slowly (as long as the surface of leaves is small), then more quickly, and with a large density of biomass it increases slowly once again because of the decline in photosynthesis under conditions of considerable shade for lower layers of leaves. Thus, the relation $\Pi_a(B_a)$ in a certain interval of change in the biomass has the form of an $S$-shaped curve. With very high values of the biomass, when additional expenditure on respiration is not fully compensated for by photosynthesis, $\Pi_a$ begins to decline. The relation $\Pi_a(B_b)$ is determined by the action of herbivorous animals in promoting the productivity of plants with a given biomass, by stimulating an additional growth of plants by browsing a portion of them.

The productivity of herbivores $\Pi_b$ usually depends on their biomass $B_b$ in a manner that is qualitatively similar to the relation $\Pi_a(B_a)$, because for a very small and also very large number of animals per unit area, the ratio of productivity to biomass declines, mainly due to the reduction of the birth rate coefficient. It is evident that the productivity $\Pi_b$ depends on the biomass of the plants $B_a$ and increases as the latter grows. A similar influence is exerted on $\Pi_b$ for a given value of $B_b$ by the biomass of carnivores $B_c$, which in the main destroy sick and old animals, thus contributing to an increase in the birth rate coefficient.

The productivity of carnivores $\Pi_c$ changes as their biomass $B_c$ increases in a manner that is qualitatively similar to the relation $\Pi_b(B_b)$ as $\Pi_c$ increases with the growth of the biomass $B_b$.

The consumption of plant and animal biomass by parasites is usually proportional to their own biomass, although this dependence is sometimes stronger than the one described as direct proportionality, since an increase in the density of populations facilitates the transportation of parasites from one organism to another.

By representing Equations (5.15)–(5.20) in a quantitative form it is possible to study the dynamics of populations of organisms belonging to an ecological system. As a result of such an analysis it is possible to draw the following conclusions concerning the conditions that are characteristic of many ecological systems.

1. For values of biomass for each group of organisms that are constant over time (the stationary state of an ecological system) there is often more than one solution to the equations described above, which corresponds to different possible magnitudes of biomass of the different components of the ecological system. However, not all of these solutions correspond to a stable regime of the ecological system, when small deviations in the biomass of one of the groups of organisms from its value for a stationary state bring about a return to the same state.

2. Ecological systems are to some extent stable in relation to deviations in the values of biomass of each component of the system from their values for a stationary state. This stability is determined by the interval over which changes in the biomass of the system's components take to return to a stationary state. For changes in the biomass that exceed this interval, restoration back to the stationary state does not occur, and one or more components of the system are destroyed.

3. The stability of ecological systems including all the above-mentioned components is usually higher than that of the systems within which certain groups of are missing.

Since trophic interrelations (and also non-energy ecological relations) of the components of ecological systems add a certain stability to the systems, these possess some traits of an integrated whole (a similar point of view has been expressed earlier by Elton, 1930). In this connection ecological systems must produce some influence on the evolution of organisms.

The number of organisms in each group forming an ecological system continually changes. Changes in the number of a particular species $n$ in time $t$ per unit area they occupy can be defined by the equation

$$\frac{dn}{dt} = \alpha n - \beta n, \tag{5.21}$$

where $\alpha$ and $\beta$ are the coefficients of birth and death rates. If $\alpha$ and $\beta$ are constant, and assuming that at the initial moment of time $n = N$, we find from Equation (5.21) that

$$n = N e^{t(\alpha - \beta)}. \tag{5.22}$$

With $\alpha > \beta$ the number of animals will increase and with $\alpha < \beta$ it will decline.

If the population declines the extinction of animals will take place shortly after the magnitude $n$ attains a critical value $n_1$, at which a small number of animals does not allow for their further reproduction. As the number of animals increases an ecological crisis will occur after $n$ has passed a critical value $n_2$, which corresponds to the maximum number of animals that the food resources of a given locality can maintain. In this case it is possible that animals will destroy the source of their food, and this will lead to the destruction of the ecological system and endanger the existence of the species it contains. In the present context the destruction of the ecological system is the annihilation of one or several main components of this system under the influence of external factors, and also changes in the relations among the main components of the system that lead to the same result. The groups of organisms that survive the destruction of the former ecological system may then form, either themselves or together with organisms from other systems, a new ecological system.

Taking these considerations into account, let us find from Equation (5.22) the duration of existence of a population of animals for both the first and the second case:

$$t_1 = \frac{1}{\alpha - \beta} \ln \frac{n_1}{N}; \qquad t_2 = \frac{1}{\alpha - \beta} \ln \frac{n_2}{N}. \tag{5.23}$$

These formulas may be rewritten in the following form:

$$g_1 = \frac{1}{(\alpha/\beta) - 1} \ln \frac{n_1}{N}; \qquad g_2 = \frac{1}{1 - (\alpha/\beta)} \ln \frac{N}{n_2}, \tag{5.24}$$

where $g_1 = \beta t_1$ and $g_2 = \beta t_2$ correspond to the number of successive generations of animals from the initial moment of time to the moment of extinction.

It is evident that under natural conditions fluctuations in the birth and death rates are inevitable, and as a result, their average values over a long time interval do not fully coincide with each other. Certain differences in their magnitudes will also take place under the influence of fluctuations in the external conditions that are either favourable or unfavourable for a given species of animals.

It follows from Equation (5.24) that if the difference between the birth and death rates is very small, the existence of any population may come to an end. Considering that under natural conditions, $\ln N/n_1$ and $\ln n_2/N$ as a rule are less than 10, then even a 1% difference between the birth and death rates will lead to the extinction of that population over a period of time not exceeding the life of several hundreds of generations. Since such interval of time is much smaller than the average existence time of a species in the course of its evolution, it is evident that under natural conditions the birth and death rates are somehow regulated, otherwise the prolonged existence of populations would have been impossible. Thus, the prolonged existence of populations is ensured by the dependence of the birth and death rates on the density of populations and also by the interaction of organisms belonging to ecological systems.

We point out that the terms of Equations (5.12)–(5.20) describing the biomass of different components of the ecological system can be expressed in the number of relevant organisms. For this purpose it is necessary to divide these terms by the average biomass of each organism considered. Afterwards we may take into account in a quantitative form the effects of the relationships promoting the stabilization of the size of populations of organisms entering the ecological system.

The most important relationships of this kind are represented by Equations (5.15)–(5.20). They include a reduction in the productivity of excessively numerous populations because of insufficient nutrition, an increase in the death rate of these populations as a result of the activities of parasites and predators and so on. The reverse regularities determine the dynamics of small populations.

As a result of the influence of ecological factors, a successful struggle for existence of living organisms is determined to a great extent by their ability to adapt to conditions that are favourable for maintaining the stability of ecological systems.

Our considerations concerning the dynamics of animal populations may be supplemented with the results of other empirical and theoretical studies of this problem.

It has long been known that considerable seasonal changes occur in the sizes of animal populations with short life expectancy (less than decades). For instance, the number of highly reproductive rodents in middle latitudes is frequently tens of times greater towards the end of the warmer half of the year than at its beginning. Still more variable are the seasonal numbers of some insects.

In addition to seasonal variability in the number of animals, there are also annual and long-term fluctuations in the size of populations. An understanding of long-term fluctuations in the numbers of animals, as has been treated earlier, is of vital importance in examining the causes of the extinction of different animal species.

Of the empirical studies investigating the variability of animal numbers we need to mention the monographs of Severtsev (1941) and Lack (1954). Severtsev collected a vast amount of information on variations in the number of animals in different groups of vertebrates and made certain inferences that they are related to specific features of the adaptive evolution of these groups. Severtsev emphasized that the relationships between the reproduction of mammals, birds and reptiles and the length of their life are of inverse character and found quantitative forms of these relationships. Using mainly data on birds, Lack has come to the conclusion about the ecological factors affecting the birth and death rates of animals. These books and many other studies consider the consequences of a rapid decrease in the number of animals as a result of climatic factors, of epizootic and other causes.

In studies of fluctuations in the number of animals particular attention has been paid to the mutual relation between the dynamics of the populations of predators and the animals they prey on that belong to the same ecological systems. Already in the 1920s it was found, through the analysis of empirical data, and theoretically, that this relation often leads to periodic fluctuations in the number of predators and the animals they pursue, these fluctuations lagging in time for the indicated groups of animals.

Of considerable interest are the results of laboratory investigations of the sizes of populations of animals started by the works of Gause (1934). Such experiments made possible the examination of quantitative dependences of variations in the size of animal populations of one or several species of organisms in terms of ecological factors and mutual relations of different organisms.

In recent years many detailed theoretical studies have investigated the dynamics of populations with respect to age factors and the genetic characteristics of the populations. Among these studies are the book edited by Poluektov (1974), the monograph of Svirezhev and Logofet (1978), and others. Investigations of the dynamics of animal populations are of great importance in developing well-grounded plans for fishing, optimizing hunting, and combating agricultural pests.

CHAPTER 6

# HISTORY OF THE BIOSPHERE

## 6.1. Origin of the Biosphere

The emergence of the biosphere, marked by the first appearence of living organisms, occurred at an early stage of our planet's history. The exact time when the biosphere came into being is difficult to determine because records of primitive living forms preserved in ancient sediments are indistinct and rare in the modern Earth's crust. Due to recent progress in palaeontological studies it has been possible to increase the age of the earliest organisms from an estimated 2—3 billion years to 3.5—3.8 billion years. Since the possibility of finding evidence of more ancient organisms still remains, the dates given above refer to the lower boundary of the real age of the biosphere. However, from analysing the history of the formation of the Earth as a planet, we can suggest that the real age of the Earth's biosphere cannot exceed the available estimates by very much.

It is assumed that the Earth, like other planets, was formed from a gas-dust cloud as a result of gravitational condensation of individual particles which aggregated into a celestial body, the size of which increased as new particles joined it by gravity. It is rather probable that the Earth's mass also grew as a result of impacts of more or less large-size external bodies formed through the same process.

In forming the Earth, the deep layers were gradually heated by the radiogenic decay of elements such as uranium and thorium. In addition, a considerable amount of heat was liberated when the Earth contracted by gravity and when interstellar matter fell on the Earth's surface. The heating of the Earth contributed to its differentiation to several geospheres.

The central geosphere is a nucleus composed of heavier elements at very high temperatures and pressures. The nucleus is surrounded by mantle consisting of less heavy elements but also heated to high temperatures. The Earth's crust is composed mainly of light elements, exuded upward from the mantle.

From data on the long-living isotope content of radioactive elements it has been found that the most ancient rocks in the Earth's crust are about 3.4—3.8 billion years old. At the same time the age of the Earth is assumed to be about 4.5—4.7 billion years. However, it is difficult at present to find records of the rocks formed throughout the first billion years of the Earth's history. There are various assumptions explaining the causes of this phenomenon. For example it is likely that in the earliest part of the Earth's history an intense bormbardment of relatively large bolides prevented the formation of a stable crust on the Earth's surface.

Impact craters can be seen distinctly on the planets close to Earth (Mercury and Mars) as well as on our satellite, the Moon. The question of meteoroid impacts on the Earth is discussed in Chapter 8 in detail. Here we will only consider the origin of the Moon — one of the large (particularly compared with the planet Earth) satellites revolving around the planets in our solar system. From astronomical observations is has been established that, owing to tidal forces, the Moon gradually moves away from the Earth, assuming that they were very close to each other several billion years ago. Since the Moon is usually considered to have become the Earth's satellite as a result of its capture by gravitational forces, this hypothesis does not rule out the possibility that, under other conditions of their approach, the Moon and the Earth could collide. Such an event would lead to the destruction of a large extent of the Earth's crust. It is fairly probable that similar collisions of celestial bodies occurred rather frequently in the early stages of the Earth's formation.

In modern studies it has been found that the first records of living organisms occur very close to the time at which the most ancient rocks of the Earth's crust were formed. This observation deserves much attention. The carbon isotope ratios in strongly metamorphosed sedimentary rocks in West Greenland, of age 3.8 billion years, enable the assumption that autotrophic organisms influenced the formation of these rocks. The remnants of micro-organisms have been detected in Western Australian rocks aged at 3.5 billion years. In sediments from South Africa the remnants of blue-green algae dating from about 3 billion years ago have been discovered.

Thus, it can be assumed that the first organisms appeared soon (from geologic point of view) after the formation of the Earth's crust. However, since the Earth's crust could have been formed and destroyed before this time, the possibility exists that life on Earth may have appeared (or appeared and then disappeared temporarily) before the formation of a stable crust.

The origin of life on Earth is not yet completely understood. The hypotheses concerning this question are divided into two groups: the first includes the assumptions of the transfer of primary organisms from space onto the Earth, and the second — the supposition that they emerged on the Earth.

The first group attracts at present less attention mainly because of a lack of necessary data for discussing this suggestion. The assumption of transfer of primary organisms to the Earth by meteorites was shared by many prominent scientists of the 19th century including Lord Kelvin and Helmholtz. At the beginning of the 20th century the hypothesis of Arrhenius, concerning microscopic spores that spread in space by pressure of the light emitted by stars, was widely known.

Later the meteoroid hypothesis was refuted because investigations of meteorite matter did not find living organisms or even biogenic organic substances. The assumption that spores could be transferred from space was considered unfounded because microorganisms are sensitive to ultraviolet radiation. The following conclusion, formulated most clearly by Oparin, has been frequently drawn: "... we must discard completely the idea that the germs of life flew to us, on the Earth from somewhere outside of the Earth ..." (1957). However, such conclusion cannot be considered valid. The number of meteorites investigated is negligibly small in comparison with the number that may have impacted on the Earth's surface for ten or a hundred million years, i.e. the time interval comparable with the duration of the epoch when life originated on Earth.

The results of studying the known meteorites could be important if they had confirmed the fact that organisms are transferred in the Universe. However, the absence of such transfer by this means cannot be proven because the number of meteorites investigated is insufficient. The inference that microorganism transfer in space is impossible, drawn during experiments exposing modern bacteria to ultraviolet irradiation, is also unconvincing. Such experiments may be conclusive only if they yield a positive result which confirms the possibility that some microorganisms can endure certain doses of radiation. A negative result does not solve the question since it is based on experiments with modern bacteria that are the product of billions of years of evolution. In the course of this evolution, bacteria could lose the mechanism of adaptation to intensive ultraviolet irradiance since this adaptation is not necessary on the modern Earth as the ozone screen attenuates the ultraviolet radiation.

Let us quote the objections from the afore-mentioned book by Oparin against the assumptions about the possible mechanism of life transfer in space: "... to uphold the theory under consideration, even such 'hypotheses' have recently been proposed as bringing the germs of life on the surface of the Earth by astronauts, the intelligent beings travelling in a spacecraft around the universe who once visited our planet. However, such assumptions resemble rather a science fiction than a serious scientific hypothesis." This opinion seems so evident to Oparin that he has not corroborated it by any scientific arguments.

Soon after the publication of Oparin's monograph spacecraft could approach various planets in the Solar System. When preparing for space flight to the Moon particular attention was paid to preventing the transfer of the Earth's microorganisms, in order to exclude the event that Oparin considered possible only in science fiction novels. Although the era of space flight began not very long ago, we now have the technical means that permit the transfer of microorganisms to any celestial body within and even outside the confines of our solar system through the possibility of preserving viable microorganisms or spores on the spacecraft for a very long time. As such probability exists, it has recently been proposed that a 'directed panspermia' (i.e. the intentional transfer of the simplest organisms (protozoa) to our planet by spacecraft constructed by extraterrestrial intelligent beings, could contribute to the origin of life on Earth. Arguments against the probability of this hypothesis are considered in Chapter 10; however, it cannot be completely discarded given the present state of knowledge.

Let us note that the above-mentioned example of an extremely categorical conclusion about the impossibility of transferring organisms through space is typical of the erroneous inferences that easily appear in the epoch of the scientific—technological revolution. In particular, it should be remembered that not long before the first atomic reactors were built, one of the founders of nuclear physics, Lord Rutherford had expressed the opinion that the investigations in this field would never be of practical significance. Similar cases confirm the necessity of being prudent in negation of assumptions when the ways in which science will advance are frequently difficult to predict.

Returning to the hypothesis of an extraterrestrial origin of life, we conclude that at present there is no proof of life transfer to Earth from other celestial bodies or of the impossibility of such transfer. This question will be considered in more detail in Chapter 10.

The second group of hypotheses about the origin of life on the Earth associates this event with converting inanimate matter into animate matter on the Earth itself. Since antiquity, similar proposals have been made by many thinkers. In the 19th century this idea was supported by outstanding scientists of that time; Lamarck, Huxley, Tyndall, Darwin and, particularly, Haeckel. In one of his letters Darwin (1887) expressed an idea which is recognized as very important in modern investigations of the origin of life. He wrote that, in a warm reservoir containing a mixture of ammonium and phosphate, light, heat, electricity and so on, albumen, capable of being subjected subsequently to more complicated transformations, is produced. At present such a substance would be eaten or absorbed. This could not have taken place in the period before the appearance of living things.

The hypothesis about the origin of life on the Earth was developed in detail in the studies of Oparin. In his first monograph Oparin (1924) proposed that the chemical composition of the ancient atmosphere was of great importance for the origin of life. This atmosphere, in Oparin's opinion, had a reducing character and consisted of hydrogen, water vapour, ammonia and carbon compounds (methane and cyanogen). Oparin supposed that life on the Earth originated approximately two billion years after the formation of the planet and believed that such a long interval of time was necessary for carrying out the process of chemical evolution in the course of which complex organic molecules were synthesized. These molecules gave birth, at first, to protobionts (pre-cell systems) and then to protocells which possessed the properties of animate matter. Similar views were expressed subsequently by Holdane (1929), Bernal (1951) and other scientists.

Although the assumption by Oparin that the stage of prebiological (chemical) evolution could last up to two billion years is not corroborated by the afore-mentioned data of contemporary investigations of the Earth's history, this does not produce insuperable difficulties for substantiating his idea, since the duration of this stage cannot be determined *a priori* and it could be far shorter than the time interval indicated by Oparin. The belief of Oparin that the first organisms were anaerobic has been indirectly confirmed by experiments in which many organic substances, important in the composition of living organisms, were synthesized under conditions of a reducing medium.

Among such investigations are the experiments made in 1953 by Miller and Urey who used a mixture of methane, hydrogen, ammonia and water vapour for modelling the ancient atmosphere. Running an electric spark discharge through this mixture they were able to synthesize four amino acids that are essential components of the protein molecule, as well as a number of other organic compounds. In subsequent investigations, other energy sources were used (ultraviolet radiation, heat, ultrasound, etc.) which reproduced the natural conditions of the energy regime in the epoch of the origin of life. In these investigations many organic substances that occur in living organisms were synthesized.

The widespread opinion that the ancient atmosphere was reducing in nature is not shared by all investigators (refer to Chapter 7). In this connection, several studies have considered the possible mechanism of the origin of life in an atmosphere consisting mainly of carbon dioxide and water vapour.

One of the major difficulties in the experimental studies on the origin of life is associated with the necessity of reproducing the emergence of a genetic and protein synthesizing mechanism based on the activity of deoxyribonucleic acid (DNA) molecules. These contain the information used in synthesizing proteins and transmitted in the course of an organism reproduction. It should be noted that in spite of the great complexity in the mechanism of genetic information transmission, this mechanism is found in even the simplest organisms belonging to the group of procaryotes.

Although questions relating to the formation of genetic apparatus in ancient living organisms has received particular attention in the various investigations, they remain unsolved. Because of this it has not yet been possible to create under laboratory conditions an autonomous organism which can reproduce itself by energy and matter exchange with the environment. The solution of this problem would be an important argument in favour of a local origin of life on the Earth, although even in this case we would not have negated the possibility of life being transferred to the Earth from interstellar space.

It is probable that the ancient biosphere of the Earth (i.e. a zone of spread of primary organisms) occupied a very limited space compared with the modern biosphere and, possibly, for a long time it consisted of a group of small 'islands of life' where primitive chemotrophic organisms found favourable conditions for extracting energy and food from the mineral substratum surrounding them. It is unclear where these 'islands of life' were placed. Possibly, they were located in shallow water bodies or at the hydrosphere—lithosphere boundaries.

The most important stage in developing the ancient biosphere was the formation of phototrophic organisms which could use solar radiation energy coming to the Earth's surface. The formation of phototrophic organisms expanded the life zone and led first to slow and then to more rapid accumulation of oxygen in the atmosphere and surface waters. This resulted in the appearance of aerobic organisms, which represented another great event in the history of the biosphere.

## 6.2. Geochronologic Scale and the History of Organisms

In studying the history of the biosphere, it is necessary to use the geochronologic scale showing the sequence and duration of the main stages of the Earth's history. The geochronologic scale used here has been constructed over a long time period and difficulties in reconstructing geological history inevitably result in a lack of coincidence among the different geochronologic scales proposed by different authors. The task of constructing a generally accepted geochronologic scale was put before national and international commissions in which leading specialists in stratigraphy took part. Because of these activities, the geochronologic scales used in modern investigations are much more similar to each other although some particular differences still occur.

The geochronologic scale is based on data describing successive changes in the character of faunas, floras and lithogenesis found in studying the upper layers of the Earth's crust. In the past, the age of various stages of the Earth's history was determined very approximately on the basis of indirect evidence. In the last several decades absolute age data, obtained by geochemical methods of analysing geological formations, have been used to determine the age of these stages.

The main geochemical method is radiological, based on studying the decay of long-lived radioactive elements such as uranium, thorium, potassium and so on. In particular, the age of a number of potassium minerals can be estimated from data on the argon content in them — argon being a decay product of radioactive potassium.

The radiocarbon method is used to determine the age of comparatively young formations not exceeding several tens of thousands of years. It is based on measuring relative amounts of the radioactive isotope $^{14}C$ in remnants of organisms. The quantity of this isotope gradually decreases after the death of organisms and has a half-life of about 6000 years.

Table 6.1 represents the geochronologic scale used in this book. In this table, the names of eras and of periods incorporated into the three last eras are given. The epochs of various periods are indicated only for the

Tertiary. Let us note that the Tertiary is frequently divided into two parts or periods: the Palaeogene including Palaeocene, Eocene and Oligocene; and the Neogene incorporating Miocene and Pliocene. In studies of the U.S. scientists the Carboniferous is divided into two periods: the Mississipian, an earlier one; and the Pennsylvanian, a later one.

TABLE 6.1

Geochronologic scale

| Era | Period | Epoch | Duration of the period or era ($10^6$ yr). | Absolute age of the onset of the period ($10^6$ yr) |
|---|---|---|---|---|
| Cenozoic | Quaternary | Holocene<br>Pleistocene | 2 | 2 |
| | Tertiary | Pliocene<br>Miocene<br>Oligocene<br>Eocene<br>Palaeocene | 64 | 66 |
| Mesozoic | Cretaceous | | 66 | 132 |
| | Jurassic | | 53 | 185 |
| | Triassic | | 50 | 235 |
| Palaeozoic | Permian | | 45 | 280 |
| | Carboniferous | | 65 | 345 |
| | Devonian | | 55 | 400 |
| | Silurian | | 35 | 435 |
| | Ordovician | | 55 | 490 |
| | Cambrian | | 80 | 570 |
| Proterozoic | | | 2030 | 2600 |
| Archean | | | 900 | 3500 |

The upper part of the geochronologic scale (Palaeozoic, Mesozoic and Cenozoic eras) is the Phanerozoic Eon, the lower part is the Cryptozoic Eon (the Precambrian). Although the Cryptozoic Eon makes up about five-sixths of the whole geologic history of the Earth, the available evidence about this time is not so comprehensive compared with data on the history of the later Phanerozoic Eon. It is assumed that the Archean began more than 3500 million years ago and lasted about 900 million years. The Proterozoic started 2600 million years ago, its duration is about 2000 million years. Brief information concerning the history of organisms throughout the basic time intervals shown in geochronologic scale is given below. It should be noted that this information is somewhat inaccurate because of inevitable incompleteness of palaeontological evidence.

The time of appearance of new organisms estimated by palaeontologic data corresponds usually to the time of their more or less wide distribution rather than their emergence. Similarly, the conclusion about the extinction of some groups of organisms in many cases may indicate only a sharp reduction in their area of inhabitation and in the size of populations.

### 6.2.1. THE PRECAMBRIAN

As has already been noted, the first indications of biologic activity date from 3.8 billion years ago. It is assumed that amongst the ancient living organisms were bacteria and blue-green algae belonging to the group of prokaryotes (unicellular organisms lacking a nucleus).

During the Precambrian various primitive plants developed (animals came after plants). Fossil rocks including their earlier forms are very rare because the organisms had no skeletons or other organs that could produce distinct imprints in the palaeontologic record. Identification of the remnants of Precambrian animals presents great difficulties

Sedimentary rocks more than three billion years old contain microstructures considered to be the imprints of protocells, assuming that these preceded the appearance of real biological structures. In the period 2.5—3 billion years ago, the stromatolites (carbonate sediments consisting of thin wavy plates) are found. The stromatolites were the products of the life activity of blue-green algae and bacteria. Microfossils, among which, some investigators believe are found the earliest eukaryotes (organisms with a nucleus), date from 2 to 2.5 billion years ago.

A considerable diversity of animals including a number of multicellular forms appeared at the late Precambrian (the Vendian period) as first revealed by the Soviet palaeontologist B. S. Sokolov in 1950. The Vendian period lasted for about 110 million years. Along with the more ancient organisms (blue-green algae and others), numerous forms of organisms without skeleton including those similar to modern sea-pen, ancient annelids and organisms similar to medusa occurred at this time. One should mention the discovery of the remnants of an animal with a segmented body resembling the trilobites that appeared later, but lacking their two-sided symmetry (*Venedia Sokolovy*).

Apparently, all (or almost all) organisms of the Precambrian and the first two periods of the Palaeozoic inhabited the hydrosphere. As the land was largely uninhabited by organisms it in essence was not a part of the biosphere.

### 6.2.2. THE CAMBRIAN

A drastic change in animate nature took place in the Cambrian and was of great significance in the biosphere's history. At this time, skeleton animals developed, among which were archaeocyathids, gastropods, sponges, radiolaria, brachiopods, trilobites and coelenterates. At this time, various molluscs were widespread and cephalopods appeared. At the end of the Cambrian, the first vertebrates (testaceous jawless fishes) evolved.

The Cambrian vegetation included blue-green algae, red algae and diverse forms of microphytoplankton.

The unique feature of the history of life in Cambrian time was the appearance of representatives of the overwhelming majority of animal phyla that comprised the basis of faunas for all subsequent periods of the Phanerozoic including the modern epoch.

### 6.2.3. THE ORDOVICIAN

In the Ordovician a rich fauna of the Cambrian hydrosphere was supplemented with many new forms. At this time there lived the representatives of most classes of marine invertebrates known at present. Throughout the Ordovician, the variety of the oceanic forms of many animal groups considerably increased, however at the end of the Ordovician, the extinction of many representatives of a number of groups (graptolites, corals, trilobites and others), took place which resulted in total decrease of the diversity of animal forms.

Many green and red algae existed in the Ordovician. It is possible that at the end of this period the first few vascular plants appeared on land.

### 6.2.4. THE SILURIAN

In the seas during the Silurian, the shell and coral forms were widespread. Trilobites occupied a noticeable place among numerous arthropods; however, the variety of their forms decreased compared with the Ordovician.

Jawless fishes (including testaceous jawless) and acanthodii were among vertebrates in the Silurian. At the end of the Silurian plants began spreading on land. These were mainly psilophytes, vascular herbaceous plants which probably grew only in regions with the most humid climate.

### 6.2.5. THE DEVONIAN

The Devonian marine invertebrates had various forms, among which were numerous ammonites, foraminifera, brachiopods and corals. Many forms of testaceous fishes inhabited the seas. The dissemination in the Devonian of both main groups of fishes (cartilaginous and bony) was of great importance for the development of successive faunas. These fishes, besides inhabiting the oceans and seas, began gradually to inhabit fresh water bodies.

In the Devonian the plants spread widely on the continents and the land animals appeared. At this time the vegetation cover on the continents embraced ever increasing areas and many groups of higher land plants, in particular, the original ferns and gymnosperms first emerged. By the end of the Devonian, vegetation on the continents was far richer than at the beginning of the period. This is particularly pronounced in the late Devonian when the first real gymnosperms evolved.

In the late Devonian, amphibians (stegocephals) originated from cross-opterygians and gave birth to the diverse vertebrate group of animals which inhabited the land later on.

### 6.2.6. THE CARBONIFEROUS

In the Carboniferous period, the dissemination of plants and animals on the continents continued. Vast continental areas were covered at this time by high-stem forests of club mosses, ferns and calamites. Throughout the greater part of Carboniferous time, a considerable portion of land experienced conditions of abundant moisture, which made it possible for the accumulation of organic matter from the Carboniferous forests in swampy soil. In this manner the coal deposits, characteristic of the present day, were formed. Carbon also accumulated in water bodies, mainly from the matter of numerous green algae.

The Carboniferous seas were inhabited by various forms of foraminifera, coelenterates, molluscs, ammonites, nautiluses, brachiopods, pearl-worts, arthropods, echinoderms and other invertebrates. Only few representatives of the earlier prosperous trilobite group survived. The diversity of major groups of fishes that appeared in the Devonian increased.

Great changes occurred in the animals inhabiting the continents. In the mid-Carboniferous, the reptiles first emerged. However, their diversity at this time was not great. Numerous insects existed on land among which were gigantic forms. The penetration of animals into air space was a new step in expanding the boundaries of the biosphere.

### 6.2.7. THE PERMIAN

In the early Permian a great continental glaciation, which first appeared in the late Carboniferous, took place over a considerable part of all southern continents (known as Gondwanaland). This produced, besides a cooling in the glaciation zone, an arid climate in a number of continental regions. This is discussed in detail in Chapter 7. An increase in the aridity of some regions is recorded in the Carboniferous, the process of aridity being considerably strengthened during the Permian. Such changes in the environmental conditions are apt to exert great influence on animate nature. In particular, the arid climate probably aided in widely distributing the reptiles, the diversity of which rapidly increased. Of particular importance was the appearance of a progressive group of mammal-like reptiles (Therapsida). Many groups of earlier diverse amphibians, including most of stegocephals, became extinct during the Permian. The rich marine fauna of the Permian decreased sharply at the Permian–Triassic boundary. This striking scale in the extinction of animals was one of the great events in the biosphere's history. The extinction at the end of the Permian period embraced a number of groups of coelenterates, ammonites and other marine invertebrates. Trilobites that existed since the beginning of the Phanerozoic were extinct before the termination of the Permian. The question of mass extinctions (critical epochs of geologic history) is discussed in Chapter 8.

The history of the vegetation during the Permian period is characterized by a rapid dissemination of gymnosperms among which the conifers were very important. Vast areas of the continents with comparatively humid climate were covered by forests consisting mainly of gymnosperms.

### 6.2.8. THE TRIASSIC

As mentioned above, the end of the Palaeozoic is characterized by disappearance of the many ancient groups of animals which were substituted by new groups typical of the Mesozoic era. Although the late Permian extinctions played an important role in this process it was only one link of a chain of events causing the substitution of many Palaeozoic forms by Mesozoic forms. The extinction of typical Palaeozoic forms took place throughout the greater part of the Permian and the entire Triassic. This contributed to a considerable decrease in the diversity of living organisms, particularly of animals. During the Triassic, woody club mosses, ferns and ancient conifers gradually disappeared from the land vegetation. At the same time, new groups of plants, such as cycads, ginkgo and others, expanded. The animal world on the continents included miscellaneous groups of reptiles. However, throughout the Triassic the number of genera of most progressive mammal-like reptiles (Therapsida) drastically decreased. Their place was occupied mainly by representatives of more primitive Sauropsida. In the mid-Triassic, the first dinosaurs appeared. This group became widely distributed subsequently. At the end of the Triassic, one of the remaining groups of mammal-like reptiles (theriodonts) gave birth to the mammals.

The variety in the groups of Triassic marine animals was reduced compared with the Permian period (brachyopods, foraminifera, sea lilies and others). At the same time, the higher bony fishes (teleosts) which subsequently spread widely, first appeared.

### 6.2.9. THE JURASSIC

The animal world in the Jurassic period changed noticeably due to the great success of reptiles in occupying new ecological niches. For the first time in the Earth's history giant animals appeared on the continents — herbivorous dinosaurs (*Diplodocus, Brontosaurus*, and others) and carnivorous dinosaurs (*Ceratosaurus* and *Allosaurus*). Another big step in the development of animate nature was the occurrence of flying vertebrates — pterosaurs (belonging to the group of archosaurs) and birds originating from small dinosaurs. The seas were inhabited by reptiles which often reached large sizes (ichthyosaurs and plesiosaurs). The land vegetation included a diverse flora of gymnosperms, ferns and many other ancient plants.

It should be noted that although the transition from the Triassic to the Jurassic was accompanied with the extinction of many animals, the number and variety of the surviving groups increased considerably.

### 6.2.10. THE CRETACEOUS

On the continents in the Cretaceous period, angiosperms became widely distributed and included magnolias, oaks, bays and platans. In some regions these groups supplanted the previously widespread ferns and some primitive gynmosperms. In the Cretaceous, reptiles continued to prosper and still, included the giant forms of herbivorous and carnivorous dinosaurs. Some flying reptiles reached very large sizes.

Throughout the Cretaceous the diversity of mammals increased, among which appeared the first placental animals. However, the mammals of this period were comparatively few and small animals, much smaller than the large land reptiles. During the Cretaceous, the progressive evolution of birds took place, although their numbers remained comparatively small.

The rich fauna of marine invertebrates at this time included some gigantic forms, among which occurred some species of belemnites and ammonites. The bony fishes were very miscellaneous and their diversity rapidly increased. The Cretaceous seas were also inhabited by large reptiles (ichthyosaurs, plesiosaurs and mosasaurs).

Although some groups of the Cretaceous fauna disappeared before the termination of this period, it was only at the end of the Cretaceous that the widespread extinction of many forms of animals, as well as a number of groups of marine plankton occurred. This extinction is treated in Chapter 8.

### 6.2.11. THE TERTIARY

The extinction of all the groups of dinosaurs, flying reptiles and big marine reptiles at the end of the Cretaceous period greatly affected the evolution of the animal world. This extinction led to the liberation of many ecological niches which were rapidly occupied by mammals and birds. In the Palaeocene and Eocene, almost all of now existing orders of mammals and birds were formed. However, most of the mammal families in the early Tertiary were subsequently changed for new families. At the Palaeogene—Neogene boundary, a particularly drastic renewal of mammal fauna occurred. In the early Neogene, the first representatives of such groups as bears, badgers, hyenas, pigs, antelopes, deer, oxen and sheep appeared. At this time, anthropoid apes also emerged. In the Neogene, the dissemination of existing aquatic mammals accelerated. This group included whales (Cetacea) some of which were the largest animals ever inhabiting the Earth.

In the late Neogene, climatic changes towards cooler conditions considerably impoverished high latitude faunas. However, in the tropics a rich animal world was being retained until recently.

Throughout a greater part of the Tertiary, the fauna of South America (which was isolated at this time from the other continents) included many representatives of comparatively primitive mammal groups. The fauna of Australia was also isolated and the mammals were represented mainly by marsupials.

The Tertiary period is characterized by a rapid increase in the diversity of birds which occupied new ecological niches. For some time large non-flying birds competed with ground mammals. Although several species of large non-flying birds have survived up to the present, this group as a whole was replaced by mammals. This can be seen in the present epoch. The largest non-flying birds are confined to islands where large mammals are absent (Dinornithes in New Zealand, Aepyornithes in Madagascar). However, the largest of the flightless birds were annihilated by man in historic times.

In the Tertiary flora, the angiosperms were the dominant plants although gymnosperms were still widespread. In the Eocene, the specific flora of humid forests first appeared and is still retained in tropical latitudes in the present epoch. In the Miocene, herbaceous steppe plants appeared in regions with insufficient moisture.

### 6.2.12. THE QUATERNARY

The Quaternary period is much shorter than all previous periods and great changes in fauna and flora have not taken place. The peculiar reasons that justify the separation of Quaternary time into an individual period are, firstly, the development of land and sea glaciations which constantly existed in high latitudes and from time to time advanced to the middle latitudes, and, secondly, the development of man whose activity exerted an ever-increasing influence on the biosphere.

Taking into account the latter, many scientists call the Quaternary period the Anthropogene. This name, however, is not completely logical, since representatives of the group of hominids and possibly of the genus *Homo* appeared as long ago as in the Pliocene, i.e. before the onset of the Quaternary period. Along with this, man's impact on the biosphere became noticeable only at the end of the Quaternary period during the time of the last

Quaternary glaciation and in the Holocene, i.e. throughout the last tens of thousands of years.

The effects of the ice ages on the vegetational cover and on animals were mainly expressed in the shift of geographical zones to lower latitudes when glaciers advanced and in their return to higher latitudes when glaciers retreated. At the same time, specific flora and fauna evolved in the regions close to the ice sheets. Vegetational cover in these regions combined the features of tundra and steppe. Among the animals, conspicuous place belonged to now extinct mammoths and woolly rhinoceros as well as to musk-oxen, reindeer, polar fox and others.

The evolution of modern man (*Homo sapiens sapiens*), occurring about 40 000 years ago, was the greatest event in the Quaternary period. The activity of modern man apparently resulted in the extinction of many large-sized animals all over the continents (except the Antarctic) prior to the Holocene (i.e. up to 10 000 years ago). A number of large-sized and smaller animals were annihilated by man during the Holocene, this process being continued at present. The relations between man and other living organisms are treated in Chapter 9.

In conclusion, let us note the most significant events in the biosphere's history which relate to the evolution of living organisms. Firstly, the origin of life on the Earth. Secondly, the appearance of phototrophic plants. Thirdly, the dissemination of animals. All these events occurred in the Precambrian. The fourth event is the 'evolution explosion' in the animal history of the Early Palaeozoic when, during the Cambrian and Ordovician, almost all now existing phyla of animals spread. The fifth great change in the biosphere's structure is related to the expansion of vegetational cover on the continents. It took place mainly throughout the Devonian. As plants inhabited new environments, their more progressive forms emerged. The appearance of plants on the continents made it possible for the origin of land animals and flying animals as well. Adaptation of animals to miscellaneous conditions of life on the continents was a prolonged process in the course of which new progressive groups of animals arose.

It is important to emphasize that in the second half of the Phanerozoic the most important changes in animate nature took place on the continents. In the oceans (particularly, in the open ocean), a large number of the living organisms are from representatives of ancient groups. This is still so today.

Considering populating the continents with animals to be the sixth great event in the history of the biosphere, the seventh notable event will be the appearance of modern man.

The question of the causes which stipulated the occurrence of the last three events is discussed in Chapters 8 and 9.

## 6.3. Lithosphere and Hydrosphere

### 6.3.1. THE LITHOSPHERE

As noted in Chapter 1, the biosphere consists of three major areas inhabited by organisms — the lithospere, the hydrosphere and the atmosphere. All these areas are of abiotic origin and all of them are modified to a certain extent by living organisms. The atmosphere has been most affected by organisms and at the same time exerts a considerable influence on the evolution of animals and plants. This is discussed in Chapters 7 and 8.

Changes in the hydrosphere and, particularly, in the lithosphere also depend on abiotic factors. However, this influence is more essential in comparison with the atmosphere and therefore, the hydrosphere and the lithosphere may be considered the abiotic factors of the biosphere.

In Chapter 1 a brief description of soils as the part of the lithosphere inhabited by living organisms is presented. It should be pointed out that although in deeper layers of the lithosphere organisms are rare, to understand the history of the biosphere it is necessary to have information about past changes in the lithosphere as a whole. Studying the lithosphere's evolution permits us to elucidate the distinctive features of the formation of the continents and oceans, to investigate the relations between organisms and abiotic factors when developing the sedimental shell of the Earth and to detect impact basins of meteorites that greatly influenced the biosphere's history.

Previously the lithosphere was called the Earth's crust, i.e. the outer rigid shell of the Earth that overlies the Mohorovicic discontinuity in which seismic gradient increases spasmodically from top to bottom.

The Earth's crust is divided into two parts: the continental and the oceanic crust. The former includes the sedimentary layer 20—30 km thick, the middle or granitic layer 10—40 km thick and the lower basaltic layer 10—70 km thick, the latter comprising a thin sedimentary layer less than 1 km thick, the middle layer consisting of various rocks 1—2.5 km thick and the lower or basaltic layer about 5 km thick. In addition to these two major types of the crust, there are various intermediate variants.

The products of organisms make up a considerable part of lithosphere deposits, many of which (e.g. limestone, coals, dolomites and others) being of biogenic origin. Many abiogenic sedimentary rocks are also produced due to the activity of organisms taking part in the process of weathering of the initial rocks. Sediments cover the greater part of the Earth's surface, their total mass being about 10% of the Earth's crust mass. About 70% of the total amount of sediments are located on the continents and continental slopes and 30% on the bottom of the oceans.

Below the crust, there is the upper mantle 850—900 km thick. It incorporates several layers, the upper (substrate) being frequently combined with the crust in the common term 'lithosphere'. The mantle's matter can be determined from data on the conditions of the propagation of seismic waves and by other indirect methods. Probably, it consists mainly of olivine.

The thermal regime of the lithosphere is characterized by a lapse rate, i.e. the temperature increase with depth. This lapse rate is associated with the vertical heat flux from deeper layers to the Earth's surface caused by the radioactive decay of long-lived isotopes of a number of elements. The value of this flux is much smaller than the solar radiation flux coming to the Earth's surface, therefore the heat flux does not influence global climate. The temperature at a depth of 100 km is assumed to be approximately 1000 °C under average conditions. This temperature is smaller than the mean fusion temperature of the upper mantle matter (which somewhat exceeds the corresponding fusion temperature at the Earth's surface because of high pressure reaching 50 kbar). In some regions, the temperature rises faster with depth, therefore at depths of several tens of kilometres the magma chambers emerge. This often corresponds to volcanically active zones.

It is usually believed that the most stable elements of the Earth's continental crust are the ancient platforms over which the strata of sedimentary rocks lie. Between ancient platforms there are more mobile geosynclinal facies characterized by high tectonic activity, intensive deformation of rocks and volcanism.

The comparatively thin crust of the oceans is divided into vast, relatively flat, and stable plates and into the tectonic mobile zones of oceanic rift belts which considerably differ from geosynclinal fold belts of the continents. The zones where the oceans are crossed by mid-oceanic ridges with rift valleys are characterized by high volcanic activity.

Beginning in the 1960s, the tectonic hypothesis of continental movement (i.e. the drift of the continents relative to each other and to the poles) has become widespread. Attention has long been given to the fact that the shapes of the coastal lines (or the shelf zone boundaries) of several continents separated by vast oceanic areas are similar. This similarity is particularly noticeable between the coastal lines of Europe and Africa in the East and the Americas in the West of the Atlantic Ocean. To explain this it has been proposed that all these continents were some time in the past, part of a common continent Pangaea. These parts have subsequently shifted horizontally, separated and moved away for considerable distances. The explorer A. Wegener developed this hypothesis most thoroughly.

The Wegener hypothesis did not gain recognition at first because of the lack of a convincing explanation of the physical mechanism causing the displacement of the continents. His assumptions that these displacements can be attributed to the effects of centrifugal forces and Earth tides caused by the gravitational action of the Moon and the Sun, were rejected by quantitative calculations which showed that these forces were insufficient to displace the continents. Subsequently Wegener changed his views concerning the causes of continental migration and suggested that this phenomenon should be explained by the action of subcrustal convective currents. The hypothesis of subcrustal currents also did not gain widespread acceptance at that time.

In 20 or 30 years after the death of Wegener, in the Greenland expedition of 1930, the opinions of most geologists about the tectonic hypothesis of mobilism gradually changed. Extensive results from palaeomagnetic investigations determining latitudes and azimuths of different regions for successive time intervals of the geological past, contributed to this change. Palaeomagnetic results agreed well with the suggestion of considerable continental drift, although its causes remained unknown.

In the 1960s, the idea of ocean-floor spreading was advanced. This has turned out to be essential in studying the mechanism of continental drift. In studies by Dietz (1961) and Hess (1962) it was assumed that as a result of upward migration of molten rocks in the regions of submarine ridges, new crust was constantly formed, causing the spreading of oceanic floor. This made North America move away from Europe. The spreading of the Atlantic Ocean is compensated for by contraction of the Pacific Ocean area where in marginal regions the crust subducts into the mantle due to convective processes which in some regions extrude the mantle matter into the upper layers of the lithosphere and in other regions produce its downwrapping. The basic ideas of plate tectonics formulated by McKenzie and Parker (1967), and Morgan (1968) are closely related to the idea of oceanic-floor spreading. These authors believed that the lithosphere enclosed stable plates changing little under the influence of tectonic processes. These plates are separated by mobile layers which are considerably affected by tectonic activity. It is assumed that lighter plates overlie a more plastic asthenosphere consisting of basalts and heavier rocks. The convective motions in the asthenosphere slide apart the plates in a direction away from submarine ridges. As a result, some plates expand while others, suppressed by the first, contract their areas. The idea of plate tectonics allows one to imagine more distinctly the changes in the structure of the Earth's surface in the geological past. Based on the material collected and analysed when developing the idea of mobilism, many investigators suggested that in the Palaeozoic, about 570—235 million years ago, two supercontinents existed: Gondwanaland (which divided later into South America, Africa, Madagascar, India, Australia and Antarctica) and Laurasia (which formed North America, Greenland, Europe and the greater part of Asia). Gondwanaland was situated nearer to the South Pole and Laurasia to the North Pole.

The splitting of Laurasia, which was associated with the formation of the Atlantic Ocean, probably occurred in the late Mesozoic, about 100 million years ago. The expansion of the Atlantic Ocean still continues today.

Although the idea of plate tectonics won wide acceptance, it is not yet universely adopted. Some researchers only partially adopt it while others support the concept of a stable position of continents and oceans during hundreds of millions of years (fixism).

In addition to possible shifts of continents relative to each other the life conditions of organisms could be affected by motions of the Earth poles. The assumption that the geographical poles changed considerably in the past is based mainly on palaeomagnetic data. From this some specialists believe that during the past 500—600 million years the north magnetic pole has shifted from central areas to northwestern regions of the Pacific Ocean (at about 285—230 million years ago) and then via northeastern Asia to its modern position. It is probable that in the past the Earth's rotational poles were not far from the magnetic poles as occurs in the present epoch. However, this assumption is not proven, and the displacement of geographic poles in the past is not fully understood.

Of great importance in studying the history of the biosphere are the data on the structure of the sedimentary shell of Earth. Most detailed information concerning this problem is obtained in studies by Ronov including his book *Sedimentary Envelope of the Earth* (1980). As Ronov notes, the sedimentary shell covers about 80% of the land surface, while the remaining 20% is covered by crystalline rocks. Sedimentary stratum of the continents consists mainly of the Phanerozoic and Upper Proterozoic rocks. More ancient sediments are comparatively rare.

Sedimentary rocks of the oceans are considerably younger on average, as compared with the continental sediments, usually being confined to the Cretaceous and Cenozoic sediments. Most ancient oceanic sediments date back only to Upper Jurassic times. The mean thickness of the sedimentary cover for the Earth as a whole is about 2.2 km changing in individual regions from 0 to 20 or 30 km. The components of the sedimentary cover consist basically of the rocks formed on the Earth's surface as a result of weathering, material deposited from the atmosphere and hydrosphere and rocks produced by volcanoes.

Of special interest are the carbonate rocks in the sedimentary layer the formation of which is connected with carbon cycle in the atmosphere and hydrosphere. In most ancient deposits (Archean time), there were very small amounts of carbonate rocks. These increased in early Proterozoic times and, in particular, in late Proterozoic and in the Palaeozoic when dolomites were important in these rocks. Later, Mesozoic and Cenozoic limestones became the major component of carbonate rocks.

The data on changes of organic carbon by Ronov are of great importance for studying the history of the biosphere. The amount of organic carbon is basically defined by the mass of scattered organic carbon which exceeds by approximately 200 times that in coals, petroliferous shales, petroleum and other caustobioliths. The relative amount of organic carbon in the total mass of carbonate sediments increased with time from values of about 5% in the upper Proterozoic to 30% in the Cenozoic. This change corresponds to a considerable increase in the role of photosynthesizing plants in the formation of carbonate sediments. Changes in organic carbon quantities in sediments of different age reflect the variation in the rates of oxygen created and released into the atmosphere during photosynthesis. This subject is treated in Chapter 7.

Comparing the volumes of volcanogenic and carbonate rocks generated for various time intervals during the Phanerozoic, Ronov has found that these volumes are closely related to each other. This conclusion confirms the important role of volcanic activity in the global carbon cycle.

In the first section of this chapter the existing ideas concerning the mechanism of the Earth's formation during the course of gravitational condensation of the primary gas-dust cloud have been mentioned. There are grounds to believe that the processes caused by this mechanism were important not only for early history of the lithosphere and biosphere but also for far later time.

It is assumed that for the formation of a planet of Earth's type through the processes described earlier, it requires about 100 million years. As noted in the review of the problem of planet formation (Wetherill, 1978), at that time, many comparatively large celestial bodies (planetoids) impacted on the surface of the growing Earth. At the same time in the zone of formation of Earth's group of planets, rather big bodies reaching the size of modern large asteroids and the Moon were produced from the same material. At the beginning of this chapter the possibility has been mentioned that such bodies bombarded the surface of the young Earth.

Studies of impact basins on the Earth's surface and on Earth's group of planets show that although the intensity of bombardment gradually decreased, it has lasted throughout the entire history of the solar system up to the present time. In this connection it can be proposed that the process of Earth formation has not yet finished.

Asteroids and comets were the bodies impacting on the surface of the Earth, after free passage through the atmosphere and oceans. The possibility of bombarding the Earth's surface by the Appolo group asteroids whose orbits are close to that of the Earth is particularly great. At present, several dozens of such asteroids are known; undoubtedly, their actual number is much greater. Approximate calculations show that in the modern epoch, collisions of asteroids of not less than 1 km in diameter with the Earth can occur nearly once every 250000 years and of not less than 10 km in diameter once every 100 million years (Pollack, 1982).

In studies on the effects of meteoroid impacts on the Earth's surface structure, some information is obtained concerning changes in the frequency of such impacts throughout the history of the solar system (*Geology of Astroblems*, 1980). From data on the frequency of crater formation on the Moon during various periods of its history we may conclude that for the last billion years this frequency decreased by 100 to 1000 times when compared with the highest level about four billion years ago. However, such change does not mean the cessation of crater formation in the present epoch.

Since the gravity on Earth is considerably greater than on the Moon, it may be believed that the impacts of comparatively large-sized external objects can produce greater deformations on the Earth's surface compared with lunar craters. To detect these deformations on the Earth's surface presents a considerable difficulty because of their smoothing in the course of the processes that occur in the atmosphere and hydrosphere. Despite a far greater complexity in searching for impact craters (astroblems) on the Earth's surface compared with the Moon, Mars and Mercury, about a hundred astroblems up to 100 km across have been discovered at present. Probably in the future far larger meteoroid impact craters on the Earth's crust will be found. In this connection a round structure of 2800 km in diameter has been discovered in North America in investigating the anomalies of gravity field (Simon, 1980). In this paper, the assumption that such large deformations of the Earth's crust produced by meteorites could be of importance for the formation of continental plates merits consideration.

The possible effects of impacts by comparatively large asteroids and comets on the biosphere is discussed in Chapter 8.

### 6.3.2. THE OCEANS

Almost all the mass of the hydrosphere water is concentrated in the oceans. Water evaporated from their surface is transferred by air flows to the continents and provides their moistening. The history of the oceans, as well as the history of the lithosphere, has only partially been elucidated.

It is assumed that the World Ocean is mainly formed of water vapour released in degassing the mantle during the Earth's history. This vapour, being condensed in the atmosphere, precipitated as liquid water on the Earth's surface where gradually increasing reservoirs were formed. There are different points of view concerning the time interval during which the accumulation of a greater portion of oceanic water occurred. In a review by Schopf (1980), the difficulty of solving this question is noted and the assumption is made that oceanic waters were generated mainly during the first two billion years of the Earth's history.

The existence of ocean during a very long period of time proves that throughout the greater part of the Earth's history the temperature of its surface did not exceed the boiling point of water which at modern atmospheric pressure is about 100 °C. In the distant past the mass of the atmosphere could have been less than today, which would lower the boiling point and narrow the temperature interval in which the liquid ocean could exist.

At the same time, with the process of degassing of the mantle, the sources of juvenile waters may have significantly increased the mass of water in the biosphere.

It should be pointed out that a considerable difference occurs between changes in water volume in the oceans and in their total area. A comparatively small amount of water would be sufficient to entirely cover a completely plain Earth. When increasing the difference between the mean level of land surface and the mean depth of oceans, the area of ocean can decrease, not only with constant volume of oceanic waters but even when the volume is increased.

Studying the changes in the Earth's relief during the Phanerozoic shows that with the rhythmic variations in the continental area there was a tendency to increasing the area of land due to an increase in the difference between the land and sea-floor levels. This resulted in a shrinkage of the ocean area during the Phanerozoic (Klige, 1980).

Noticeable variations in ocean level took place during the continental glaciations when the oceans lost a portion of water in the formation of glaciers. In the epochs of greatest Quaternary glaciations, the ocean level was approximately 200 m lower than during epochs without ice. This resulted in the connection of previously separated continents and islands, which considerably influenced the spread of many land animals and plants. The present continental glaciation, concentrated almost entirely in the Antarctic, is equivalent to a lowering of the ocean level by about 90 m (an exact estimation of this level is difficult because of appreciable differences in assessing the mass of Antarctic ice).

In the course of ocean formation, the ocean waters were gradually enriched by various mineral compounds mentioned in Chapter 1. An important source of these compounds was the process of land erosion in which exposed rocks were destroyed or dissolved by land waters, air and also by living organisms after life developed on land.

In the course of chemical weathering many magma rocks were eroded forming clay and other loose materials. In addition, physical weathering also contributed much to the formation of loose material from bedrock. Such weathering was particularly intensive in regions with drastic temperature variations and abundant precipitation. A considerable amount of mineral substances also entered the ocean from submarine volcanoes.

The appearance of living organisms created the necessary prerequisites for an essential change in the composition of oceanic water. The release of free oxygen by autotrophic plants led to oxidizing many compounds dissolved in ocean waters. Oxygen dissolved in ocean water made it possible for water-aerobic organisms to exist.

Some water organisms possess the ability to concentrate mineral and organic substances contained in water in shells or other solid parts of their bodies. These solid parts in many cases were deposited after the death of animals and formed thick strata of sedimentary rocks on sea floor.

Changes in the lithosphere and hydrosphere throughout the history of the biosphere partially occurred as a result of organism activity. These changes in turn exerted profound influence on this activity. In some cases this influence was direct, since many organisms inhabited the hydrosphere and upper layers of the lithosphere where an energy exchange and circulation of different matter take place between the organisms and the environment.

It is known, in particular, that the chemical composition of oceanic water is similar to the chemical composition of animal blood. This is indicative of great antiquity of existing chemical composition of the oceans and confirms the long existence of the ancestors of modern forms of animals in oceanic waters.

In cases when the influence of the hydrosphere and lithosphere on organisms was indirect, it also could be essential. For instance, the structure of the World Ocean and of the continents affected global climate which influences all the living organisms. The history of the atmosphere is even more closely connected with the history of organisms and is considered in Chapter 7.

CHAPTER 7

# CHANGES OF THE ATMOSPHERE

## 7.1. Chemical Composition of the Atmosphere and the Life Activity of Organisms

In the geological past both the chemical composition of the atmosphere and its physical state, i.e. climate, varied. The question of the chemical composition of the atmosphere of the past is covered in Sections 1 and 2 of this chapter and the question of climatic change, in Sections 3 and 4.

The evolution of the atmospheric chemical composition was considered mainly in geological and geochemical studies. The results obtained in previous investigations of this problem are discussed in a number of reviews (Garrels and Mackenzie, 1971; Rutten, 1971; Cloud, 1974; Holland, 1980; etc.).

As seen from the available studies, the history of the Earth's atmosphere can be divided into three major stages.

The primary atmosphere was chemically similar to the now existing atmospheres of large planets. This atmosphere contained a considerable amount of light gases — hydrogen and helium. At present these elements are kept by gravity in the atmospheres of Jupiter and Saturn, while the Earth's atmosphere, because of smaller mass and therefore insufficient gravitational forces, lost them long ago. Probably, the primary atmosphere also contained some noble gases (argon, neon, etc.).

The secondary atmosphere was formed gradually as a result of degassing of the heated Earth's layers. For a long time this atmosphere did not contain a noticeable amount of oxygen. It consisted of water vapour, carbon dioxide, nitrogen and several other gases.

The question of the presence in the pre-oxygen atmosphere of considerable quantities of carbon monoxide, ammonia and methane is open to discussion (Gribbin, 1982 and others). Earlier it was proposed that life could arise only in an atmosphere rich in reducing gases. In the last few years some authors have assumed that, since gaseous volcanic ejections always consisted mainly of carbon dioxide and water vapour, the major component of the pre-oxygen atmosphere created in degassing the Earth was carbon dioxide. This idea is corroborated by the fact that the atmospheres of our neighbouring planets Venus and Mars contain basically carbon dioxide. This assumption represents an explanation of the warm climatic conditions in the remote past when the solar constant was considerably less than the modern one (ref. to Section 4 of this chapter).

More or less noticeable amounts of oxygen began accumulating in the atmosphere about two billion years ago after photosynthesizing plants appeared. There are grounds to believe that the secondary atmosphere did not contain a noticeable amount of oxygen. Among them, according to Cloud, are the following conclusions: (1) chemical evolution that resulted in creating living matter can occur only in an oxygen-free atmosphere; (2) the presence of iron- and uranium-containing compounds in sediments more than two billion years ago which could not have been formed in a medium with a considerable amount of atmospheric oxygen; (3) only small quantities of oxides of various chemicals in these sediments (Cloud, 1974).

The last stage in the history of the atmosphere is the epoch of accumulating oxygen. The presence of oxygen in the atmosphere made it impossible for the continued existence of easily oxidized gases. The ancient atmosphere, therefore, consisted of the same major components that are characteristic of the modern atmosphere (nitrogen, oxygen, water vapour, carbon dioxide, and some noble gases). Among the indicated components, oxygen and carbon dioxide are of the greatest importance for living organisms. Moreover, the amounts of oxygen and carbon dioxide in the atmosphere themselves depend considerably on the life activity of these organisms. It might be thought that the present-day state of the biosphere and its evolution throughout the last two billion years were, to a considerable extent, determined by the relations between the organisms and atmospheric gases, mainly carbon dioxide and oxygen. In accordance with this, to study the evolution of the biosphere, information is necessary concerning variations in the quantities of atmospheric carbon dioxide and oxygen in the geological past.

As mentioned above, the process of creating organic matter is mainly based on the absorption by autotrophic plants of carbon dioxide from the atmosphere and hydrosphere.

At present carbon dioxide is one of the relatively small components of the atmosphere. The modern atmosphere contains approximately $2.6 \times 10^{12}$ tonnes of carbon dioxide, which represents about 0.035% of the entire atmospheric air (volume percentage). Much more carbon dioxide is contained in the hydrosphere (basically in ocean waters) in which approximately $130 \times 10^{12}$ tonnes of carbon dioxide is dissolved. A continuous exchange of carbon dioxide takes place between the atmosphere and the hydrosphere by molecular and turbulent diffusion.

Measurements of carbon dioxide concentration in the atmosphere show that it varies little both geographically and with height within the troposphere. The relative stability in the content of carbon dioxide in the atmosphere compared with water vapour is explained by smaller changes in the capacity of sources and sinks of carbon dioxide on the Earth's surface and by the absence of any substantial dependence of carbon dioxide content on temperature. Observational data indicate that the concentration of carbon dioxide increases slightly near the equator and declines in high latitudes. These changes are attributable to the higher solubility of carbon dioxide in the cold waters of high latitudes in comparison with the warm waters of the tropics. As a result of this, at high latitudes the atmosphere loses a part of its carbon dioxide, which is dissolved in oceans. From there surpluses of carbon dioxide are carried by cold deep streams to low latitudes where it is returned to the atmosphere. The magnitude of the flow of atmospheric carbon dioxide between the equator and the North Pole that is produced by this mechanism is approximately $2 \times 10^{10}$ tonnes per year (Bolin and Keeling, 1963).

There is a constant exchange of carbon dioxide between the atmosphere and hydrosphere on the one hand, and between living organisms and the lithosphere on the other.

The major components of the carbon dioxide cycle are determined by biological processes. During the life activity of autotrophic plants and of heterotrophic organisms (in particular, during respiration) carbon dioxide is almost completely returned into the atmosphere and hydrosphere and only a comparatively small quantity of carbon is deposited in the lithosphere.

A certain quantity of carbon dioxide annually enters the atmosphere from the depths of the Earth's crust as a result of volcanic eruptions, from mineral springs, etc. Carbon dioxide is expended in the course of the weathering of silicate rocks and in the formation of various carbon compounds.

Thus, there are two cycles within the carbon dioxide balance in the atmosphere: a biological cycle and a geological one, and within each there is an inflow and expenditure of carbon dioxide. Available data indicate that the biological components of the yearly carbon dioxide cycle are substantially larger than the geological ones.

From the data on plant productivity cited earlier, we find that the annual expenditure of carbon dioxide on photosynthesis at present amounts to about $3 \times 10^{11}$ tonnes. A similar volume of carbon dioxide is formed through the process of respiration and as a result of the decomposition of living organisms. A far smaller amount of carbon dioxide comes from the Earth's crust — about $10^8$ tonnes. The same, in the order of magnitude, quantity of carbon dioxide is expended in different geological processes (Plass, 1956; Müller, 1960; Lieth, 1963, and others).

Living organisms, their life activity products and the lithosphere contain large quantities of carbon extracted from the atmosphere and the hydrosphere as carbon dioxide. The quantity of carbon in the lithosphere corresponds to approximately $2 \times 10^{17}$ tonnes of carbon dioxide, of which the main part is bound in carbonate rocks (Vinogradov, 1972).

While these estimates relating to the components of the carbon dioxide balance and to the reserves of carbon in different environments are very rough, they lead to certain conclusions about the rates of carbon dioxide circulation.

The average time of renewal of carbon dioxide in the atmosphere is approximately ten years. The time during which carbon is accumulated in the lithosphere is very long and of the same order of magnitude as the history of the biosphere. There are grounds to believe that, in the geological past, the concentration of carbon dioxide in the atmosphere differed substantially from its present value. This suggests that the total of all forms of inflow of carbon dioxide into the atmosphere and of its losses was not always equal to zero.

It may be concluded, however, that the difference of this total from zero over every year represents a very small part of the absolute value of the main components of the carbon dioxide balance in the atmosphere. Thus, if we assume that during the last million years the concentration of carbon dioxide in the atmosphere has declined by one-half, this will correspond to a change in its magnitude per year by only 0.001% of the annual amount of photosynthesis.

As mentioned above, oxygen and carbon dioxide are involved in the biotic cycle, which includes the interaction of these gases with living things and their waste products. Oxygen in the biological cycle is generated by photosynthesis of autotrophic plants and consumed in the oxidation of organic material. According to the data given above, the annual output of dry organic matter via photosynthesis appears to be about $2 \times 10^{11}$ t, of which about three-quarters is produced by continental vegetation cover and about one-quarter in the ocean. This process yields $2 \times 10^{11}$ t of oxygen annually.

Nearly the entire mass of oxygen originating from green plants is assimilated by the mineralization of organic matter. The difference between oxygen production and consumption in the biotic cycle corresponds to the amount of organic carbon which accumulates in sedimentary rocks.

The relatively small differences between production and consumption of oxygen and carbon dioxide in the biotic cycle were very important in the evolution of the atmosphere, since, for long-term time intervals, they are comparable with the main components of the balances of these gases.

Variations in the atmospheric oxygen amount over a given period of time correspond to the difference between production (which is determined by the total input and output in the biotic cycle as well as by photodissociation of water vapour in the atmosphere) and consumption (mainly by oxidation of various minerals and gases coming from the Earth's crust).

Undoubtedly one of the most important features of the atmospheric oxygen balance in Phanerozoic times was the dominance of $O_2$ production over consumption as a result of autotrophic plant activity. The problem of how oxygen from the dissociation of water vapour molecules by solar ultraviolet radiation affects this balance is far more complicated.

Quantitative estimations of photodissociation carried out in available studies have basic differences. For instance, Berkner and Marshall (1965a, b, 1966, etc.) contend that, without photosynthesis, the amount of $O_2$ produced by water vapour photodissociation could not exceed 0.1% of the modern value. Of great importance for this estimate is taking into account the 'Urey effect' which means that a very small quantity of oxygen absorbs a considerable portion of ultraviolet radiation. As a result, the screening effect of $O_2$ makes it impossible for an appreciable mass of it to accumulate.

There are also some sources (Byutner, 1961; Brinkman, 1969) with higher estimates for the water vapour photodissociation effect, comparable to the results for photosynthesis.

Most authors of contemporary surveys are convinced that only very small concentrations of atmospheric oxygen can be maintained by photodissociation, since biological and geological processes in the epoch preceding the appearance of autotrophic plants reveal an almost complete absence of $O_2$ in the ancient atmosphere. From this standpoint, photosynthesis seems to be the basic source of atmospheric oxygen during the Phanerozoic.

## 7.2. Chemical Composition of the Atmosphere in Phanerozoic Time

Information on oxygen income in the course of the biotic cycle during different periods can be obtained from the data concerning organic carbon in sedimentary rocks.

Carbon dioxide enters the atmosphere and the hydrosphere from the Earth's crust to a considerable extent from volcanic eruptions during which both $CO_2$ and CO are ejected, the latter being converted into $CO_2$ by interaction with atmospheric $O_2$. In addition to carbon dioxide expenditure on the formation of organic carbon, of great importance is its outgo on the weathering of different minerals and on production of carbonate rocks. The estimation of free atmospheric and hydrospheric carbon dioxide consumption during various periods can be based on the total $CO_2$ mass in sedimentary rocks of different ages.

The data necessary for studying atmospheric evolution in Phanerozoic time have been obtained in Ronov's investigations of sedimentary and igneous rocks on the continents (Ronov, 1959, 1964, 1976, etc.). These data, slightly improved compared with preceding publications, are shown in Table 7.1. Here, results will be cited that have been obtained in the study of the evolution of the atmospheric chemical composition in the Phanerozoic carried out by the author, Ronov and Yanshin (Budyko, 1977a; Budyko and Ronov, 1979).

According to Table 7.1, Phanerozoic sediments contained $7.3 \times 10^{21}$ g of organic carbon, which corresponded to the formation of about $19.5 \times 10^{21}$ g of oxygen. Since the indicated data cover only the continental surface, the actual emission of $O_2$ during Phanerozoic time exceeded the indicated estimates because of photosynthesis occurring in the ocean. As mentioned earlier, the total amount of photosynthesis in water reservoirs in the modern epoch is considerably less than that on the continents. The available data show that the organic carbon mass in oceanic sediments over the last 150 million years is far less than that on the continents (Ronov, 1982).

TABLE 7.1

The rate of formations of continental sediments ($10^{21}$ g per million years)

| Stratigraphic interval | Absolute age (millions of years) | Igneous rocks | $CO_2$ total in sedimentary rocks (without organic carbon) | Organic carbon in sediments |
|---|---|---|---|---|
| Lower Cambrian | 490—570 | 0.36 | 0.45 | 0.009 |
| Middle Cambrian | | 0.18 | 0.32 | 0.005 |
| Upper Cambrian | | 0.14 | 0.32 | 0.003 |
| Ordovician | 435—490 | 0.33 | 0.27 | 0.009 |
| Silurian | 400—435 | 0.19 | 0.22 | 0.005 |
| Lower Devonian | 345—400 | 0.62 | 0.24 | 0.003 |
| Middle Devonian | | 1.43 | 0.66 | 0.018 |
| Upper Devonian | | 1.76 | 0.68 | 0.024 |
| Lower Carboniferous | 280—345 | 1.02 | 0.70 | 0.022 |
| Middle and Upper Carboniferous | | 0.27 | 0.30 | 0.010 |
| Lower Permian | 235—280 | 0.95 | 0.61 | 0.009 |
| Upper Permian | | 0.35 | 0.23 | 0.005 |
| Lower Triassic | 185—235 | 0.32 | 0.18 | 0.004 |
| Middle Triassic | | 1.10 | 0.34 | 0.004 |
| Upper Triassic | | 0.85 | 0.28 | 0.009 |
| Lower Jurassic | 132—185 | 0.33 | 0.31 | 0.016 |
| Middle Jurassic | | 0.38 | 0.39 | 0.033 |
| Upper Jurassic | | 0.31 | 0.40 | 0.028 |
| Lower Cretaceous | 66—132 | 0.67 | 0.35 | 0.023 |
| Upper Cretaceous | | 0.65 | 0.44 | 0.018 |
| Palaeocene | 2—66 | 0.31 | 0.20 | 0.012 |
| Eocene | | 0.37 | 0.31 | 0.020 |
| Oligocene | | 0.17 | 0.08 | 0.017 |
| Miocene | | 0.35 | 0.18 | 0.019 |
| Pliocene | | 0.32 | 0.09 | 0.014 |

Since the highest rate of assimilation in the hydrosphere generally occurs in shallow reservoirs within the continental shelf, the data on continental organic carbon sediments is believed to roughly reflect the dynamics of $O_2$ production on a global scale during the Phanerozoic. Considering that the present atmosphere contains $1.2 \times 10^{21}$ g of oxygen, one may conclude that more than 90% of the total oxygen production in the Phanerozoic was consumed in the oxidation of mineral compounds.

The variation of atmospheric oxygen with time, $dM/dt$ is determined by the equation

$$\frac{dM}{dt} = A - B, \tag{7.1}$$

where $A$ and $B$ are oxygen production and consumption rates. The oxygen production rate, $A$, for different periods of time can be calculated from the data on organic carbon in sediments. To estimate the oxygen consumption rate, $B$, by oxidation of mineral matter, we should bear in mind that the rate was not constant but was dependent on the atmospheric $O_2$ mass, tending to zero when there was no $O_2$ in the atmosphere and increasing with the expansion of the $O_2$ mass. Taking this into account we can assume $B = \alpha M$, where $\alpha$ is a proportionality coefficient.

Assuming that $M = M_0$ at $t = 0$, we find from (7.1)

$$M - \frac{A}{\alpha} = \left( M_0 - \frac{A}{\alpha} \right) e^{-\alpha t}. \tag{7.2}$$

The $O_2$ variations during the Phanerozoic can be estimated by proceeding from Equation (7.2) and using the data on organic carbon presented in Table 7.1. In this estimation, by assuming a certain amount of oxygen at the beginning of the early Cambrian, it is possible to calculate variations during this time interval. In the subsequent calculation the value obtained for the end of the Lower Cambrian is assumed to be the initial oxygen amount in the Middle Cambrian, etc. Thus all necessary values of the parameter $M_0$ are obtained by the calculations, except the first one, for the early Phanerozoic.

It can be shown that the initial oxygen amount assumed has only a slight effect on the calculated results. For example, if one uses the two extreme assumptions that the oxygen amount at the beginning of the Cambrian is either zero or the modern value, the results practically coincide for all the subsequent periods since the Ordovician.

Taking this into consideration, we assume that, during the Lower Cambrian, $dM/dt$ was considerably less than the $O_2$ production in the biotic cycle. Therefore, throughout this interval, $M = A/\alpha$.

To calculate $O_2$ variations, it is necessary to determine $\alpha$. We suppose that this parameter varied with time, increasing in the epochs of active volcanism and orogenesis. Since it is difficult to take these variations into account, we confine ourselves to a mean value of $\alpha$ that can be obtained by different methods based on the available empirical data. For instance, $\alpha$ is $3.1 \times 10^{-8}$ yr$^{-1}$ if the $O_2$ mass obtained for the end of the Phanerozoic is equal to its present-day value.

The substantiation of this method for determining the $O_2$ mass includes a hypothesis that the above data on organic carbon deposition are representative of their global values. To see whether the results are dependent on that kind of supposition, we can carry out a similar computation considering the data on sediments for each time interval to be proportional to global values but representing only a small portion of it. In this case, the rate of $O_2$ variation will be low compared to its production or consumption rate and the $O_2$ mass $M'$ appears to be proportional to the organic carbon mass for the appropriate period.

To define variations in atmospheric carbon dioxide, we assume, by analogy with the computation carried out for $O_2$, that they depend on the total carbon dioxide being fixed in sedimentary rocks. We assume also that this value approaches zero when the atmospheric carbon dioxide concentration is negligible and increases with its rise. Hence, it is natural to suppose that the carbon mass in sediments over a unit time interval is roughly proportional to the atmospheric carbon dioxide concentration.

In this case the proportionality factor can be found by comparison of the present-day carbon dioxide concentration and the amount of carbon dioxide in sediments at the end of the Quaternary period, the latter obtained by extrapolation of the variations during the Tertiary from the data in Table 7.1.

Calculated variations in $O_2$ and $CO_2$ during the Phanerozoic are presented in Fig. 7.1. The curves $M$ and $M'$ correspond to $O_2$ fluctuations calculated by the previous two methods. As seen from these curves, they are rather similar. This shows that the assumptions concerning the global representativeness of the data used have an insignificant effect on the results of the calculation.

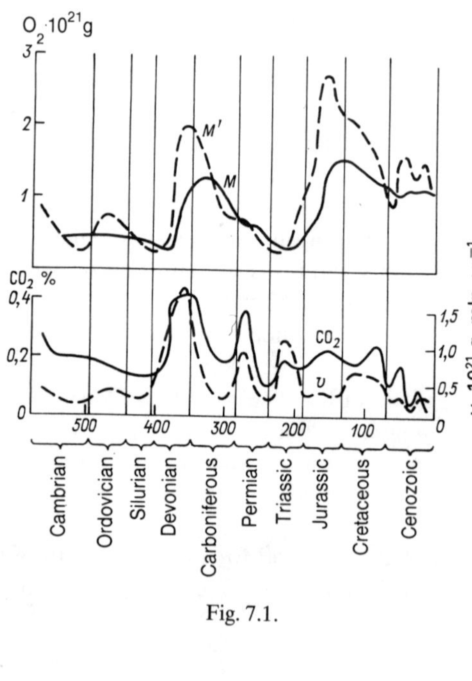


Fig. 7.1.

It is seen from the figure that the amount of atmospheric oxygen in the early Phanerozoic represents about one-third of its present amount. Throughout the Phanerozoic the $O_2$ mass was increasing, but unevenly. The $O_2$ mass increased abruptly for the first time in the Devonian and Carboniferous periods, when it reached its present-day level. At the end of the Palaeozoic the $O_2$ mass decreased and in the Triassic it approached the level of the first half of the Palaeozoic. In the mid-Mesozoic a second sharp rise in $O_2$ concentration took place. Afterwards it decreased noticeably.

It is worth noting that the increase in $O_2$ in the Devonian and the Carboniferous was accompanied with the expansion of continental vegetation cover. A considerable decrease in $O_2$ in the Permian and most of the Triassic is possibly linked to the arid conditions spreading over the continents at that time.

A high $O_2$ content of the atmosphere in the second half of the Mesozoic and the Cenozoic, probably depended on an increased productivity of autotrophic plants caused by the expansion of their progressive forms.

The data on variations in $O_2$ presented in Fig. 7.1 can be compared with the results of a later calculation based on the information concerning sedimentation of organic carbon both for the continents and for the oceans (Fig. 7.2). This comparison shows that a consideration of the global balance of organic carbon had comparatively little effect on the results of determining the changes in atmospheric $O_2$ concentration during the Phanerozoic.

Returning to the analysis of the data in Fig. 7.1, we may note that, as seen from the carbon dioxide curve describing variations in carbon dioxide mass, throughout most of the Phanerozoic the carbon dioxide concentration varied within a range of 0.1—0.4%. In the late Mesozoic the carbon dioxide concentration started to decrease gradually. The process was uneven,

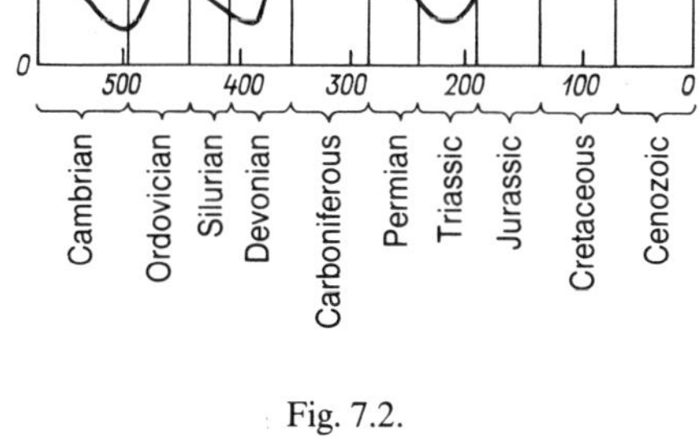


Fig. 7.2.

repeatedly giving rise to temporary increases of carbon dioxide mass. The last decrease in carbon dioxide concentration took place in the Pliocene. As a result, the present concentration of carbon dioxide has reached an unprecedentedly low level, several times lower than the Phanerozoic mean concentration.

To clarify the causes of fluctuations in atmospheric carbon dioxide content, we should study data on igneous rock formation per unit time interval. These data are presented in Fig. 7.1 in the form of curve $v$. Since variations in igneous rock mass are considered to correspond to volcanic activity fluctuations, we conclude that great changes in volcanicity took place during Phanerozoic time. As shown in Fig. 7.1, volcanic activity varied rhythmically, its highest peaks occurring approximately every 100 million years.

According to Tikhonov *et al.* (1969), who studied the theory of thermal convection in the lithosphere, this process is characterized by rhythmical upheavals of heated substances to the Earth's surface at intervals of about 100 million years. This conclusion has been confirmed by the empirical data.

Comparison of the curves $v$ and carbon dioxide in Fig. 7.1 shows that fluctuations in volcanicity undoubtedly influenced the carbon dioxide content of the atmosphere. The highest peaks of volcanic activity were accompanied with increases in carbon dioxide. It is interesting to note that comparatively small fluctuations in volcanicity during the Cenozoic also affected carbon dioxide content. The decreasing tendency of volcanic activity that prevailed since the mid-Cretaceous and corresponded to a decrease in carbon dioxide was of great importance in the evolution of the atmosphere in the late Phanerozoic.

In spite of the indicated common features of the variations in igneous rock formation and carbon dioxide content, there are certain differences between them. This is probably explained by some inaccuracy in the data used in the calculations, but mainly by the fact that the mass of igneous rocks is only an indirect indicator of carbon dioxide entering the atmosphere from the Earth's crust.

We can compare our data on the evolution of the composition of the atmosphere during the Phanerozoic with some concepts concerning the nature of this development that have been presented in a few modern studies. It should be borne in mind that these suppositions are based on indirect considerations that have not been susbtantiated by quantitative calculations.

Surveys of atmospheric evolution are frequently based on the previously cited works of Berkner and Marshall, who supposed that, in the early Phanerozoic, $O_2$ content was 1% of its contemporary value, in the Devonian period it rose to 10%, and in the Carboniferous it reached a maximum of 2—3 times its present-day value.

It is worth noting a hypothesis relating to $O_2$ variations in the post-Carboniferous period presented by Berkner and Marshall (1965a). A diagram in this paper shows that, after the $O_2$ concentration reached its maximum in the Carboniferous, it decreased in the Permian to a small fraction of the modern concentration. Then, the $O_2$ mass started increasing, and in the Mesozoic it approached the present-day level.

If we compare this supposition with our results we shall see that they agree reasonably well in a qualitative way. As for the quantitative estimates of $O_2$ fluctuations, those of Berkner and Marshall appear to be greatly overestimated. It should also be mentioned that the quantity of $O_2$ assumed in their work for the beginning of Phanerozoic time is strongly underestimated (by a factor of $\sim 30$). This error is easily explained since Berkner and Marshall made their estimates with the assumption that there were no multicellular organisms in the Precambrian period. From their point of view, this fact accounted for by the prevalence in the atmosphere of anaerobic processes that proceed with a small quantity of oxygen. As shown by present-day studies, a variety of multicellular organisms existed in the late Proterozoic. This invalidates the assumption of Berkner and Marshall.

Some authors (Rutten, 1971; Cloud, 1974) using the data of Berkner and Marshall discarded the hypothesis about very small $O_2$ concentration in the early Phanerozoic, although their estimates of this concentration are also too low. Cloud (1974), in his article on the origin of the atmosphere published in the *Encyclopaedia Britannica*, has emphasized that great oxygen mass changes undoubtedly took place during Phanerozoic time, but their nature is still unknown.

An opinion about atmospheric carbon dioxide fluctuations expressed long ago states that carbon dioxide concentration was gradually decreasing during Phanerozoic time. Some authors overestimated the range of carbon dioxide mass variations because they considered carbon dioxide one of the main air constituents in the early Phanerozoic. More correct estimates of carbon dioxide fluctuations were presented by Rutten (1971), who believed that the carbon dioxide concentration in the Phanerozoic was never more than 10 times the modern level. This supposition agrees well with the results of our study.

We should like to mention here Garrels' concept of the evolution of atmospheric carbon dioxide, which is most fully set forth in his Russian booklet (Garrels, 1975). In this work he emphasizes that, in the Phanerozoic, the lowest carbon dioxide concentration was no less than one-third of the contemporary value because autotrophic plants could not exist with lower levels of carbon dioxide. Though the upper limit of carbon dioxide partial pressure is difficult to determine, it probably did not exceed 0.1 atm. In this connection, Garrels refers to other authors who believe that the upper limit of carbon dioxide concentration exceeded the present-day one by no more than a factor of 100. Further, Garrels holds that the possible range of atmospheric carbon dioxide variation was not so wide and concludes that, during the Phanerozoic, the carbon dioxide content of the atmosphere varied within the 0.01—0.09% range. This conclusion is similar to the above statement that carbon dioxide concentration changed in the Phanerozoic by approximately 10 times. However, our estimates of the lower and upper limits of these changes exceed those of Garrels.

One of the first studies of the evolution of the atmospheric chemical composition was carried out by Grigoriev (1936) who believed that the amount of carbon dioxide and oxygen in the past atmosphere changed, the variations in the former being comparatively greater than in the latter. Grigoriev pointed out that, with increasing carbon dioxide income to the atmosphere from the depths of the Earth's crust, the growth in atmospheric carbon dioxide concentration cannot be considerably decreased by a rise in biomass of living organisms or by the absorption of carbon dioxide in the ocean. This study emphasizes that variations in carbon dioxide concentration affected climatic conditions of the past and poses the question of the possible importance of variations in atmospheric oxygen for the evolution of organisms.

Among the studies of variations in oxygen in the past atmospheres which have recently appeared, one by Sochava and Glikman (1973) should be considered. This work was one of the first steps in investigating the effects of fluctuations in oxygen content on the process of evolution.

Quantitative calculations of variations in the chemical composition of the atmosphere for the entire history of the Earth have been made by Hart (1978). Unlike our studies, Hart has not used empirical data on the composition of sediments in his calculations of oxygen and carbon dioxide variations. As a result, some of his conclusions appear to be unrealistic. For instance, assuming that in the Phanerozoic carbonates were not produced at the expense of the degassing of the lithosphere, Hart arrives at the conclusion that the atmospheric carbon dioxide concentration did not vary at that time. Since most of the carbonates were actually formed from gases ejected by the lithosphere in the Phanerozoic, this conclusion cannot be considered correct.

In studying climatic changes, the results of calculations of the carbon dioxide concentration in the geological past are of great importance. We think that the above data on carbon dioxide variations, as a rough approximation, reflect the actual evolution of the composition of the atmosphere. This supposition is based on a few independent grounds. First, the highest carbon dioxide concentrations obtained by our calculations occurred mostly during the epochs of highest volcanicity. Second, the results given below on climatic change caused by carbon dioxide fluctuations agree reasonably well with the empirical data on climatic conditions in the past. Third, the mean carbon dioxide concentration obtained for the Phanerozoic seems to approach the optimum concentration for photosynthesis. This confirms the ideas that autotrophic plants adapted to an atmospheric chemical composition that was richer in carbon dioxide than the present one. Fourth, the range of carbon dioxide variations obtained by our calculations agrees well with earlier suppositions, based on the features of sediment formation, as to how that concentration could vary.

## 7.3. Climates in the Geological Past

### 7.3.1. METHODS FOR STUDYING PAST CLIMATES

Information on the climates of remote epochs has been obtained by the examination of data concerning natural conditions in the past. Since sediment formation, rock weathering, the formation of water reservoirs and the existence of living organisms were dependent on atmospheric factors, data relating to these processes allow the evaluation of climatic conditions of the relevant periods of time.

The interpretation of data pertaining to natural conditions in the geological past in order to estimate the climate regime presents great difficulties, some of them crucial. Among these difficulties is, in particular, the necessity of applying the principle of uniformitarianism. Here, this corresponds to the assumption that, in the past, relationships between the climate and other natural phenomena were the same as at present. Although this approach is open to argument, the variety of natural processes that depend on climate enables the results of past climate reconstruction to be verified independently with the help of various palaeogeographical indicators. The most general regularities of climatic conditions in the geological past, as established in palaeogeographical investigations, are undoubtedly reliable, whereas some individual results are open to question and require further examination.

In addition to palaeogeographical data, information on palaeotemperatures obtained by isotopic analyses of organic remains appears to be very helpful in the study of ancient climates. It is difficult to evaluate the accuracy and to properly interpret these data, but these problems are gradually being overcome with improvements in the methods for palaeotemperature studies.

In studying climates of the past, information on sedimentary phenomena, geomorphological evidence, and data on fossil flora and fauna are used.

The first approach is based on the well-known relations between lithogenesis and climatic factors. For example, chemical weathering of rocks by the decomposition of unstable minerals is more rapid in hot, humid climates. When the climate is hot and dry, chemical weathering is less intensive and rocks are largely weathered by wind and temperature fluctuations. Chemical weathering is even more retarded in a cold climate where physical weathering, which does not affect chemically unstable minerals, dominates.

The extent and composition of deposits are strongly dependent on moisture conditions. The volume of sedimentary deposits is usually small in dry areas and increases in humid regions, where alluvial deposits prevail.

Since coal formations are closely related to climatic conditions, data on fossil coal can be used for reconstructing past climates. However, the interpretation of these data is fraught with difficulties because, at different periods, the relationship between coal accumulation and climate varied to a great extent, depending on the nature of the vegetation that formed the coal. For example, many Devonian coal beds were formed in a dry climate, while the coal formations of the Carboniferous period were usually connected with more humid climatic conditions.

Data on the geographical distribution of deposits of limestone and dolomites, as well as on the salt deposits that are so essential for reconstructing the climatic conditions of ancient water reservoirs, are also used in studying climates of the past.

In some cases, data on the sediment composition are used for estimating seasonal variations in the climate. For example, varved clays around the margins of continental glaciations are examined for this purpose. During the summer melting of glaciers, water streams carried away a great deal of rough debris, while in cold seasons, far less fine-grained argillaceous substance was deposited in the same area. The stratified structure of varved clays allows the length of their formation period to be evaluated.

Geomorphological evidence is widely used for studying the atmospheric precipitation regime and ice cover formation. For instance, data on the position of the ocean coastline enable one to estimate the water lost or gained by the ocean in the formation or melting of continental ice sheets.

Surface relief variations due to glaciations are a major indicator of glacier development. In addition, important evidence regarding past climates comes from data on snow levels in mountains, which depend on the temperature and precipitation regimes.

Data on ancient lakes and river valleys may give certain information concerning moisture conditions. For instance, impressions of numerous lakes and rivers in present deserts are indicative of great changes in the moisture conditions in these regions.

Information on fluctuations in the levels of such land-locked reservoirs as the Caspian Sea allows the water inflow to be evaluated and, hence, changes in precipitation over the basins of the feeding rivers may be estimated.

Data on fossil soils are used for studying moisture and thermal conditions in past times. Traces of permafrost are valuable in reconstructing zones of cold climate.

The nature of erosion processes is strongly dependent on climatic conditions, in particular, on moisture conditions, so such information is also useful in palaeooclimatic studies.

The geographical distribution of living organisms, particularly vegetation which is vitally dependent on climatic conditions, can be of great importance in the study of past climates. This method provides quite reliable results for the not-so-distant past when plants were similar to modern species and their distribution was evidently affected by climatic conditions in much the same way as at present. Great advances have been achieved in recent decades following the development of fossil pollen and spore analysis, which allows the composition of a regional vegetation cover to be defined. This method has proved to be efficient in the reconstruction of the epochs when plants were similar to modern species, but extrapolation of the relationships between modern plants and meteorological factors to more ancient times presents greater difficulties, which restrict the use of plant distribution data in palaeoclimatic studies.

We should mention here data on annual tree rings, which are also used in the study of climatic change. Variations in the annual growth layers of a tree reveal short-term climatic fluctuations, whereas their structure is indicative of general climatic conditions (for example, trees growing in carbonaceous marshes show little inter-annual width variability and/or poorly defined rings, reflecting no seasonal climatic variations).

Information on fossil fauna is more difficult to regard as diagnostic of past climates, since the geographical distribution of animals is generally not as strongly dependent on climate as that of vegetation. However, these data are a valuable addition to other methods used in the study of climate change. As was the case with plants, the data on the distribution of animals are interpreted under the assumption that climatic effects on the vital activities of the relevant organisms were similar to those existing at present.

A palaeotemperature method based on the $^{18}O$ isotope content of the fossil remains of aquatic animals is of particular importance in palaeoclimatology. It has been established that the $^{18}O/^{16}O$ isotope ratio in mollusc shells and other remnants of marine organisms depends on the temperature of the environment in which the animals lived. Since the middle of our century the method of measurement of palaeotemperatures has been widely applied in studies of climatic conditions of the recent past and of hundreds of millions of years ago.

In evaluating the results of these methods for studying climatic variations, it should be recognized that, except for a relatively short contemporary period, almost all the available information on past climates concerns the air temperature regime near the Earth's surface, the surface temperature of the land, the temperature of water reservoirs, and, to a lesser degree, the moisture conditions on the continents.

Palaeogeographical data can also supply information on some other climatic elements. For example, in some cases the shape of fossil dunes and barkhans can be indicative of the direction of the prevailing winds. But these data are not so widespread as the above-mentioned.

### 7.3.2. PRECAMBRIAN CLIMATES

Until recently there has been lack of sufficiently reliable information on climatic conditions in the Precambrian — the time covering five-sixths of the duration of the Earth's history. Although some progress has been achieved in studying Precambrian climates, the available concepts are far more schematic than data on Phanerozoic climates.

Three facts are of considerable importance for estimating the climatic conditions of Precambrian time: (1) the presence of living organisms on the Earth's surface not less than 3.8 billion years ago; (2) the occurrence of large water bodies throughout the greater part of the Precambrian; and (3) the advance of glaciers over vast areas during some epochs. These facts show that, in the Precambrian, over the whole period the temperature of most of the Earth's surface was within the rather narrow range of 0 to 50 °C favourable for the existence of ancient organisms. Only occasionally did the temperature of part of the Earth's surface drop below 0 °C.

Some results of determining palaeotemperatures from oxygen-isotope data for the Precambrian are given by Sagan and Mullen (1972). These data correspond to temperatures of 280—340 K and they enable one to infer that, during the greater part of the Precambrian, the surface temperature tended to decrease. Possible causes of this decline are discussed in the next section of this chapter.

Information on glaciations in the Precambrian presented by Frakes (1979) and other authors is of great importance. It is difficult to interpret this information because an accurate determination of the age of Precambrian glaciations is almost impossible. The first large glaciation of the Precambrian is assumed to have occurred about 2.3 billion years ago. This is confirmed by sediments in North America, Africa and Australia characteristic of the ice ages. It is possible, however, that glaciations in these regions were not simultaneous and reflect climatic changes separated by hundreds of millions of years.

A large glaciation that occurred at the end of Precambrian time and lasted over the period of 950—600 million years ago has been studied most comprehensively. The best evidence of this glaciation identifies three stages with the average ages of about 940, 770 and 615 million years. Each of these stages continued for several tens of millions of years. This glaciation spread over a considerable part of Australia, several regions in Africa, Asia, northwest Europe, Greenland, Spitsbergen and, possibly, South America.

The relative positions of ice sheets during this glacial epoch are difficult to establish because of insufficient knowledge of the relief at that time. Possibly, a part of glaciation embraced the high-mountain regions. However, the considerable number of glaciers show that they were interrelated and represented the parts of large ice sheets which, as for the Quaternary glaciations, covered vast lowlands.

In a number of studies, the position of large ice masses at the end of Precambrian time has been found from palaeomagnetic data. Although these studies are not yet completed, they show that a considerable part of the ice was located in the low latitudes of Gondwanaland continents. However, the hypothesis that the records of ice sheets of the late Precambrian reflect the glaciation of the entire planet at that time seems improbable because of the difficulty of maintaining life on the Earth under such conditions.

### 7.3.3. THE CLIMATES OF THE PALAEOZOIC AND THE MESOZOIC

Sinitsyn (1965, 1966, 1967, 1970 and 1976) was the first to thoroughly describe climatic variations in the Phanerozoic by analysing data on lithogenesis, floras and faunas of the geological past. Sinitsyn constructed palaeoclimatic maps for various periods of the Phanerozoic and their epochs that cover the territory of Europe, extratropical Asia and North Africa. Much new information based on palaeotemperature data has been recently obtained concerning Phanerozoic climates. Unfortunately, most later investigations refer to restricted areas and do not contain palaeoclimatic maps embracing large fractions of the globe.

Data on the Phanerozoic climate conditions given below are based on the monographs by Sinitsyn, by Frakes (1979) and other authors. In these studies it is found that, at the beginning of the Palaeozoic, climate was warmer than during the ice ages of the late Precambrian. Data on the Cambrian glaciation in northwest Africa (which was, at that time, close to one of the poles) only cover a limited area and are of doubtful validity. More clear evidence is available for the glaciation at the Ordovician–Silurian boundary, records of which have been discovered in Africa, West Europe and North Asia. However, in this case, most of the valid data come from northwest Africa which makes it difficult to determine the margins of ice cover. It is possible that this glaciation was restricted to the polar zone.

In the second half of the Palaeozoic, in the late Carboniferous and early Permian periods, a far greater glaciation occurred. This glaciation spread over vast areas of Gondwanaland continents and left particularly numerous tracks in Australia and Africa, as well as in South America, the Antarctic and India. In Australia it appears that the glaciation reached its maximum area at the beginning of the Permian period when it occupied approximately half of the modern territory of the continent. The glaciation in Africa (the dating of which is imprecise) embraced the greater part of south Africa. The ice sheets in South America were located in the northeast parts of the continents which was connected to South Africa when these two continents were part of Gondwanaland.

The study of the late Palaeozoic glaciation was important in clearing up the question of continental drift. Modern investigations suppose that this glaciation started in the west of Gondwanaland (the south of South America). Moving gradually to the east it reached Australia. Such a drift corresponds to a change in the position of the South Pole during the glaciations.

The available data show a comparatively small climatic change in the Northern Hemisphere during the Palaeozoic. Far more drastic changes in climatic conditions took place throughout the late-Palaeozoic glaciation in Gondwanaland. Undoubtedly, this glaciation was accompanied with cooling over the entire planet, the largest temperature drop being in the glacier zone and in adjoining regions. Although the thermal conditions in the Palaeozoic can be estimated only approximately, there are grounds to believe that, during epochs without glaciations, the climate was very warm in low and warm in high latitudes. When ice masses began advancing the climate continued to be warm in all regions other than the glacial zones and adjoining regions.

The moisture conditions on the continents changed considerably in the Palaeozoic. In particular, vast territories became arid in the late Silurian and in the first half of the Devonian. After the humid climatic conditions of the end of the Devonian and the first half of the Carboniferous, aridity spread again in the second half of the Carboniferous, increasing further in the Permian.

The climatic conditions in the Mesozoic era were comparatively stable. Over the most part of the Earth, climate was very warm with a comparatively small temperature decrease in high latitudes where seasonal temperature variations took place.

The early Triassic climate was moderately warm and, by the mid-Triassic, the temperature increased noticeably. This warming continued for a considerable time period. Vast arid zones of the late Palaeozoic widened even more in the Triassic. In the mid-Triassic, the zone of dry climate is thought to have reached its greatest area over the entire Phanerozoic. The contraction of this vast aridity zone began in the late Triassic and continued in the Jurassic. The available records of palaeotemperatures for the Jurassic period are sometimes inconsistent and seem to be frequently inaccurate. Nevertheless, an estimate of 14 °C for the maximum temperature of the surface water layers at 75°S is of interest, since this temperature is more than 10 °C greater than today's temperature. This estimate agrees with palaeogeographical data which indicate a warm climate in high latitudes in the Mesozoic.

Beginning with the Cretaceous period, the amount of data on past climatic conditions increases sharply. In particular there is a large increase in the body of palaeotemperature data determined by the isotope method. In the Lower Cretaceous a small temperature increase from 15 to 17 °C occurred in the depths of the oceans. In the Upper Cretaceous this temperature noticeably decreased to approximately 11 °C. Surface water temperatures were higher than at present and reached about 28 °C at middle latitudes in the mid-Cretaceous. Palaeobotanic data indicate a tendency to warming throughout the Jurassic period and in the early Cretaceous. This changed into cooling in the late Cretaceous. Frakes draws the conclusion that the mean temperature in the mid-Cretaceous was 10—15 °C higher than the present one, the meridional temperature gradient being half the modern one.

At the end of the Mesozoic era a relatively short-term temperature decrease of 3—5 °C is recorded, after which, in the early Palaeocene, a warming of 1—3 °C occurred (Savin, 1977).

### 7.3.4. THE TERTIARY CLIMATES

A cooling trend dominated during the Tertiary. This cooling was global, although the minimum temperatures are recorded in extratropical latitudes and, particularly, in the polar zones. The Tertiary cooling was not permanent: it was, on several occasions, interrupted by more or less long-term warm epochs which eventually yielded to new, still stronger coolings.

The general pattern of variations in the thermal regime during the Tertiary is seen from data in Fig. 7.3. In this figure we show the temperature differences for various epochs of the Tertiary and for the present time obtained from three studies. The first refers to the west of North America and is based on palaeobotanic evidence (Axelrod and Baily, 1969); the second is an integration of isotope data from deep sea carbonate sediments in the oceans of the Southern Hemisphere (Shackleton and Kennett, 1975); and the third uses isotope data from the remnants of marine molluscs from the southern part of the North Sea (Buchardt, 1978). The results of these studies are presented in Fig. 7.3 as curves 1—3 respectively. These curves show the warming that began in the Palaeocene and which reached a maximum in the Eocene. In the early Oligocene the temperature decreased sharply and then, in the Miocene, rose again. As seen in data from two of the above-mentioned studies, one more temperature increase occurred in the Pliocene, but it was not long-lived and gave way soon to drastic cooling.

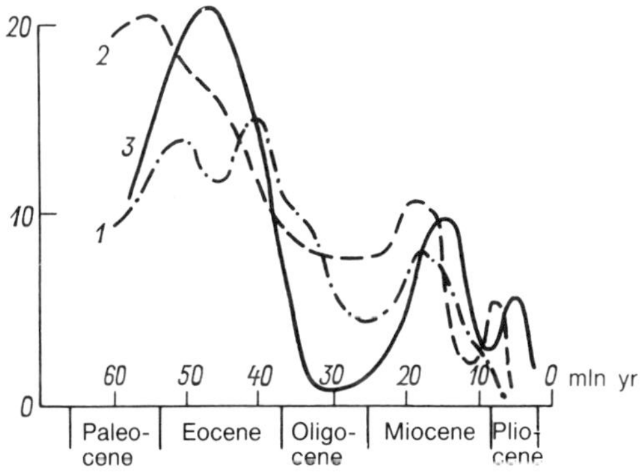


Fig. 7.3.

It is important that the maximum temperatures of the three warmings in the Tertiary decreased progressively. When the data in Fig. 7.3 are smoothed, the long time scale Tertiary cooling trend becomes apparent. Such a cooling trend was inferred in earlier investigations of Cenozoic climatic conditions, in particular, in Sinitsyn's studies mentioned above.

Recently it has been discovered that, along with climatic variations lasting for millions of years, there have been numerous shorter fluctuations lasting thousands, tens of thousands and hundreds of thousands of years. Such variations were most noticeable at the late Tertiary (Zubakov and Borzenkova, 1983).

The available data show that the amplitudes of warming and cooling were 4-5 times greater in high latitudes than in the tropics. This made it possible for high latitude ice sheets to develop during the epochs of the strongest coolings. The area of the Antarctic ice mass (which may have existed in the Eocene as individual mountain glaciers) considerably increased in the Oligocene. By the end of the Oligocene, this ice sheet reached sea level. At the end of the Miocene the boundaries of the Antarctic ice sheet seem to be wider than the present ones, with the ice volume greatly exceeding the modern volume. In the Northern Hemisphere the first continental glaciations appeared in Alaska during the Miocene, and in Greenland and in Iceland during the late Pliocene.

### 7.3.5. CLIMATE OF THE QUATERNARY

The climatic conditions of Quaternary time have been studied more thoroughly than any earlier climates. Among the numerous works devoted to natural conditions and climates of the Quaternary we mention here studies by Gerasimov and Markov (1939), Schwarzbach (1950, 1968), Flint (1957), Zeuner (1959), Fairbridge (1967), Velichko (1973) and Zubakov and Borzenkova (1983).

The main part of Quaternary period, except for the last relatively short interval called the Holocene, corresponds to the Pleistocene. During this time climatic conditions differed greatly from those in the preceding Mesozoic and Tertiary periods, when thermal zones were relatively indistinct.

In the Pleistocene a cooling trend became more pronounced in middle and high latitudes and great continental glaciations developed. They advanced repeatedly, reaching the middle latitudes, and retreated again to the high latitude belt. During these continental glaciations, sea ice expanded over vast areas. Although we do not known the exact number or dates of the Quaternary glaciations, dating accuracy for events in the Pleistocene is much better than for earlier epochs.

Alpine investigations carried out in the last century gave evidence of four major European glaciations called Günz, Mindel, Riss and Würm. Later it was discovered that each of these glaciations proceeded in several stages, with intervals marked by ice recession. The Alpine glaciations were accompanied with the development of ice cover in the lowlands of Eurasia and North America, stretching over vast territories in high and middle latitudes. It has been established that ice advance and retreat took place more or less simultaneously in Europe, Asia and North America. This also applies to the ice ages in the Northern and Southern Hemispheres.

During the ice ages the continental ice cover spread farther in the zones of humid maritime climate. Under the relatively dry climatic conditions of Northern Asia the ice covered comparatively small areas. During the greatest glaciations the continental ice cover in the Northern Hemisphere reached as far south as 57°N on the average and even 40°N in some places. Its thickness in most areas was hundreds of metres and sometimes more than a kilometre. Undoubtedly, the progress of continental glaciations was accompanied with a shift of the polar sea-ice boundary to lower latitudes. This considerably increased the total extent of permanent ice cover on the planet. At each ice advance the snow line in mountain regions outside the glaciation dropped by hundreds of metres, sometimes by a kilometre or more. At the same time the permafrost zone increased considerably in area. With the development of glaciations the permafrost boundary moved a few thousand kilometres into lower latitudes.

The ice ages were marked by a 100—150 m drop in the level of the World Ocean below its modern level, a result of the development of the thick ice masses on the continents. During the warm interglacial epochs the sea level was several metres higher than that at present. The area of the ocean surface varied by a few per cent according to fluctuations in its level.

A noticeable drop in air temperature occurred over the entire globe during the ice ages. In the warm interglacial epochs the air temperature was higher than at present, although the increase above present-day levels was less in absolute magnitude than the decrease during the glacial epochs. Variations in the precipitation regime associated with glaciations are less clear. According to the available data, moisture conditions in various parts of the Earth varied in different ways under glacial conditions. This confirms that the whole atmospheric circulation changed as a result of the variation in the temperature differences between the equator and the poles. Precipitation decreased in many continental regions during the epochs of the greatest glacier advances. Evidently this was accounted for by a reduction in evaporation as a result of the temperature drop and by the fact that the ocean surface was partially covered with ice.

### 7.3.6. CLIMATE OF THE HOLOCENE

Post-Pleistocene climatic variations occurred within a comparatively short period of the Earth's history. During this time, which is often called the Holocene, we can distinguish several stages with different climatic conditions.

Many of the works on the palaeoclimatology of the Quaternary cited in the preceding subsection present some information on the climate in the Holocene.

The last Würm glaciation reached its greatest extent about 20000 years ago. In a few thousand years its area contracted considerably. The climate of the subsequent time was comparatively cold, particularly in middle and high latitudes. About 12000 years ago the climate became much warmer (the Alleröd epoch). However, a cooling trend soon developed. As a result of these climatic fluctuations, the summer air temperature in Europe varied by a few degrees.

Later on, the warming trend resumed, and 5000—7000 years ago the last remains of glaciations disappeared in Europe and North America. It was in that epoch that the postglacial warming reached its peak. Between 5000 and 6000 years ago the air temperature in the middle latitudes of the Northern Hemisphere is believed to have been about 1—3 °C higher than at present.

At the same time the atmospheric circulation apparently underwent a change. The northward shift of the polar-ice boundary was accompanied with the displacement of the subtropical high pressure belt to higher latitudes. This resulted in the expansion of arid zones in Europe, Asia and North America. In addition, precipitation increased in low latitudes over the areas now occupied by deserts. The climate in the Sahara was then comparatively humid, favouring abundant flora and fauna. Still later, a tendency toward cooling prevailed, which became especially pronounced at the beginning of the first millennium B.C. The change in thermal regime was accompanied with an alteration of the precipitation regime that brought about a gradual approach to its present state.

A significant warming trend took place during the end of the first and the beginning of the second millennium A.D. The polar ice retreated to higher latitudes allowing the Vikings to colonize Greenland and to discover the continent of North America. A subsequent cooling favoured a new advance of ice and the Norse colony in Greenland, deprived of any communication with Europe, perished.

The cooling trend that began to develop in the 13th century reached its maximum in the early 18th century and was accompanied with the expansion of mountain glaciers. For this reason it is called the Little Ice Age. Then, the warming started again and the glaciers retreated. The climate during the second half of the 18th century and the 19th century differed comparatively little from contemporary conditions.

Of great importance in understanding the physical mechanism of contemporary climatic change is the study of climatic variations over the last century, when most of the continental surface has been monitored by a network of permanently operating meteorological stations.

The greatest climatic change within the period of instrumental observations started at the end of the 19th century. It was marked by a gradual rise in air temperature at all latitudes in the Northern Hemisphere throughout the year. This warming was especially pronounced at high latitudes and during the cold seasons.

The warming intensified in the 1910s and reached its peak in the 1930s, when the mean air temperature in the Northern Hemisphere was 0.6 °C higher than at the end of the 19th century. In the 1940s the warming trend changed to a cooling trend, which has continued until recently. The cooling was not very strong and its scale did not approach that of the preceding warming.

Although the evidence of modern climatic change in the Southern Hemisphere is less certain than that for the Northern Hemisphere, there are grounds for believing that the warming also occurred in the Southern Hemisphere in the first half of the 20th century.

In the Northern Hemisphere the air temperature rise was accompanied with a contraction of polar ice, a retreat of the permafrost boundary to higher latitudes, a northward shift of the forest and tundra boundary, and other changes in natural conditions. Of vital importance was a noticeable change in the atmospheric precipitation regime during the warming. In some regions of unstable moisture conditions, precipitation diminished during the climatic warming, especially in the cold seasons. This resulted in decreased river run-off and a fall in the water level of some land-locked reservoirs. The abrupt drop in the water level of the Caspian Sea that occurred in the 1930s and was caused by the reduced run-off of the Volga River is particularly well known. In addition, the warming brought about more frequent droughts over vast continental areas in mid-latitudes in Europe, Asia and North America. This climatic change had a definite influence on the national economy of many countries.

The question of the modern climatic changes is considered in more detail in Chapter 10.

## 7.4. Causes of Climatic Change

### 7.4.1. THE STUDIES OF THE CAUSES OF CLIMATIC CHANGE

Although many hypotheses have been proposed for explaining the causes of climatic change in the geological past, until now there has been no generally accepted opinion of the physical mechanism of climatic change for the modern epoch and the geological past. Because of the absence of an accurate climatic theory and the lack of specific observational data, conflicting hypotheses about the causes of climatic changes were not widely accepted and were subject to many objections. Considerable progress in this field has been made recently, particularly during the last decade. As a result, we now have a much better understanding of the physical mechanisms of climatic change.

In studies of the causes of climatic change, emphasis is placed on the climatic effect of carbon dioxide concentration. Over 100 years ago Tyndall pointed out that, since atmospheric carbon dioxide, together with water vapour, absorbed longwave radiation in the atmosphere, variations in carbon dioxide concentration could result in climatic fluctuations (Tyndall, 1861). Later Arrhenius (1896, 1903) and Chamberlin (1897, 1898, 1899) studied the question of the climatic effect of atmospheric carbon dioxide. They supposed that variations in carbon dioxide concentration could have caused the Quaternary glaciations.

Arrhenius studied the absorption of radiation fluxes in the atmosphere. He suggested a numerical model for determining the temperature near the Earth's surface depending on the properties of the atmosphere. Using this model, Arrhenius found that a 2.5- to 3-fold increase in carbon dioxide concentration would raise the air temperature by 8—9 °C, and a 38—45% decrease would lower the temperature by 4—5 °C.

Based on geological evidence, Arrhenius stated that the mass of $CO_2$ in the present atmosphere is only a small fraction of that which existed in the past, much of which was absorbed from the atmosphere by the formation of carbonate rocks. In this connection, Arrhenius concluded that the carbon dioxide concentration in the atmosphere could vary over a wide range. These variations, he believed, exerted a considerable influence on air temperature, sufficient for glaciations to arise or to disappear.

The geological aspect of this problem was studied by Chamberlin. Considering the carbon dioxide balance, Chamberlin stated that the amount of carbon dioxide coming from the lithosphere varied noticeably depending on the level of volcanic activity and other factors. The expenditure of carbon dioxide in geological processes also changed appreciably; in particular, in accordance with the surface area of the continents exposed to atmospheric erosion. As the area increased, the loss of carbon dioxide by weathering grew as well. Chamberlin asumed that the glaciations resulted from the intensive process of orogenesis and elevation of continents, which led to an increase in the erosion basin and the surface area of weathered rocks, and to a decrease in atmospheric carbon dioxide concentration. To corroborate this assumption, Chamberlin carried out some calculations but these cannot be considered a complete quantitative model of the process.

It was later established that, within the 13—17 μm carbon dioxide radiation absorption band, an overlap occurs with part of the absorption band of water vapour. This decreases the influence of carbon dioxide concentration variations on the thermal regime. Considering this effect, Callendar (1938) obtained smaller values than those of Arrhenius for the temperature variations at the Earth's surface. Using Callendar's data, a doubling of the carbon dioxide concentration increased the air temperature by 1.5 °C, and the effects of variations of carbon dioxide concentration on temperature decrease with increasing concentration.

Möller (1963) also studied the problem of the effect of carbon dioxide on the atmospheric thermal regime. He determined that, when air temperature varies, the absolute humidity changes, whereas the relative humidity is more or less constant. An increase in the absolute humidity as the temperature rises increases the absorption of longwave radiation in the atmosphere, further raising the temperature. Möller found that the temperature increase near the Earth's surface caused by doubling the carbon dioxide concentration with constant absolute humidity is 1.5 °C, whereas, with constant relative humidity, this value is several times larger, Möller also mentioned that the effect of carbon dioxide on the thermal regime could be compensated for by comparatively small variations in the absolute humidity or cloudiness.

The first thorough investigation of the dependence of the air temperature on the carbon dioxide concentration of the atmosphere was carried out by Manabe and Wetherald (1967). In their work they noted the inaccuracy of the calculations by Möller, who had estimated variations in the thermal regime of the atmosphere using only data on the heat balance of the Earth's surface without considering the atmospheric heat balance as a whole. Manabe and Wetherald computed the vertical temperature distribution in the atmosphere, taking into account the absorption of longwave radiation by water vapour, carbon dioxide and ozone. They used a vertical distribution of the relative humidity taken from empirical data. It was assumed that the temperature distribution was determined by conditions of local radiation balance if the lapse rate did not exceed 6.5 °C $km^{-1}$. This value was considered a maximum because of the limiting effect of convection on the growth of the vertical gradient. Manabe and Wetherald found that, under average cloud conditions with constant relative humidity, doubling the carbon dioxide concentration raised the temperature at the Earth's surface by 2.4 °C. Reduction to half the present carbon dioxide level lowered the air temperature by 2.3 °C.

In a later study, Manabe and Wetherald (1975) computed the carbon dioxide effect on climate using a three-dimensional model of the atmospheric general circulation, taking into account water exchange, heat exchange in the atmosphere, in the oceans, and on the continents (with idealized topography), and the feedback between snow and ice cover and the atmospheric thermal regime. They found that doubling the carbon dioxide concentration raises the mean air temperature at the Earth's surface by 2.9 °C. Similar estimates have been obtained in a number of recent studies that have used various simplified models of climate theory.

The relationship between the atmospheric thermal regime and the level of carbon dioxide concentration is of importance in understanding climatic changes. The assumption of the author that a decrease in atmospheric carbon dioxide concentration exerted an essential influence on climatic cooling in the late Cenozoic era (Budyko, 1974) was supported by data on the evolution of the atmospheric chemical composition in the Phanerozoic

(Budyko, 1977a; Budyko and Ronov, 1979). From these data, it was concluded that variations in the carbon dioxide concentration caused the lowering of the mean air temperature that began in the late Mesozoic era and continued in the Cenozoic era.

A French mathematician Adhèmar (1842) was the first to elucidate the causes of the Quaternary glaciations and to point out that changes in the Earth position relative to the Sun could result in glacier advance. This concept was treated comprehensively by Milankovich (1920, 1930, 1941). To explain climatic changes during the Quaternary period, Milankovich used the calculated results of the secular change of three astronomical elements: the Earth's orbital eccentricity, the inclination of the Earth's axis to the ecliptic plane, and the time of precession of the equinoxes due to the precession of the Earth's axis of rotation (i.e. the moments during the year when the Sun crosses the celestial equator and the distance between the Sun and the Earth is the shortest). All these elements vary with time because of the effect of the Moon and the other planets on the Earth's movement: the period of change in the eccentricity is ~92000 yr; in the inclination of the axis, ~40000 yr; and in the precession of the equinoxes, ~21000 yr.

Although oscillations in these elements do not noticeably influence the amount of solar radiation coming to the outer boundary of the atmosphere of the Earth as a whole, they do alter the total radiation received by various latitudinal zones in different seasons of the year.

Milankovich developed an approximate theory of radiation and the thermal regime of the atmosphere, from which it followed that the surface air temperature in the warm period of the year, at middle and high latitudes, can vary by several degrees, depending on the relationships among the above astronomical elements. Such a variation usually coincided, to a greater or lesser degree, with a temperature change of opposite sign in the cold season. However, as is known from empirical investigations, the glaciation regime depends mainly on the temperature of the warm season (when ice melts). The thermal conditions of the cold season exert a far smaller influence on ice cover.

By using calculated results for variations in the astronomical elements for the Quaternary period, Milankovich assumed that the periods of temperature decrease during the warm season in the zone where most glaciation developed (60—70° latitude) corresponded to the glacial epochs, and the periods of temperature rise to the intervals between them.

Köppen was the first to compare Milankovich's results and climatic conditions during the Quaternary period. He found good agreement between Milankovich's conclusions and the history of Quaternary glaciations (Köppen und Wegener, 1924). Similar comparisons were carried out by other authors, but their conclusions proved to be contradictory. For this reason, the question of the climatic effect of astronomical factors has generated lively discussion.

To determine the causes of Quaternary climatic changes, one can use semi-empirical theories of the atmospheric thermal regime (Budyko, 1968; Sellers, 1969) through which it is possible to quantitatively account for the effect on the thermal regime of air temperature—polar ice feedback.

During discussions on the causes of the Quaternary glaciations, several attempts were made to construct numerical models of the development of glaciations which would make it possible to relate changes in the area of polar ice cover to the changing position of the Earth in relation to the Sun. In a number of cases these attempts led to the conclusion that the influence of astronomical factors on the development of glaciations was unimportant. This was due to the fact that the models used did not adequately take into account the ice cover—atmospheric thermal regime feedback. When this feedback was considered, it was shown that orbital changes exerted an essential influence on the development of the Quaternary glaciations (Budyko and Vasishcheva, 1971; Budyko, 1972; Berger, 1973). In these works, semi-empirical models were used for calculating the atmospheric thermal regime. Later, a similar conclusion was obtained using more general models of climate theory (Suarez and Held, 1976).

Of great importance for determining the causes of the Quaternary glaciations was an empirical investigation which established a close agreement between the intervals of decreasing total summer radiation at high latitudes and an improved chronology of Quaternary glaciations (Hays *et al.*, 1976).

In addition to the effects of Earth-orbit changes on climate, variations in volcanic activity, resulting in fluctuations in atmospheric carbon dioxide content and transparency of the atmosphere, greatly affect the climate. The first of these mechanisms of climatic change is discussed earlier and the second, pertaining mainly to modern climatic changes, is treated at the end of this chapter.

Besides the indicated factors, climatic conditions in the remote past were influenced by changes in the shape of the continents, in continental and sea-floor relief and in the position of the poles. The dependence of climatic conditions on these factors is, as yet, imperfectly understood. The question of physical relations of climatic changes with other factors is considered further more thoroughly.

### 7.4.2. EFFECTS OF HEAT INCOME ON THE THERMAL REGIME

To explain climatic changes, it is necessary to have reliable estimates of the climate sensitivity to variations in the major climate-forming factors. Among these, the heat income to the Earth—atmosphere system is a most important factor. The amount of heat income can vary due to increasing or decreasing either the astronomical solar constant (i.e. the solar energy flux at the top of the atmosphere) or the meteorological solar constant (i.e. the heat flux at the top of the troposphere which can change with constant solar radiation because of variations in the aerosol mass in the lower stratosphere that attenuates the shortwave radiation flux). At the same time, the heat influx to the climate system may depend on variations in the chemical composition of the atmosphere, in particular, on the increase or decrease in carbon dioxide quantity that influences the greenhouse effect in the atmosphere.

To evaluate the sensitivity of the thermal regime to variations in heat income, the parameter $\Delta T_1$ is frequently used. This corresponds to the increase or decrease of the mean air temperature near the Earth's surface produced by a 1% change in the solar constant. Table 7.2 presents values of $\Delta T_1$ obtained by different methods.

TABLE 7.2

The sensitivity of the thermal regime to heat income variations ($\Delta T_1$)

| Simplified climate models | | | General circulation models | | | Modern climatic variations | | | Satellite observations of outgoing radiation | | |
|---|---|---|---|---|---|---|---|---|---|---|---|
| 1. | 1967 | 1.2 °C | 4. | 1975 | 1.5 °C | 5. | 1969 | 1.1 °C | 7. | 1975 | 1.1—1.4 °C |
| 2. | 1968 | 1.5 °C | | | | 6. | 1977c | 1.2 °C | 8. | 1976 | 1.45 °C |
| 3. | 1979 | 1.4 °C | | | | | | | | | |

*Note*: Estimates: 1 — Manabe and Wetherald (1967); 2, 3, 5, 6, 7 — Budyko (1968, 1979, 1969, 1977c and 1975); 4 — Wetherald and Manabe (1975); 8 — Cess (1976).

Manabe and Wetherald (1967) were the first to realistically evaluate the sensitivity of the thermal regime to heat income variations for the globe as a whole. They took into account an important feedback among the air temperature, absolute humidity, and outgoing longwave emission, thus considerably changing the value of $\Delta T_1$ compared with previous estimates obtained without considering this feedback.

Another method of estimating $\Delta T_1$ can be applied, based on Budyko (1968) and mentioned above, in which an empirical formula relates the value of outgoing longwave emission to the air temperature near the Earth's surface. This equation was introduced into a numerical model of the atmospheric thermal regime that allowed us to take into account the influence on air temperature of the feedback from the polar ice position. This effect operates mainly over long time periods; therefore, it can be neglected when computing $\Delta T_1$ for present-day climatic changes the duration of which does not exceed several decades. The $\Delta T_1$ value obtained in Budyko (1968) is 1.5 °C. When the parameters of the formula for outgoing emission were improved by using satellite data, $\Delta T_1$ turned out to be equal to 1.4 °C (Budyko, 1979).

Until recently, it has not been clear to what extent we can neglect the air temperature—cloudiness feedback when estimating the sensitivity of climate. Variations in cloudiness lead to changes in the amount of absorbed solar radiation and outgoing longwave emission.

This question has been treated in a number of studies. In the first of these it was proposed that the effect on global mean air temperature of a decrease in absorbed radiation due to the increase in albedo with increasing cloudiness, is compensated for by the influence of the decrease in outgoing longwave radiation (Budyko, 1971; Schneider, 1972). However, the empirical data used for the analysis of this question in these studies were insufficient to either prove or disprove the proposal.

The presence of such a compensation was suggested by studies where satellite data on outgoing longwave radiation were compared with those on the temperature of the lower air layer and on cloudiness. This comparison, carried out for various seasons, showed the absence of any large influence of variations in cloudiness on the global mean temperature of the lower air layer (Budyko, 1975; Cess, 1976).

This problem was also investigated by Wetherald and Manabe (1980) on the basis of an atmospheric general circulation model. They reached a similar conclusion.

Thus, models which fail to account for variations in cloudiness can still yield realistic estimates of the sensitivity of global mean air temperature to fluctuations in heat income.

Table 7.2 indicates that values of the $\Delta T_1$ parameter can be derived by several other independent methods. Among these are calculations by general circulation models (estimate 4), the analysis of empirical data on modern climatic changes (estimates 5 and 6), and the comparison of satellite data on outgoing emission with temperature data for the lower air layer (estimates 7 and 8). All these methods produce similar results, which gives us confidence that estimates of $\Delta T_1$ are realistic.

To estimate the effects of carbon dioxide on climatic conditions, three independent methods are currently used (see Budyko *et al.*, 1983). The first of these methods is connected with the application of climate models. In 1967 Manabe and Wetherald obtained a realistic estimate of the dependence of the temperature of the lower air layer on the carbon dioxide concentration on the basis of radiative—convective model of the vertical distribution of temperature. Later, the dependence of the mean latitudinal temperature of the lower air layer on the carbon dioxide content was obtained using simplified energy-balance climate models (Budyko, 1974). Great opportunities for studying the climatic effects of carbon dioxide have appeared as a result of using general circulation models. Studies of this type have given information on the relationships between a number of the elements of the meteorological regime (air temperature, precipitation, evaporation, soil moisture, etc.) and the carbon dioxide concentration. In a number of studies, variations in the air temperature, preciptation and other meteorological elements with increasing carbon dioxide concentration have been calculated for various latitudes and longitudes, making it possible to construct maps of the indicated variations.

The second method for evaluating the climatic effects of carbon dioxide is based on the results described in paragraph 2 of this chapter, i.e. on the considerable excess of atmospheric carbon dioxide concentration in the geological past compared with the present. For example, in the Neogene (2—25 million years ago), the carbon dioxide concentration was 2—4 times higher than at present. Since other climate-forming factors in the Neogene (the structure of the Earth's surface, the astronomical solar constant) differed comparatively little from those in modern conditions, it might be thought that the difference between the Neogene and present-day climates depends mainly on fluctuations in carbon dioxide content. This approach has been used in constructing maps of variations in the air temperature, precipitation and other climatic elements due to a doubling of carbon dioxide concentration for the greater part of the U.S.S.R. (Budyko *et al.*, 1978). These maps are based on Sinitsyn's palaeoclimatic data. Later, to clear up the relationship between climate and the carbon dioxide content of the atmosphere, special studies concerning climatic conditions in the Neogene have been carried out. The information from these studies has been compared with model-based calculations of $CO_2$-induced variations in air temperature and precipitation (Budyko and Yefimova, 1981; *Climatic Effects* . . . , 1982). The agreement between these two independent estimates of variations in the elements of the meteorological regime has confirmed the realistic character of these estimates.

The third method for determining the relationship between climate and carbon dioxide concentration is based on the data on present-day climatic change. Since the end of the 19th century, the carbon dioxide concentration in the atmosphere has increased by approximately 15—20% due to man's economic activity (ref. to Chapter 10). These calculations show that such a change in the chemical composition of the atmosphere should result in a rise of the mean surface air temperature by approximately 0.5 °C. In analysing the empirical data of the last century it turned out to be possible to distinguish air temperature variations due to a carbon dioxide increase from those caused by atmospheric transparency fluctuations (Budyko, 1977c; Vinnikov and Groisman, 1981). This made it possible to study, in detail, the relationship between the climate and the carbon dioxide concentration using the third independent method.

It should be mentioned that, while the first two methods can be used to evaluate climatic variations corresponding to an increase in the mean surface air temperature by several degrees, the third method allows us to determine only the variations in the meteorological regime whose absolute values are comparatively small and correspond to an increase in the global mean air temperature by no more than 0.5 °C from its mean value for the last century. It is evident that clarification is needed on the agreement between these variations and results calculated using the first two methods.

Since each of the methods for determining the climate sensitivity to carbon dioxide variations in the atmosphere is approximate, the possibility of applying three independent methods is of considerable importance in justifying any definite conclusions concerning this sensitivity. A summary of the basic data from each method, showing the air temperature increase due to a doubling of the carbon dioxide concentration compared to its preindustrial value, is given in Table 7.3.

TABLE 7.3

Air temperature variations with doubled carbon dioxide concentration

| Simplified climate models | General circulation models | Modern climatic variations | Climatic variations in geological past |
|---|---|---|---|
| 1. 2.4 °C | 8. 2.9 °C | 16. 3.3 °C | 18. 3.5 °C |
| 2. 2.5—3.5 °C | 9. 3.9 °C | 17. 2.0—3.0 °C | 19. 3.4 °C |
| 3. 2.0—3.2 °C | 10. 3.5 °C | | 20. 2.8 °C |
| 4. 3.3 °C | 11. 2.0 °C | | |
| 5. 4.4 °C | 12. 2.0 °C | | |
| 6. 3.2 °C | 13. 3.0 °C | | |
| 7. 3.5 °C | 14. 2.4 °C | | |
| | 15. 2.5 °C | | |

*Note*: 1, 8, 13 — Manabe and Wetherald, 1967, 1975, 1980; 3 — Augustsson and Ramanathan (1977); 4 — Ramanathan *et al.* (1979); 5 — Kondratiev and Moskalenko (1980); 6 — Hameed *et al.* (1980); 7 — Mokhov (1981); 9, 10 — Hansen *et al.* (1979); 11 — Manabe and Stouffer (1980); 12 — Dymnikov *et al.* (1980); 14 — Wetherald and Manabe (1981); 15 — Bryan *et al.* (1982); 2, 16, 18, 19 — Budyko (1974, 1977c, 1979, 1980); 17 — Vinnikov and Groisman (1981); 20 — *Climatic Effects*. . . (1982).

The tabulated results of model calculations of the parameter defining the climate sensitivity to carbon dioxide varitaions ($\Delta T_c$) are divided into two groups: the first includes the results based on simplified climate models, the second includes those based on general circulation models. The values of $\Delta T_c$ in the table are not quite homogeneous. In determining some of them, data for the Earth as a whole or for the Northern Hemisphere alone have been used (estimates 1, 3—7, 16, 17); several estimates have been obtained from temperature variations averaged over latitude (2, 8, 12); and others have been based on data on the spatial distribution of temperature variations for the Northern Hemisphere (or its greater part, or for the whole planet with real or schematic topography: 9—11, 13—15, 18—20). In some cases calculations based on climate models were carried out considering the annual variations of meteorological elements (estimates 11 and 14); in other cases the calculations were based on the values of meteorological elements averaged over the year. Some of the theoretical calculations take account of the feedback between the thermal regime and the albedo of the Earth's surface which reflects variations in the area of the polar sea ice (estimate 2, 8, etc.); other calculations do not consider this feedback. It should be pointed out that all the feedbacks are automatically taken into account in the empirical estimates of the parameter $\Delta T_c$.

The majority of $\Delta T_c$ estimates in Table 7.3 were obtained directly for a doubled carbon dioxide concentration. Certain estimates from the corresponding studies for other values of carbon dioxide increase have been re-calculated for a doubled carbon dioxide concentration assuming that the dependence of temperature difference on the carbon dioxide concentration is logarithmic.

According to the data in Table 7.3, the mean value of $\Delta T_c$ is equal to 3.0 ± 0.5 °C. The same value of $\Delta T_c$ is considered the most reliable by the commissions of the U.S. National Academy of Sciences (*Carbon Dioxide and Climate*, 1979, 1982).

It should be mentioned that the results of calculations based on the most detailed climate models give us slightly smaller values of $\Delta T_c$, in the range of 2.0—2.4 °C (estimates 11 and 14), while more schematic models of general circulation yield about 3 °C (estimates 8 and 13). In this connection, it is possible that more detailed climate models, which require additional parametrization whose accuracy is limited, do not necessarily increase the reliability of the global estimates of climate sensitivity to variations in climate-forming factors.

The heat income absorbed by the climate system can also vary even with unvarying values of the astronomical and meteorological solar constants. In particular, variations in ocean area influence the Earth's albedo. Albedo decreases as the area of water surface increases, other things being equal. This effect can be taken into account by using the formulas for determining the Earth—atmosphere system albedo.

Noticeable changes in heat income can occur with variations in snow and ice cover, since these also affect the Earth—atmosphere system albedo. These changes, however, are not an independent climate-forming factor, they represent the albedo—surface air temperature feedback considered in many contemporary climate theories.

A redistribution of heat income within the Earth—atmosphere system can be accomplished by changing the marine current system through changes in continental and sea floor topography. Although the question of climatic effects of these factors has been discussed in a number of studies it cannot be considered well understood. It is possible that the change in heat transfer by currents influences only, other things being equal, the thermal regime of individual regions and does not modify the global mean air temperature (Budyko, 1980).

In studying the effects of carbon dioxide on climate, mean latitudinal air temperature variations may be estimated by methods of climate theory or by empirical data. Both methods reach the conclusion that these variations usually increase in high latitudes. In Fig. 7.4 quantitative data on the latitudinal dependence of $CO_2$-induced temperature change are presented for mean annual conditions of the Northern Hemisphere. Curves 1—3 are based on calculations using general circulation models (1 — Manabe and Stouffer (1980); 2 — Wetherald and Manabe (1981); 3 — Manabe and Wetherald (1980); curve 4 is based on palaeoclimatic data (*Climatic Effects* . . . , 1982); 5 — is based on data on modern climatic change (Vinnikov and Groisman, 1982). Similar curves for January (a) and July (b) are shown in Fig. 7.5, the palaeoclimatic data being depicted for the latitudes south of 70°N. Mean temperature distributions based on all above data are displayed in Fig. 7.6.

As seen from Fig. 7.6, in high latitudes the carbon dioxide-induced warming is twice the global mean warming and four times more than the low latitude warming for annual average conditions. In winter the temperature differences are still greater.

From data on Figs. 7.4—7.6 the conclusion can be drawn that all the methods for determining the relative temperature changes give more or less similar results for the latitudes south of 60°N. For higher latitudes the empirical methods yield a stronger sensitivity of the thermal regime to increasing carbon dioxide concentration compared with the results obtained from climate models. This difference may reflect a strong dependence of the model results for high latitudes on the parametrization of the genesis of polar sea ice which may influence the predicted air temperature considerably. This parametrization in the available models is probably not wholly satisfactory, as is indicated by, for example, the incomplete agreement between theoretical and empirical estimates of the modern sea ice regime. In this connection curves 4 and 5 are believed to show more realistic estimates of temperature variations for high latitudes compared with the results of climate models.

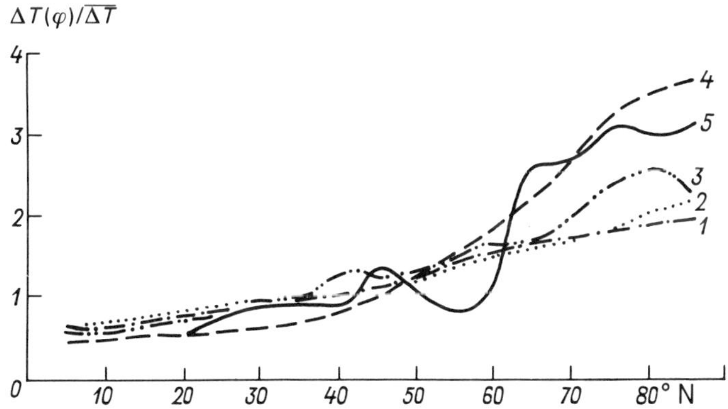

Fig. 7.4.

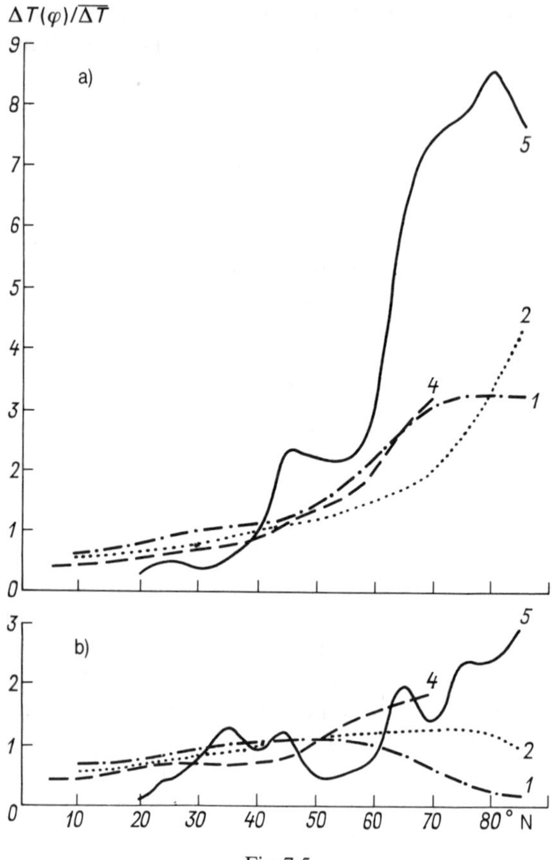


Fig. 7.5.

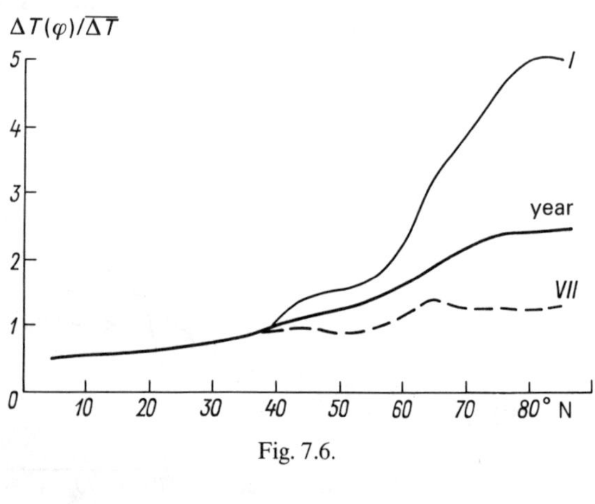


Fig. 7.6.

Figs. 7.4—7.6. $\Delta T(\varphi)/\overline{\Delta T}$ is a relative change in the global mean air temperature.

Besides data on variations in the mean latitudinal air temperatures, we can determine, by applying the above-mentioned methods, the spatial patterns of temperature changes in the lower layers of the atmosphere due to increased carbon dioxide concentration. For example, by using general circulation models global maps have been constructed showing the temperature differences for doubled and quadrupled carbon dioxide concentrations. Maps of temperature variations due to a doubling of carbon dioxide concentration have also been compiled using data from palaeotemperature records and data on the relatively small warming in modern epoch.

### 7.4.3. THE WATER CYCLE EFFECTS OF TEMPERATURE VARIATIONS

Any rise or fall in global mean air temperature will result in a change of precipitation pattern. To estimate this relationship over the entire Earth, the following considerations may be used. We assume that evaporation from the Earth's surface is

$$E = A(q_s - q), \tag{7.3}$$

where $A$ is the coefficient of proportionality, $q_s$ the specific humidity of air, saturated by water vapour at the temperature $\theta_w$ of the evaporating surface; and $q$ the specific air humidity at the level of meteorological observations. This formula is accurate for evaporation from the water surface which makes up more than 85% of the total evaporation from the entire Earth's surface, and it is approximate for evaporation from a moist land surface, comprising most of the difference between total evaporation and evaporation from the water surface.

Since relative humidity varies comparatively little over the oceans, let the value of $1 - q/q_s = \beta$ be assumed constant. In this case, we find the ratio of $\partial E/\partial \theta$ (the derivative of evaporation with respect to air temperature) and $E$ (evaporation):

$$\frac{1}{E}\frac{\partial E}{\partial \theta} = \frac{1}{q_s}\frac{\partial q_s}{\partial \theta_w}\frac{\partial \theta_w}{\partial \theta}. \tag{7.4}$$

In order to determine the value of $\partial \theta_w/\partial \theta$ we use the heat balance equation of the Earth's surface for the whole Earth,

$$R = LE + P, \tag{7.5}$$

where $R$ is the radiative balance, $L$ the latent heat of evaporation, and $P$ the turbulent heat flux from the Earth's surface to the atmosphere. This equation can be rewritten as

$$R = LA(q_s - q) + c_pA(\theta_w - \theta), \tag{7.6}$$

where $c_p$ is the specific heat of the atmosphere at constant pressure. Differentiating (7.6) with respect to $\theta$, we find that

$$\frac{\partial \theta_w}{\partial \theta} = \frac{1}{1 + (L/c_p)\beta(\partial q_s/\partial \theta_w)}. \tag{7.7}$$

Here, we have assumed that, while temperature varies, the radiative balance of the Earth's surface remains constant (actually, it varies slightly).

From (7.4) and (7.7) we have

$$\frac{1}{E}\frac{\partial E}{\partial \theta} = \frac{\frac{1}{q_s}\frac{\partial q_s}{\partial \theta_w}}{1 + \frac{L}{c_p}\beta\frac{\partial q_s}{\partial \theta_w}}. \tag{7.8}$$

Using (7.8) we can calculate the dependence of evaporation on global mean air temperature. From observational data we assume $\beta = 0.2$, and, using (7.8), find that a 1 °C increase in mean air temperature causes a 4% increase in evaporation. Because the total global precipitation is equal to the total evaporation, this dependence also characterizes the relationship between the total precipitation and air temperature.

The dependence derived here can be compared with the results obtained by Wetherald and Manabe (1975). They used a numerical climate model that considered the atmospheric general circulation. This approach enabled them to avoid some simplifications used in the above calculation. In particular, Wetherald and Manabe did not assume the relative humidity to be constant when temperature varied. They found that global evaporation (and precipitation) changes by 3% when temperature changes by 1 °C. This result does not differ greatly from that obtained using the simplified calculation.

It should be noted that the assumption that $\beta$ is independent of temperature over the continents is not accurate, because the potential evaporation from land increases with a rise in temperature. In this case, if the inflow of moisture from the oceans remains constant, the mean moisture content of the upper soil layer decreases with the increase of evaporation. This results in a reduction of relative humidity. As indicated above, warming leads to decreased horizontal transfer of water vapour in middle latitudes, an effect which would amplify the above-mentioned process. Warming, therefore, would result in reduced precipitation over a considerable part of the continents. Brückner (1901) was the first to detect this effect. Further studies have found a negative correlation between temperature and precipitation in the warm half-year for the Eurasian continent (Drozdov and Grigorieva, 1963) and in various seasons over the greater part of the U.S. territory (CIAP, 1975).

Various methods can be used for determining the effects of carbon dioxide concentration change on the latitudinal precipitation distribution over the continents. Figure 7.7 shows the distribution of difference in annual precipitation sums in various latitudinal zones of the continents for a doubled

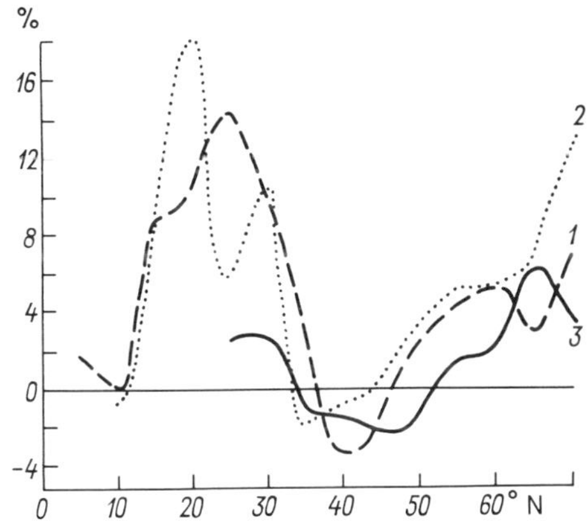


Fig. 7.7.

carbon dioxide concentration above its initial value close to 0.03%. Curve 1 in Fig. 7.7 represents the results obtained by the atmospheric general circulation model (Manabe and Stouffer, 1980), curve 2 is based on palaeoclimatic evidence (*Climatic Effects* . . . , 1982), and cruve 3 on the empirical data on modern changes in climate (Vinnikov and Groisman, 1982). The values of precipitation variations taken from these studies are reduced to a 1 °C increase in the mean global surface air temperature. Good agreement of all curves in Fig. 7.7, indicating the reality of the results obtained, deserves our attention.

As seen from this figure, global warming causes a considerable increase in the average precipitation amount in the 10—30°N zone and north of 50°N. In the 30—50°N zone the amount of precipitation tended to decrease. Of significant importance is the question concerning the distribution of changes in sums of precipitation on the continents for specific geographical regions. At present such data have been obtained in the above-mentioned palaeoclimatic studies and in investigations of modern climatic changes. For further study of this problem it would be desirable to obtain such data using climate models.

### 7.4.4. CLIMATIC CHANGES IN PRECAMBRIAN TIME

In considering the question of the causes of climatic changes during various epochs of the geological past let us dwell, first, on the climate genesis of Precambrian time.

As noted in the preceding section, although there are not much data on Precambrian climate, there are grounds to believe that, throughout the late Precambrian, the thermal regime differed only slightly from the modern one. In the mid-Precambrian the climate may have been a little warmer than today. Climatic conditions in the Precambrian were considerably affected by variations in the solar constant which at that time was smaller than at present. Schwarzschild (1958) has drawn the conclusion that, during the evolution of the Sun, its diameter decreased and its luminosity increased by 60% over the period of Earth's existence. In later studies the increase in the Sun's luminosity was found to be smaller. Now this value is usually assumed to grow by approximately 5% per billion years (Newman and Rood, 1977).

As indicated above, the mean surface air temperature increases by approximately 1.4 °C with a 1% rise in the solar constant. This means that, other things being equal, a billion years ago the mean temperature should have been smaller than the modern one by 7 °C, and two billion years ago by 14 °C. Since semi-empirical climate theory shows that a 5—10% decrease in the solar constant should cause glaciation of the entire Earth, it is obvious that the assumption of 'other equal things' for Precambrian atmosphere is erroneous.

The question of causes of far higher temperatures during the Precambrian compared with those obtained in the calculations mentioned here, has been discussed repeatedly. Sagan and Mullen (1972) proposed that a warm climate in Precambrian time can be attributed to the presence of ammonia in the ancient atmosphere which intensified the greenhouse effect. This hypothesis is disputable, since not all researchers believe that it is possible to have considerable amounts of such an unstable substance as ammonia even in the oxygen-free atmosphere of the early Precambrian. The presence of noticeable amounts of ammonia in the oxygen atmosphere that emerged long before the onset of the Precambrian is still less probable.

Warm climatic conditions of the remote past can be explained by a greater amount of carbon dioxide in the atmosphere compared with the modern epoch (Budyko, 1974) and by a smaller value of the Earth's albedo which could have been close to the albedo of the Moon, approximately equal to 0.07.

Chapter 3 is concerned with calculations of the mean surface air temperature for conditions when the atmosphere consists of gases that are practically transparent to shortwave and longwave radiation. A calculation using Stefan—Bolzmann's formula shows that this temperature is 3 °C if albedo equals 0.07 with the modern value of solar constant.

Assuming that 4.5 billion years ago the solar constant was 22.5% less than at present, we find that the mean Earth's surface temperature in this case must have been close to −28 °C. If the Earth was not heated at that time by internal thermal sources, such conditions may have prevailed when the primary atmosphere existed. Later, in degassing the Earth's crust, the atmosphere was enriched in water vapour and carbon dioxide, which led to a considerable increase in its temperature. Then, the growth of temperature fell somewhat because of increasing albedo, the value of which gradually approached its modern magnitude.

One may believe that, even in the early Precambrian after the appearance of an atmosphere consisting mainly of carbon dioxide and water vapour, the formation of water bodies, where carbonate sediments were later deposited, began due to water vapour condensation.

As the rate of income of gases from the Earth's crust to the atmosphere decreased, the loss of carbon dioxide due to the formation of carbonate sediments exceeded its output due to degassing of the Earth's crust. This resulted in a reduction of atmospheric carbon dioxide content. As the findings presented in Section 2 of this chapter show, carbon dioxide concentration in the early Cambrian amounts to several tenths of a per cent of the atmosphere volume, i.e. it was noticeably less than the carbon dioxide quantity existing in the atmosphere billions of years ago. Throughout the Phanerozoic, atmospheric carbon dioxide concentration continued to decrease.

Undoubtedly, the decrease in carbon dioxide concentration in the Precambrian atmosphere was uneven. As seen from the data for Phanerozoic time treated in Section 7.2, volcanic activity varies over wide limits, and the carbon dioxide concentration increases when it intensifies and decreases when it reduces. Sharp reductions in carbon dioxide concentration seem to occur during the Precambrian glaciations mentioned in the preceding section. The onset of glaciations decreased the loss of carbon dioxide due to carbonate sedimentation because of the freezing of some water bodies, reduced weathering of the rocks covered with ice, etc. This first retarded a reduction in atmospheric carbon dioxide concentration and then led to its increase. It might be thought that in the Precambrian, the periods of lowered volcanic activity changed into maximum far greater than those in the Phanerozoic. These maxima were accompanied with a drastic carbon dioxide increase and (in the second half of Precambrian time) by some elevation in oxygen content of the atmosphere due to the growth in productivity of autotrophic plants. The maxima of carbon dioxide mass in the Precambrian atmosphere were higher than those in the Phanerozoic, whereas the increase in the amount of oxygen was considerably smaller because of insufficient productivity of primitive photoautotrophs. Nevertheless, an increased content of oxygen in the Upper Precambrian should have exerted a pronounced effect on living organisms, in particular, on the emergence at that time of multicellular organisms.

### 7.4.5. CLIMATIC CHANGES IN THE PHANEROZOIC

The mean air temperature near the Earth's surface during Phanerozoic time depended mainly on variations in atmospheric carbon dioxide concentration, on the growth of the solar constant value, and on fluctuations in the Earth's albedo. We now discuss the results of calculations of the effects of these factors on the mean surface air temperature (Budyko, 1981). In this calculation it has been assumed that the temperature increases by 2.5 °C with a doubling of the carbon dioxide concentration. We also assume that the temperature changes logarithmically with increasing carbon dioxide concentration.

Table 7.4 shows the differences of mean surface air temperatures between the past and present ($\Delta T$) pertaining to various periods of the Phanerozoic. In calculating these differences, the dependence of $\Delta T$ on changes in atmospheric carbon dioxide mass, on a rise in solar radiation and on fluctua-

TABLE 7.4

Variations in mean air temperature compared with the contemporary epoch

| Time interval | Absolute age (million years) | $CO_2$ (%) | $-\Delta S$ | $-\Delta\alpha$ | $\Delta T$ (°C) |
|---|---|---|---|---|---|
| 1 | 2 | 3 | 4 | 5 | 6 |
| Lower Cambrian | 545—570 | 0.27 | 0.028 | 0.018 | 7.5 |
| Middle Cambrian | 515—545 | 0.19 | 0.026 | 0.015 | 6.2 |
| Upper Cambrian | 490—515 | 0.19 | 0.025 | 0.015 | 6.2 |
| Ordovician | 435—490 | 0.16 | 0.023 | 0.017 | 6.4 |
| Silurian | 400—435 | 0.13 | 0.021 | 0.015 | 5.4 |
| Lower Devonian | 376—400 | 0.14 | 0.019 | 0.016 | 6.1 |
| Middle Devonian | 360—376 | 0.40 | 0.018 | 0.015 | 9.9 |
| Upper Devonian | 345—360 | 0.41 | 0.018 | 0.015 | 10.1 |
| Lower Carboniferous | 320—345 | 0.42 | 0.016 | 0.016 | 10.6 |
| Mid-Upper Carboniferous | 280—320 | 0.18 | 0.015 | 0.010 | 6.5 |
| Lower Permian | 255—280 | 0.37 | 0.013 | 0.009 | 9.2 |
| Upper Permian | 235—255 | 0.14 | 0.012 | 0.016 | 7.1 |
| Lower Triassic | 220—235 | 0.11 | 0.011 | 0.015 | 6.2 |
| Middle Triassic | 210—220 | 0.20 | 0.011 | 0.014 | 8.2 |
| Upper Triassic | 185—210 | 0.17 | 0.010 | 0.014 | 7.7 |
| Lower Jurassic | 168—185 | 0.19 | 0.009 | 0.015 | 8.3 |
| Middle Jurassic | 153—168 | 0.23 | 0.008 | 0.016 | 9.3 |
| Upper Jurassic | 132—153 | 0.24 | 0.007 | 0.015 | 9.5 |
| Lower Cretaceous | 100—132 | 0.19 | 0.006 | 0.015 | 8.8 |
| Upper Cretaceous | 66—100 | 0.26 | 0.004 | 0.016 | 10.2 |
| Palaeocene | 58—66 | 0.12 | 0.003 | 0.015 | 7.6 |
| Eocene | 37—58 | 0.19 | 0.002 | 0.014 | 9.1 |
| Oligocene | 25—37 | 0.055 | 0.002 | 0.014 | 4.5 |
| Miocene | 9—25 | 0.11 | 0.001 | 0.012 | 7.0 |
| Pliocene | 2—9 | 0.055 | 0 | 0.008 | 3.6 |

tions in the Earth's albedo have all been taken into account. The values of carbon dioxide concentration, the differences between the past and modern solar radiation ($\Delta S$) expressed in fractions of the modern solar constant, the differences of the Earth's albedo $\Delta\alpha$ expressed in fractions of unity, are also presented in Table 7.4. In determining the value of $\Delta\alpha$ the relationship between albedo and variations in ocean area, and the effects on albedo of ice cover during the Permian-Carboniferous and Neogene glaciations, have been considered. The value of $\Delta T$ has been assumed to be equal to the sum of the three temperature differences depending on variations in carbon dioxide concentration, in solar radiation and albedo.

From the data in the Table 7.4 it follows that, throughout the Phanerozoic up to the Eocene inclusive, the mean surface air temperature was 5.4—10.6 °C higher than at present, i.e. it ranged between 20.5—25.5 °C. A noticeable decrease in temperature occurred in the Oligocene, the Pliocene, and at the Pliocene-modern epoch boundary, in the Quaternary period.

The temperature variations shown in Table 7.4 can be compared with data from many empirical studies of the past climates. For the late Mesozoic and for the Cenozoic similar comparisons have already been carried out for temperature calculations which took account of fluctuations in carbon dioxide concentration (Budyko, 1980) and in the Earth's albedo (Budyko and Yefimova, 1981). These showed a close correlation with mean air temperatures in the geological past obtained by independent methods. Table 7.5 shows the results of a comparison of the $\Delta T$ values from Table 7.4 with similar values of $\Delta T'$ derived from Sinitsyn's data. This comparison has taken into account differences in the geochronological scales used in compiling Table 7.4 and in Sinitsyn's study.

TABLE 7.5

Variations in the mean air temperature compared with the present epoch

| | Pliocene | Miocene | Palaeocene—Eocene | Cretaceous |
|---|---|---|---|---|
| $\Delta T$ (°C) | 3.6 | 7.0 | 8.4 | 9.5 |
| $\Delta T'$ (°C) | 4.8 | 6.0 | 8.2 | 11.0 |

As seen from Table 7.5, the mean discrepancy between the values of $\Delta T$ and $\Delta T'$ for the indicated time intervals does not exceed 1 °C.

The data on variations in $\Delta T$ with time given in Table 7.4 can be compared with the mean palaeotemperatures obtained from data on various geographical zones. The results of this comparison for the Cenozoic era and the Upper Cretaceous are displayed in Fig. 7.8.

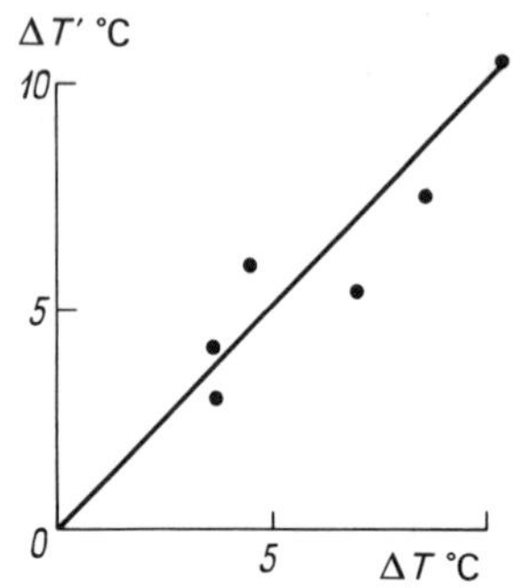


Fig. 7.8. $\Delta T$, empirical data; $\Delta T'$, calculated results.

Using data from empirical studies of climatic changes for the whole Phanerozoic obtained in studies of Yasamanov (1980, etc.), it can be found that, as a rule, they correlate well quantitatively with the data in Table 7.4. The above materials confirm that variations in the mean air temperature in the Phanerozoic depended mainly on variations in carbon dioxide concentration and, to a lesser degree, on changes in solar radiation and the Earth's albedo.

As indicated above, with increasing carbon dioxide concentration the amount of precipitation over the Earth's surface rises. This is attributed mainly to increasing mean surface temperature which in turn determines the saturation specific humidity. With rising carbon dioxide concentration mean air temperature also increases. Since relative humidity usually changes rather little, the warming results in increasing mean absolute humidity. Consequently the amount of water vapour transferred by vertical fluxes into upper troposphere layers rises, leading to increasing precipitation.

At the same time, in individual regions of the Earth the precipitation amount can increase or decrease with increasing carbon dioxide concentration depending on variations in the atmospheric circulation and on other factors. In calculating precipitation patterns, as well as in determining variations of air temperature due to a considerable increase in atmospheric carbon dioxide concentration, we may use both detailed climate models and palaeoclimatic data from epochs with a higher atmospheric carbon dioxide content.

The results of calculating the change in latitudinal mean annual precipitation over the Northern Hemisphere continents due to a doubling of carbon dioxide concentration over the pre-industrial level are shown in Fig. 7.9. Curve 1 represents the data of Manabe and Wetherald (1980). Curve 2 is based on the data of Sinitsyn for Eurasia. Curve 3 gives the mean latitudinal values derived from the data of Muratova and Suyetova who, using palaeobotanical evidence, calculated the annual sums of precipitation in ten regions in middle latitudes of Eurasia and North America for the Middle Pliocene.

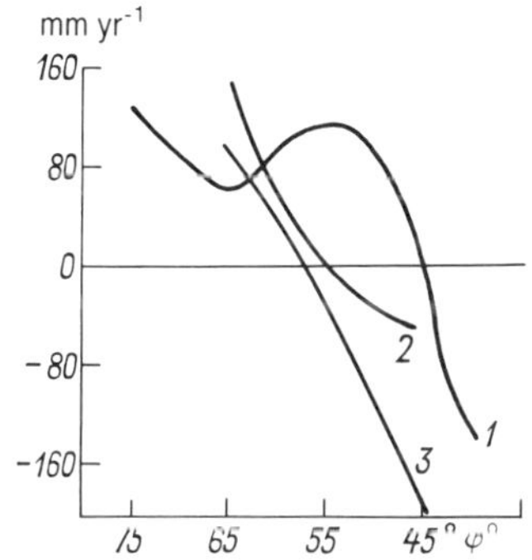


Fig. 7.9.

As seen from Fig. 7.9, the results of these three studies agree satisfactorily. They all indicate an increase in total precipitation in higher latitudes and a decrease in lower latitudes. These findings confirm the data presented in Fig. 7.7.

The cause of decreasing precipitation in the 40—50°N zone seems to be associated with a movement to the north of the high-pressure subtropical belt over which little precipitation is observed. In higher latitudes precipitation increases due to the greater moisture content of the warmer atmospheric air.

Variations in total precipitation over the continents during the Neogene depended mainly on the character of atmospheric processes. Before the Neogene, the location of drifting continents relative to the equator and the poles was of considerable importance in determining the moisture regime over the continents since these locations determined their position in relation to the stable high- and low-pressure zones. Apparently, epochs existed (for example, in the Permian and Triassic) when a considerable portion of the continental surface was in the high pressure zone, thus causing more arid conditions on the continents.

In summary, we consider the question of the relationship between large glaciations and the thermal regime of the entire Earth. As seen from Table 7.4, first in the Oligocene and then in the Pliocene, the mean air temperature was lower than during all the preceding epochs of the Phanerozoic. At present it is still lower. This agrees with the fact of the development of Cenozoic glaciation first in the Southern Hemisphere and then in high (and, from time to time, in middle) latitudes of both hemispheres.

The second largest glaciation of Phanerozoic time occurred in the late Carboniferous and early Permian periods. From Table 7.4 it is clear that, in the Middle—Upper Carboniferous, the temperature was considerably lowered compared with the Lower Carboniferous, whereas in the Lower Permian it was rather high. The agreement between variations in the mean temperature and the formation of another Palaeozoic glaciation which occurred in the late Ordovician—early Silurian is somewhat better. From Table 7.4 one can see that the air temperature in the Ordovician was not very high, and in the Silurian temperatures were the lowest over the entire Phanerozoic up to the Oligocene.

In comparing data on mean air temperatures from Table 7.4 with the history of past glacial epochs one should take into account three circumstances. First, the reliability of the data in Table 7.4 reduces as the distance between the epochs under consideration and the present time grows. Therefore, the data for the Palaeozoic are less accurate than for the Cenozoic. Second, temperature values for comparatively short time intervals are available only for the Cenozoic. The temperature values for preceding epochs refer to time intervals of tens of millions of years. Such intervals are longer than the time of the corresponding glacial epochs. In this connection, for example, the temperature during time of the early Permian glaciation could be considerably lower than the mean temperature for the entire Lower Permian period whose value is all that the resolution of Table 7.4 can give. Third, data for the Miocene show that a considerable (although not maximal in area) glaciation can exist even with a relatively high global mean air temperature. The cause of this is clear: such a glaciation appears if a continent is located in high latitudes in one of the hemispheres provided the atmospheric carbon dioxide concentration is not too high. From calculations using semi-empirical climate models it follows that, in this case, the temperature in the glaciation zone was further decreased because of the influence of ice-albedo feedback. Low temperatures can thereby be maintained in ice-covered regions even with warm to very warm climate in all other regions of the globe.

These considerations explain only the most general features of climate genesis for the different epochs of the Phanerozoic. For a more detailed study of this question it is desirable to use atmospheric general circulation models taking into account the real topography and other climate-forming factors over different periods of the Phanerozoic.

### 7.4.6. CLIMATIC CHANGES IN THE QUATERNARY PERIOD

The Quaternary period was preceded by a prolonged evolution of the climate towards a more pronounced thermal zonality resulting from changes in the chemical composition of the atmosphere and changes in the Earth's surface structure. This was largely expressed in a continual decline in air temperature at middle and high latitudes.

During the Neogene, climatic conditions were influenced by a reduction in atmospheric carbon dioxide concentration. This led to a decline in global average temperature by several degrees. Subsequently, polar ice caps extended and their development resulted in a further reduction of the global average temperature, particularly in temperature at high latitudes.

The appearance of polar ice caps sharply increased the sensitivity of the thermal regime to small changes in climate-forming factors. Very large oscillations in the boundaries of snow and ice cover on land and in the oceans as a result of Earth orbital changes therefore became possible. Before polar ice caps appeared, this factor had not influenced the climate substantially.

A continuing reduction in atmospheric carbon dioxide concentration contributed to greater advances of later glaciations compared with the earlier ones, even though the major influence on their scope was the combination of astronomical factors which determine the location of the Earth's surface in relation to the Sun. These factors include, as mentioned above, the eccentricity of the Earth's orbit, the inclination of the Earth's axis in relation to the plane of its orbit, and the time of equinoxes. These astronomical factors change periodically and produce changes in the quantity of radiation reaching diffrent latitudes in various seasons. The characteristic time periods over which these changes occur extend to tens of thousands of years. In order to verify the influence of astronomical factors on Quaternary climate, experiments with numerical models that calculate the positions of individual ice sheets, as influenced by external climate-forming factors, are required.

To clarify this question, use was made of the semi-empirical model of the thermal regime for annual average conditions (Budyko, 1968). This has shown that changes in the radiation regime during the last (Würmian) glaciation could have led to a southward shift in the ice cover in the Northern Hemisphere by approximately one degree latitude. This is substantially less than the shift in the ice cover that did, in fact, take place. The reason for this discrepancy is that a determination of annual average temperatures alone is insufficient to estimate the influence of orbital element changes on glaciation, since glacial regimes are largely influenced by the thermal conditions of warm seasons.

In a subsequent investigation (Budyko and Vasishcheva, 1971), studies of climatic conditions during glacial periods were based on a model describing the distribution of average latitudinal temperature in different seasons. Calculations were carried out for the position of the average boundary of polar ice during periods when astronomical factors resulted in a considerable reduction of radiation income during the warm season in high latitudes. Some of the resulting findings are presented in Table 7.6. These show that fluctuations in the radiation regime caused by changes in the position of the Earth's surface in relation to the Sun may lead to substantial changes in the climate. The corresponding calculations show that, at such times, global average temperature does not fluctuate much. Nevertheless, such modest fluctuations are accompanied with perceptible shifts in the ice margin.

The maximum changes in an average latitude reached by the Northern

Hemisphere ice cover obtained by these calculations accord well with empirical data. Thus, for example, Lamb (1964) noted that, at the height of glaciation, the average boundary of ice in the Northern Hemisphere attained 57°N latitude, a southward shift in that boundary by 15° in relation to current conditions. The corresponding value established through calculations is 12°, relatively close to the value noted by Lamb.

TABLE 7.6

Climatic change during glacial epochs

| Ice ages | $\Delta\varphi$ (°N) | $\Delta\varphi$ (°S) | $\Delta T$ (°C) |
|---|---|---|---|
| Würm III | 8 | 5 | −5.2 |
| Würm II | 10 | 3 | −5.9 |
| Würm I | 11 | 2 | −6.5 |
| Riss II | 11 | 0 | −6.4 |
| Riss I | 12 | −4 | −7.1 |

*Note* $\Delta\varphi$ (°N) = reduction in the average latitude of the boundary of polar ice in the Northern Hemisphere compared with its current position;
$\Delta\varphi$ (°S) = reduction in the average latitude of the boundary of polar ice in the Southern Hemisphere compared with its current position;
$\Delta T$ (°C) change in the average temperature in warm seasons at 65°N.

New empirical and theoretical investigations of Quaternary glaciations carried out in recent years have revealed information essential for understanding the physical mechanism of the ice ages. CLIMAP compiled extensive data on temperatures during the last glaciation, based on plankton distribution data in various ocean regions. These data have been used to construct world maps of the temperature of ocean surface waters. An important conclusion reached by these studies was that, during the glaciations, the mean temperature of the ice-free ocean surface was only 2 °C below the present (CLIMAP Project Members, 1976). This conclusion agrees reasonably well with the above results based on the semi-empirical theory of the atmospheric thermal regime.

According to empirical data, the temperature decrease was somewhat greater than the calculation results. This can be explained by an increase in the Earth's surface albedo due to the expansion of arid conditions on the continents during the glacial advance (Cess, 1978). At the same time, the development of ice cover seems to be accompanied with a decrease in carbon dioxide concentration, which also caused the air temperature to drop (see below).

Hays, Imbrie and Shackleton have compared temperature fluctuations for the last 450000 years derived from the composition of marine organic sediments in the Southern Hemisphere with the variations in radiation due to astronomical factors (Hays *et al.*, 1976). A detailed statistical analysis of temperature variations revealed clearly pronounced fluctuations with periods of 23000, 42000 and 100000 years. The first of these periods coincides with the oscillatory period for the precession of the equinoxes, the second corresponds to the oscillatory period for the inclination of the Earth's axis, and the third approaches the oscillatory period for the eccentricity of the Earth's orbit. Based on this agreement, the authors concluded that variations in the astronomical factors were the basic reasons for the successive Quaternary glaciations.

Considerable advances have recently been made in determining the chemical composition of air preserved as bubbles in the polar ice that formed during the last 30000 years. These studies have shown that the atmospheric carbon dioxide concentration varied noticeably. Such variations must have led to changes in the thermal regime. About 20000 years ago the carbon dioxide concentration was 0.02%. Later it increased considerably. Noting this, Schneider and Thompson (1981) have assumed that these changes in carbon dioxide concentration can contribute approximately half of the difference in the mean air temperature for glacial and postglacial epochs. It is possible that variations in carbon dioxide concentration are the result of one more positive feedback in the climate system, a feedback which reinforces the effects of comparatively small changes in solar radiation input studied by Milankovich. Further studies of this question require validation of the accuracy of the method for determining atmospheric chemical composition from bubbles in ice cores.

The studies by Gates (1976) and Manabe and Hahn (1977) should be mentioned among the theoretical investigations of ice age climates. These authors used general circulation models to study the atmospheric circulation during the last glaciation. Using the CLIMAP data on temperature distribution over the ocean, the authors were able to elucidate many features of the ice age climate. For instance, precipitation was found to decrease noticeably during glaciations leading to an expansion of continental aridity. This conclusion has been confirmed by an empirical study of Pleistocene conditions (Velichko, 1973).

Some authors have applied simplified climate theories to elucidate causes of the Quaternary glaciations, allowing them to avoid the use of empirical data for certain elements of the past climate. Berger (1973, 1975, 1977, 1978, etc.) carried out many investigations of this kind using a semi-empirical model of the atmospheric thermal regime and decided that astronomical factors were the main determining cause of Quaternary glaciations. Using various parametrized climate models, Suarez and Held (1976), Pollard (1978) and others have come to similar conclusions. Such studies have emphasized that a realistic description of Pleistocene glaciations can only be made with the help of climate models based on the feedback between polar ice and the thermal regime of the atmosphere. Schneider and Thompson (1979), for example, have shown that, if this feedback is neglected, temperature variations in high latitudes are underestimated in comparison with palaeoclimatic data.

Cess and Wronka (1979) have mentioned that, for the correct estimation of air temperature variations during glaciations, it is necessary to take into account a positive feedback between temperature fluctuations and the variations in the albedo of the continental surface determined by the state of the vegetation cover.

Research presenting new and more accurate calculations of radiation variations due to Earth orbital changes (Sharaf and Budnikova, 1969; Vernekar, 1972; Berger, 1973, 1976, etc.) was also of great significance in studying the origin of the ice ages.

It follows from the palaeogeographical data that, as a result of the cooling due to a decrease in carbon dioxide concentration, the polar ice cover first appeared in the middle of the Tertiary. Astronomical factors started to influence the climate significantly only after the development of large polar glaciations, since it is these that made Quaternary climatic conditions so sensitive to minor variations in the climate-forming factors. During earlier epochs, when polar glaciations were relatively small or absent, astronomical factors varied in a manner similar to that in the Quaternary, but their climatic effect was insignificant.

### 7.4.7. THE CLIMATIC EFFECTS OF VOLCANIC ERUPTIONS

In conclusion, let us dwell in more detail on the physical mechanism of climatic effects of volcanic eruptions. These effects are complex and sometimes ambiguous. As noted above, fluctuations in volcanic activity affected climate through variations in atmospheric carbon dioxide concentration caused by changes in the rate of carbon dioxide influx into the atmosphere.

At the same time, during the explosive volcanic eruptions sulphur-containing gases were ejected into the atmosphere producing aerosol particles — droplets of concentrated sulphuric acid, typically tenths of a micrometre in size. Over several weeks or months, stratospheric winds scattered these particles over a single hemisphere (if an eruption occurred in extratropical latitudes) or over the entire globe (when an eruption took place in the tropics). These particles of volcanogenic aerosol concentrate mainly in the lower stratosphere at heights of 15—30 km forming a stratospheric aerosol layer. The quantity of particles in this layer then gradually decreases, by a factor of 2 to 3 each year.

The idea that volcanogenic aerosols can affect the climate appeared in several early studies devoted to this problem. Savinov (1913), Kimball (1918) and others established that, after great explosive volcanic eruptions, an abrupt decrease in incoming solar radiation takes place over vast territories.

Figure 7.10 shows the change in the ratio of monthly mean direct radiation under a clear sky to its normal value after the eruption of the volcano Katmai in Alaska (1912). This curve has been constructed from observational data from several actinometric stations in Europe and North America. It shows that the direct radiation was reduced by more than 20% in some months (Budyko, 1971). It has been established that this decrease is caused by the spreading in the stratosphere of droplets of sulphuric acid that can remain there for up to several years producing a screen that decreases the incoming flux of solar radiation to the Earth's surface.

If the surface air temperature variations were dependent on fluctuations in direct radiation, each explosive volcanic eruption would result in a drastic cooling. As mentioned above, a 1% decrease in solar radiation causes approximately a 1.5 °C reduction in the mean surface air temperature. In this case a 20% decrease in solar radiation should lower the air temperature by 30 °C.

However, there are two physical mechanisms that attenuate the effects of direct solar radiation fluctuations on the surface air temperature. One of them is associated with the strengthening of the scattered radiation because of aerosol effects on direct radiation. Through this effect a reduction of the total solar radiation reaching the Earth's surface amounts only to one-sixth of the decrease in direct radiation. The other mechanism is the influence of the thermal inertia of the Earth—atmosphere system due, mainly, to the thermal inertia of the mixed layer of the ocean. Calculations show that this factor decreases the influence of attenuated radiation on the temperature after individual volcanic eruptions by approximately a factor of 10 (Budyko,

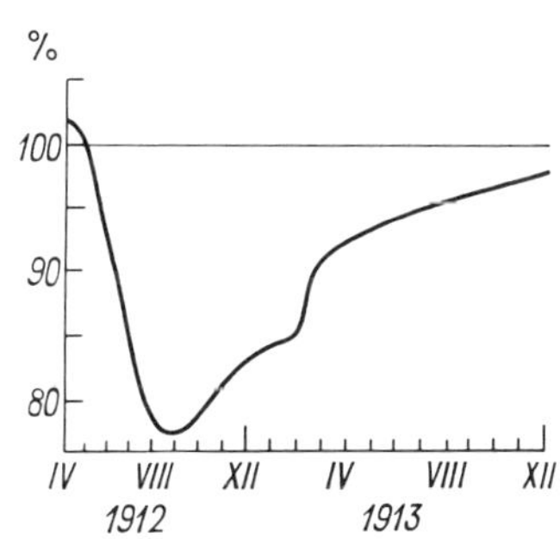


Fig. 7.10.

1971). Taking the influence of these two factors into account, the actual temperature drop due to a 20% decrease in the direct radiation flux is only about 0.5 °C. This value agrees well with data obtained at meteorological stations when observing the variations in mean air temperature after explosive volcanic eruptions (Taylor *et al.*, 1980).

Changes in the frequency of such eruptions exert a profound effect on modern climatic variations which are characterized by fluctuations in global mean temperature of several tenths of a degree. This is discussed in detail in the first section of Chapter 10.

It might be thought that, in the geological past, sets of strong explosive volcanic eruptions could lead to far larger cooling. During the last hundred years only about five strong explosive volcanic eruptions occurred. Undoubtedly, in the past, there were time intervals with far greater levels of volcanic activity. The drop in mean surface air temperature could reach several degrees if, due to the strengthening of volcanic activity caused by tectonic processes or due to chance coincidences, a set of ten or more successive explosive eruptions occurred during one or two decades. One of the reasons for this is a decrease in the effect of oceanic thermal inertia on the temperature regime for longer time scale fluctuations. If the thermal inertia effect is small, the above case could cause the temperature drop to reach about 5 °C. Since, with the growth of volcanic activity the mean intensity of explosive eruptions increases, the temperature lowering under such conditions could be still greater. One would expect that, over many millions of years, there ought to have been some periods when tens of volcanic explosions occurred in an interval of only a few decades. Such clusterings must have resulted in strong coolings. The time required to produce such a clustering (and consequent cooling) can be shown to be $10^7$—$10^8$ years (Budyko, 1971). The biospheric effects of such coolings are discussed in Section 2 of Chapter 8.

Note that, according to the point of view outlined above, a short-term increase in volcanic activity causes a cooling (due to the decrease in stratospheric transparency), whereas a long-term increase leads to warming due to the growth of atmospheric carbon dioxide concentration. Let us consider the question of the causes of such difference. We shall analyse first the relationship between surface air temperature and the stratospheric aerosol and carbon dioxide levels when the climatic system is in a steady state (i.e. when no climatic changes take place over a long time interval.)

From modern observational data one may evaluate the sensitivity of global mean temperature to fluctuations in volcanic activity. Over the last hundred years the range of variations in this temperature has been about 0.6 °C. During periods with maximum atmospheric transparency the amount of stratospheric aerosol was insignificant, so one may assume that a 0.3 °C temperature drop corresponds to the mean value of stratospheric aerosol mass over the last hundred years. Now, since the change in surface air temperature is proportional to the change in stratospheric aerosol concentration over a wide concentration range, we can use the formula:

$$\Delta T' = -0.3\,(\alpha' - 1), \tag{7.9}$$

where $\Delta T'$ (°C) is the change in mean temperature due to the departure of stratospheric aerosol mass from the mean and $\alpha'$ is the ratio of stratospheric aerosol amount to its modern average. The proportionality coefficient in (7.9) is based on the argument given above. Since this coefficient is based on temperature variations occurring over several decades, the effects of the thermal inertia of the climatic system on this coefficient are comparatively small.

At the same time the following formula can be used:

$$\Delta T'' = 3\,\frac{\ln \alpha''}{\ln 2}, \tag{7.10}$$

where $\Delta T''$ (°C) is the change of global mean surface air temperature due to any difference in carbon dioxide concentration from the modern mean and $\alpha''$ is the ratio of atmospheric carbon dioxide amount to this mean. Equation (7.10) is based on the logarithmic relationship between temperature and carbon dioxide concentration.

If the concentration ratio of sulphur-containing to carbon-containing gases in volcanic ejecta is assumed to be constant in time for mean global conditions, we may suppose that $\alpha' = \alpha'' = \alpha$.

It is obvious that the surface air temperature will be elevated if $\Delta T' + \Delta T'' > 0$ and decreased if $\Delta T' + \Delta T'' < 0$. The transition from heating to cooling occurs with $\alpha = \alpha_0$, when $\Delta T' + \Delta T'' = 0$. This condition occurs when

$$\frac{\ln \alpha_0}{\alpha_0 - 1} = \frac{\ln 2}{10}, \tag{7.11}$$

which shows that $\alpha_0$ is approximately 60, corresponding to an atmospheric $CO_2$ level of about 1.8%. Hence, under steady-state conditions with $\alpha < 60$, a high level of volcanic activity corresponds to a warmer climate and a low level to a cooler climate.

The possibility of changing the sign of this dependence by having $\alpha$ more than 60 is doubtful for a number of reasons. For such conditions, Equation (7.9) and the assumption that $\alpha' = \alpha''$ will not be accurate, since the assumed linear dependence of temperature change on the amount of gases forming the stratospheric aerosol will not hold. In this case the dependence will be weaker than direct proportionality. This is explained, firstly, by the limiting influence of coagulation processes on the growth of aerosol mass. Secondly, atmospheric optics shows that the attenuation of shortwave radiation flux is only proportional to aerosol concentration for low concentrations. Both of these factors tend to diminish the volcanic cooling effect.

It should also be borne in mind that, although the average level of volcanic activity was often higher than the modern one in the past, in the Phanerozoic, for long time intervals, it never increased by more than a factor of 20 or so. For example, from data on fluctuations in the quantity of volcanic rocks during the Phanerozoic it can be concluded that the amplitude of variations in the level of volcanism for individual geological periods or their epochs corresponds to a maximum value of $\alpha$ equal to 10. One must therefore conclude that, at a steady-state of the climate system, an increase in volcanic activity results in warming.

The problem under consideration can be better understood by the data obtained recently on the atmosphere and climate of Venus. Since the albedo of Venus is far greater than the Earth's, Venus, despite its smaller distance from the Sun, receives approximately the same quantity of absorbed radiation as the Earth. However, because of the strong greenhouse effect of the very dense atmosphere of Venus (consisting mainly of carbon dioxide), the temperature of the planet's surface reaches several hundred degrees Celsius. A continuous layer of thick sulfuric acid cloud exists in the high atmosphere of Venus, but this has clearly not compensated for the strong greenhouse heating of the lower atmosphere. This feature agrees with the afore-mentioned point of view about the relative roles of greenhouse gases and the aerosol layer.

Let us now consider the unsteady state of the climate system and estimate the climatic change arising from short-term fluctuations of volcanic activity during, for example, a year. As a result of the attenuation of solar radiation by the aerosol layer the temperature change at the end of the first year after the eruption is given by

$$\Delta T' = -0.3\,(\alpha' - 1)\beta, \tag{7.12}$$

where $\beta$ is a parameter describing the influence of the thermal inertia of the climate system ($\beta$ is close to zero for very short-term intervals and approaches the unity for long-term intervals).

An annual increase in the amount of atmospheric carbon dioxide will be equal to approximately half the carbon dioxide mass produced during the eruption (this inference is based on studies of anthropogenic changes in carbon dioxide which are discussed in Chapter 10). Thus, the amount of carbon dioxide will rise over the year from $c_0$ to $c = c_0 + \frac{1}{2}c_0\gamma(\alpha' - 1)$, where $\gamma$ is the ratio of the average carbon dioxide mass generated every year from degassing the upper mantle to the atmospheric quantity of carbon dioxide. Available data show that, in the modern epoch, the rate of influx of carbon dioxide into the atmosphere amounts to about $5 \times 10^{13}$ g yr$^{-1}$ (this value is derived by extrapolating data on the carbon dioxide influx at the late Cenozoic era). Considering the carbon dioxide mass in the atmosphere to be $2.6 \times 10^{18}$ g, $\gamma$ for one year equals about $2 \times 10^{-5}$. Now, since

$$\alpha'' = \frac{c}{c_0} = 1 + \tfrac{1}{2}\gamma(\alpha' - 1), \tag{7.13}$$

we find that

$$\Delta T'' = \frac{3\beta}{\ln 2} \ln [1 + \tfrac{1}{2}\gamma(\alpha' - 1)]. \qquad (7.14)$$

Using the above value of $\gamma$, from (7.12) and (7.14) we find that the ratio of $\Delta T''$ to the absolute value of $\Delta T'$ is $1.5 \times 10^{-4}$, i.e. a very small value.

Thus, the influence of carbon dioxide emissions on climate during an individual eruption is absolutely insignificant compared with the effects of the volcanogenic aerosol layer. The same conclusion can be drawn from (7.12)–(7.14) when estimating the climatic effect of increasing (or decreasing) volcanic activity over longer periods of time, for example, a hundred years. This conclusion confirms modern ideas on the effects of volcanic activity on the present-day, comparatively short-term natural fluctuations of climate.

For the long time intervals, however, Equation (7.13) is not accurate. For the periods of 1000 or more years, in calculating the growth of atmospheric carbon dioxide concentration it is desirable to take into account the rate of loss of carbon dioxide due to the formation of carbonate sediments.

To determine the time interval over which fluctuations in volcanic activity can result in warming, we may use the above method applied for calculating variations of carbon dioxide concentration during the geological past. The equation for the balance of atmospheric carbon dioxide has the form

$$\frac{dc}{dt} = A - B, \qquad (7.15)$$

where $dc/dt$ is the rate of change with time in carbon dioxide amount, $A$ and $B$ are the rates of gain and loss of carbon dioxide. Assuming that $B = \varepsilon c$ and letting the initial amount of carbon dioxide be $c_0$, we find from (7.15) that, at a steady state with $dc/dt = 0$,

$$A = B = \varepsilon c_0, \qquad (7.16)$$

and for an unsteady state

$$\frac{dc}{dt} = \alpha\varepsilon c_0 - \varepsilon c. \qquad (7.17)$$

From (7.17) we find that

$$\frac{c}{c_0} = \alpha - (\alpha - 1)\, e^{-\varepsilon t}. \qquad (7.18)$$

For sufficiently long time intervals the influence of the thermal inertia of the climate system is comparatively small (i.e. the parameter $\beta$ is close to unity) so we can assume that

$$\delta T' = -0.3\,(\alpha - 1) \quad \text{and} \quad \Delta T'' = \frac{3}{\ln 2} \ln \frac{c}{c_0}.$$

Now, the transition between cooling and warming occurs when $\Delta T'$ +

$\Delta T'' = 0$. From (7.18) we find, therefore, that

$$\varepsilon t_0 = \ln \frac{\alpha - 1}{\alpha - e^{0.1(\alpha - 1)\ln 2}} \tag{7.19}$$

a relation which defines the time interval, $t_0$, over which the warming influence of variations in carbon dioxide mass becomes equal to the cooling influence of stratospheric aerosol. For the time intervals greater than $t_0$ the increase in volcanic activity will yield a warming.

From (7.19) it follows that for the values of parameter $\alpha$ from 0 to 20 the value of $\varepsilon t_0$ ranges from 0.07 to 0.15. To determine the value of $t_0$ we need to estimate the parameter $\varepsilon$, which equals the ratio $B/c_0$. As indicated above, the value of $B$ for the modern epoch (derived by extrapolating data on the change of the expenditure of carbon dioxide on carbonate formation during the Neogene) is equal to about $5 \times 10^{13}$ g yr$^{-1}$. To determine $c_0$, we should take into account the fact that, for comparatively long time intervals, a change in atmospheric carbon dioxide concentration is accompanied with the change in carbon dioxide mass in the oceans. Since the amount of oceanic carbon dioxide involved in the global carbon cycle is difficult to estimate accurately, we will restrict ourselves to the case where $c_0$ is close to the present value, i.e. $2.6 \times 10^{18}$ g. In this case we find that $\varepsilon$ is about $2 \times 10^{-5}$ yr$^{-1}$. Using a representative mean value for $\varepsilon t_0$ of 0.10, we obtain $t_0$ of approximately 5000 years.

The actual value of $t_0$ is thought to be greater than this because we have neglected changes in the carbon dioxide content of the oceans. The error associated with this cannot, however, alter the conclusion that an increase in mean volcanic activity for geological periods and their epochs (i.e. for time intervals exceeding one million years) always produces a warming. Conversely, an increase in volcanic activity over time intervals less than 1000 years always results in coolings.

Thus, one can conclude that fluctuations in volcanic activity in the geological past had a considerable effect on the climate, the character of this effect depending on the duration of the fluctuations. For example, stronger volcanism over decades or centuries always resulted in cooling. This inference agrees with the results of a number of empirical studies.

To further clarify the influence of volcanism on temperature variations, Lamb's data cited in this study (1970) are of importance. In this study Lamb compared the temperature anomalies for various time intervals with an index characterizing a mean lowering of atmospheric transparency after the eruptions (the dust veil index). This index is expressed in relative units determining the atmospheric transparency effects of volcanic eruptions in relation to the eruption of Krakatau in 1883. A correlation coefficient of −0.94 has been obtained between decadal temperature anomalies for the Northern Hemisphere averaged over the last hundred years and the value of the index. This proves the close relationship between volcanic activity and temperature variations. For the last 200 years Lamb compared variations in the amount of sea ice near Iceland with the dust veil index calculated for ten-year time intervals. The correlation coefficient for these two quantities was 0.61, quite high considering the very approximate characteristics of the process in question.

For longer time intervals the volcano-climate question has been studied using data on the chemical composition of ice layers in ice sheets of Greenland and the Antarctic. In this case the amount of sulphur compounds was determined in ice layers of various age, and this amount is assumed to reflect the stratospheric volcanogenic aerosol loading consisting mainly, as noted above, of sulphuric acid droplets. The data thus obtained has been repeatedly compared with temperature variations. For example, using Greenland ice core data (Hammer *et al.*, 1980) spanning roughly the past 1500 years it proved possible to relate variations in volcanic activity with temperature fluctuations in California found from dendroclimatological data (i.e. from tree-ring width). It has been found that an increase in volcanic activity causes lower air temperature over vast areas. This empirical result confirms the ideas outlined above.

Since modern climatic change depends not only on natural fluctuations of atmospheric transparency, but also on anthropogenic factors, the question of these changes is treated in more detail in Chapter 10.

CHAPTER 8

# THE EVOLUTION OF ORGANISMS

## 8.1. Abiotic Factors of Evolution

### 8.1.1. STUDIES OF THE HISTORY OF ORGANISMS

Palaeontological studies have gathered an enormous body of information illuminating the history of the development of animals and plants for many millions of years. This information constitutes the basis of the palaeontological record, revealing a general picture of the formation and changes in successive faunas and floras.

It is important to know the completeness of the available palaeontological record. To evaluate the bulk of this information, the accessible findings concerning organisms of the past are often compared with data on modern flora and fauna. For instance, Raup and Stanley (1971) point out that there are now known more than one million species of modern plants and animals, the actual number of existing species being presumably much greater (about four or five million). At the same time only about 130 000 species of fossil organisms are known. Assuming that an average length of life of a species is about three million years and assuming that the total number of species was invariable throughout Phanerozoic time, the number of species existed within this time interval should be approximately one billion. Because the average number of species in the past was less than in our epoch this estimate is evidently much too high. However, it is in the correct order of magnitude of the number of Phanerozoic species.

It may therefore be assumed that the number of fossil species known today represent less than 0.01% of their total number. This figure seems to be very small, but the volume of information concerning life forms of the past is sufficient for identifying a considerable number of the higher taxonomic groups. There was an opinion that all the orders, as well as the higher taxonomic groups, of Phanerozoic animals are already known. However, such opinion can hardly be correct. It is highly probable that in the past the organisms existed whose structural features may have allowed us to identify them as one of the higher taxons, but because of their small number, localized distribution or low possibility of fossilization, the discovery of their remains has proved impossible. Nevertheless, it may be considered that the majority of higher taxonomic groups of animals and plants that existed in the Phanerozoic are already known.

The palaeontological record of the Precambrian is quite fragmentary. Therefore palaeontological data covering Phanerozoic time are of principal value for more or less thorough studies of changes in successive floras and faunas.

On the basis of the palaeontological record it is possible to draw genealogical schemes describing the history of the appearance and extinction of different groups of organisms. An example of such a scheme for vertebrates is given in Fig. 8.1. It should be emphasized that the construction of genealogical schemes, even for such a relatively well-known group of animals as the vertebrates, is frought with certain difficulties. This arises because of the incompleteness of the palaeontological record and the lack of information on transitory forms ('missed links'), whose appearance determines the time of the distribution of higher taxonomic groups. The inadequate palaeontological record also impedes to some extent the study of the extinction of organisms, which will be treated closely in next section.

In addition to compiling palaeontological information, the study of the history of organisms involves the elucidation of the cause of diversification and extinction of various groups of organisms in certain geologic periods. Although a large number of scientists have been concerned with this problem, it is still far from being fully solved.

One of the first scientists to discuss the causes of changes in successive faunas was an outstanding biologist at the beginning of the 19th century and the founder of palaeontology and comparative anatomy, G. Cuvier (1769–1832). In his book *Discours sur révolutions de la surface du globe* (1812) Cuvier wrote: "Life on our planet was more than once shaken by frightful events. Innumerable living creatures fell victims to catastrophes. . . . Their races vanished for ever, leaving behind only few traces, which can be hardly discerned by naturalists." Cuvier thought that from time to time catastrophes wiped out the inhabitants of certain regions, whose place was later occupied by animals and plants that survived in other areas of the globe. Contrary to a

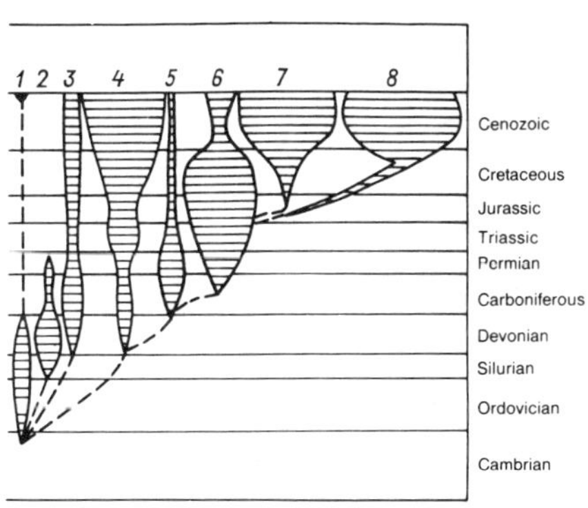


Fig. 8.1. 1, jawless fish; 2, placodermi; 3, cartilaginous fish; 4, bony fish; 5, amphibians; 6, reptiles; 7, birds; 8, mammals.

widespread opinion, Cuvier did not believe that after catastrophes new living things could appear. Therefore he supposed that the total number of species after each of the catastrophes declined.

Quite probably, Cuvier's concept sprang from the ideas of Laplace which appeared in his book *Exposition du système du monde* (1796). In this book Laplace pointed out an important principle of the movement of celestial bodies, the value of which in the history of the biosphere has been appreciated comparatively recently. Laplace noted that the probability of a collision of such celestial bodies as comets with the Earth is very low for short intervals of time (compared with the time of the Earth's existence), because the sizes of these bodies are small in comparison with the space where they travel. But for sufficiently long time intervals this probability grows considerably, which makes such a collision practically inevitable. Laplace supposed that as a result of these collisions, violent earthquakes, great disturbances of the sea and flooding of the land surface took place. He also stressed the possibility of a massive loss of life in these situations.

The ideas of Cuvier were more advanced than earlier concepts about the invariability of life forms throughout the Earth's history. Making allowance for considerable changes in the composition of successive faunas, Cuvier laid the foundations of the science concerned with the past life of organisms, i.e. palaeontology. It has been noted more than once that Cuvier's concept was to a great extent evolutionary. A particular merit of Cuvier was to regard changes in successive faunas as a determinate process, the causes of which he tried to find.

Later the accumulated palaeontological data showed that the geological past was marked, not only by the extinction of organisms, but by the appearance of new forms which had not existed earlier. In this connection the followers of Cuvier (A. d'Orbigny, L. Agassiz and others) supplemented his concept with the idea that each of the catastrophes was succeeded by a 'creative epoch' when new organisms appeared.

The ideas of Cuvier and his followers were soon rejected under the influence of the works of C. Lyell (1797—1875). In his fundamental work *Principles of Geology* (1830—1833) he substantiated the concept of a slow and gradual transformation of the lithosphere under the influence of geological factors that are still acting on the Earth's crust (a doctrine of actualism). Since this approach was sufficient to account for a great many geological processes, Lyell thought that the theory of catastrophes of Cuvier was useless.

Acknowledging the great importance of Lyell's methods for the progress of the Earth's sciences, it should be indicated that Lyell and particularly his followers overestimated the invariability of the factors influencing the biosphere and lithosphere. The overstatement is especially noticeable in the extreme form of actualism, the so called uniformitarianism, which denies any qualitative differences between geological processes in the past and at present. Later the limitations of the principles of uniformity in explaining the processes of the lithosphere's formation have been exposed quite clearly.

In contrast to Lyell, who thought that the geological processes of the past are similar to the existing ones and their intensity does not change over time, Yanshin (1961) set forth a concept that the processes forming the Earth's crust vary constantly. This idea is confirmed by the data from many investigations and is very important for understanding the biosphere's history, which is considered below.

The conclusion about the limited possibilities of using the doctrine of uniformitarianism in the study of the Earth's past has also been made in a number of other works, in particular in a series of papers published in the volume *Uniformity and Simplicity* (1967) of the Geological Society of America.

The greatest achievement of biology in the 19th century was the teaching about evolution by C. Darwin (1809—1882) and A. R. Wallace (1823—1913). Their theory of evolution by natural selection in the course of the struggle for existence and adaptation of organisms to the environment exerted a profound influence on the further development of evolutionary studies. It should be mentioned that although the theory of the origin of species was developed by Darwin somewhat under the influence of Lyell's actualistic approach, the conclusions of this theory clearly contradicted the principles of uniformity, and consequently Lyell did not immediately accepted Darwin's theory of the origin of species.

Darwin, Wallace and their followers used very little palaeontological evidence, regarding it as inadequate for detailed evolutionary studies. An exception to this was V. O. Kovalevsky (1842—1883) who begain studying the principles of evolution of animals making use of palaeontological data on horses.

At the end of the 19th and beginning of the 20th centuries the leading palaeontologists were either empirics who did not rely on evolutionary theories or partisans adhering to the ideas, which were not shared by specialists in the field of evolutionary biology. For instance there was an opinion that the development of taxonomic groups is analogous to that of individual animals and therefore it was supposed that the stages of youth, prime, gradual decline and dying are typical of each of these groups. A similar point of view was offered by some prominent palaeontologists even in the second half of the 20th century.

In the 1940s and 1950s evolutionary and palaeontological investigations became more closely connected as a result of the development of the synthetic theory of evolution combining the idea of natural selection in the course of the struggle for existence and the conclusions of the genetic studies that took place in the first half of the 20th century. Among the chief authors of the synthetic theory of evolution was a distinguished palaeontologist, G. G. Simpson (1902—1985), whose studies produced a great influence on the advancement of evolutonary palaeobiology. In his works, Simpson considered, in particular, the variability of the rate of evolution in different groups of animals.

Of considerable value in understanding the history of organisms were the works of A. N. Severtsev (1866—1936) and I. I. Schmalgauzen (1884—1963), who considered variations in the environment as the most important factor of the evolutionary processes. These studies presented an essential idea about aromorphosis, which will be considered further.

In recent decades the attention of scientists has been drawn to the problem of variability of the rate of evolution. This problem was considered by Simpson (1944) on the basis of vast palaeontological evidence. Simpson noticed that, along with very slow transformations of certain species and genera of animals that have changed very little for tens and even hundreds of millions of years, there are other groups of organisms that evolve at relatively fast rates. These fast changes, which he called quantum evolution, Simpson associated with the processes of adapting to new conditions of the environment.

Later the problem of an uneven rate of evolution attracted new attention due to the works of Eldredge and Gould (1972) and Gould and Eldredge (1977), who proposed the punctuational hypothesis of evolution. According to this idea, the evolutionary process proceeds in two ways: there are either rapid changes occurring within a limited period of time in small populations or insignificant modifications that take place within much longer time intervals (stasis). This idea was discussed by Stanley (1979), who supported the point of view that the evolution of species is largely connected with rapid and comparatively short-term changes in organisms and considered that this conclusion does not correspond to the ideas advanced by the authors of the modern evolutionary synthesis.

The ideas of punctuational evoluton have provoked some disagreement. Thus, Schopf (1981), for instance, offered reasons that questioned the wide distribution of the phenomenon of stasis. Stebbins and Ayala (1981) claimed that there are no contradictions between the punctuational hypothesis and synthetic theory of evolution and in reality both mechanisms of evolution are at work, i.e. associated with gradual transformations of organisms (the gradualistic concept) and with fast changes intermitting with long intervals of stable state of organisms (the punctuational hypothesis). It is expedient to remember Simpson's opinion (1978) that the punctuational concept is a new formulation of earlier known ideas.

One example that is worth mentioning when discussing this concept of evolution is the work of Williamson (1981) who studied the evolution of fresh water snails using data on sediments in northern Kenya. Williamson observed that new forms of snails appeared with fluctuations in lake levels within relatively short time intervals ranging from 5000 to 50 000 years. At the same time snails changed little over much longer time intervals marked by steady natural conditions. The conclusion of this work agrees fairly well with Darwin's opinion (1859) that changes in life conditions gave an impetus to an intensive variability and without them natural selection is utterly inefficacious.

There is no doubt that the rate of evolution varies considerably in many cases, particularly with a change from stable to unstable environment. At the same time it is difficult to agree that the phenomena that are considered confirming the punctuational concept greatly contradict the ideas of the synthetic theory of evolution.

The afore-mentioned collection of papers of the Geological Society of America contains an article by Newell 'Revolutions in the History of Life' (1967) which states that modern geology includes some aspects of uniformitarianism (by recognizing the preservation of the general principles determining the Earth's evolution) and certain aspects of catastrophism (by recognizing the uneven rate of the development). In the latter case, Newell pointed out that many processes in geological history proceeded very quickly, particularly the processes of extinction, which we shall discuss later.

Contemporary studies of the history of organisms usually acknowledge the importance of external factors in affecting the evolutionary process. However, there is no generally accepted view concerning the mechanism of these effects. This is because until recently, information on the external factors of evolution in the geological past has been extremely poor. It is only natural that attempts to study the influence of the environment on changes in successive faunas without sufficient factual data concerning this environment will not yield positive results.

One of the best-known examples of this arises from the idea that different stages of the evolution of animals and plants are related to tectonic processes. This idea was developed in a number of studies, including the book of Schmalgauzen *The Factors of Evolution* (1946), where, for instance, the author wrote: "It is difficult to imagine a more decisive transformation of the animal world of the Earth than the remarkable perturbation that took place at the beginning of our (the Cenozoic—M.B.) era . . . An external impetus for this perturbation was presumably intensive orogenic processes. . . . This involved also changes in climate all over the continents."

Acknowledging the importance of abiotic factors in influencing the evolutionary process raised by Schmalgauzen, it might be thought that earlier tentative ideas concerning the relationship of the tectonic processes and the vital activity of organisms were insufficient to make advances in the study of this problem. For instance, it is apparent now that changes in the relief of the continents did not produce a great influence on global climate. A stronger influence on climate may have been the effect of changes in the position of the continents with reference to the poles, but this effect would have been complex and, in some cases, not very significant.

At the same time it is known that not all climatic fluctuations resulted in abrupt changes in floras and faunas. For example, great changes in natural conditions during the Pleistocene Ice Ages influenced climate and animate nature only slightly in the tropics, while extratropical latitudes, where during glaciations noticeable climatic changes and shifts of geographical zones took place, saw neither mass extinctions nor appearances of radically new forms.

In this connection it is interesting to mention that Simpson offered in his works contradictory opinions relating to this problem. In his earlier work (1944) he wrote: "... among reptiles and mammals the proportion of orders that seem to have arisen during times of pronounced emergence and orogeny is greater than can reasonably be ascribed to chance. This is probably true of birds, also. . . ." In his later studies he viewed sceptically the possibility of establishing such relationships.

The influence of tectonic processes on the evolution of organisms is treated more fully below, where the conclusion is drawn that such an influence does take place but has a complex nature. Therefore it is difficult to reveal this influence by a simple comparison of successions of tectonic and evolutionary events.

Suppositions about the influence of other abiotic factors on evolution have been set forth repeatedly; however, in most cases they had no concrete physiological basis.

Leaving aside the role of abiotic factors in mass extinctions of organisms, which will be discussed in the next section of this chapter, it should be emphasized that the study of the effects of changes in chemical composition of the atmosphere on biological evolution deserves particular attention. The reason for these effects is that the amount of oxygen in the environment appears to be an essential factor in the vital activity of animals, and the content of carbon dioxide determines the intensity of photosynthesis of autotrophic plants.

In previous chapters it has been mentioned that there is a widespread notion that an increase in atmospheric oxygen ensured the appearance of multicellular organisms. It is also probable that after this the further growth of oxygen produced appreciable influence on evolution. In the works of Berkner and Marshall (1965a, b; 1966) and some other investigations attempts have been made to find the relationship between different stages of animal evolution and variations in oxygen content of the atmosphere. These attempts have been, however, hampered by the fact that the authors had no quantitative data on fluctuations in atmospheric oxygen obtained independently of information on the evolution of animals.

Before turning to the possibility of applying data on variations in the chemical composition of the atmosphere given in Chapter 7 to the study of general principles of the evolutionary process, let us consider the effects of the amount of oxygen contained in the environment on the vital activity of animals.

### 8.1.2. AROMORPHOSIS

The existence of living organisms is possible only if they are provided by energy influx from the environment. For the overwhelming majority of animals this flux of energy is ensured by oxidation of their organic matter through the absorption of oxygen. Since the ancient atmosphere did not contain oxygen in any considerable quantities, it has been suggested that earlier organisms were anaerobic and their existence was based on less effective, in terms of energy, reactions of disintegration of substances than aerobic organisms.

The entire history of evolution of organisms is associated with profound changes in mechanisms of their energy supply, the efficiency of these mechanisms gradually increasing in more advanced forms. As mentioned earlier, a great majority of contemporary organisms are aerobic, i.e. they use oxygen in the reaction of decomposition of glucose and related substances. By this reaction $CO_2$ and $H_2O$ are produced and energy is liberated and used by organisms.

An example of improvements of the energy supply system is modifications of blood-supply organs in different classes of vertebrates. Although a general scheme of these organs is identical in all vertebrates, the circulatory mechanism was subject to considerable changes in the course of vertebrate evolution. The heart in fish consists of consecutive chambers, and the vessels carrying venous and arterial blood are not securely isolated from each other. The auricle of amphibians is divided by a septum, which impedes the mixing of arterial and venous blood. The heart in reptiles is provided by additional septa, although only in crocodiles the flows of blood, saturated and non-saturated with oxygen, are completely divided. The most perfect circulatory systems are typical of birds and mammals, which provide the highest level of energy supply compared to other classes of vertebrates.

The respiratory organs in vertebrates also underwent similar changes. In reptiles, birds and mammals this process was accompanied with a growing dissection of the lungs, which resulted in the enlargement of their internal surface, development of a more complicated structure, and an increase in the energy income to the organism.

The level of energy supply of an organism per unit mass, all other conditions being equal, determines the possible size of the body of the organism and its physical activity. The income of trophic energy for many aerobic organisms is in a rough way proportional to the surface area of their bodies and the energy output is proportional to their mass to the power of 3/4. Because of this an increase in the body size of these organisms during the course of evolution requires more effective circulatory and respiratory systems.

Similar limitations are characteristic of the level of physical activity of animals. Other conditions being equal, the transition of animals from life in water, where they do not have to overcome gravity, to land, where gravity is a permanent load, requires a considerable increase in energy consumption. The consumption of energy also increases in the transition of land animals from walking to running and particularly to flight. It should be stressed that the energy used by different forms of movement rapidly increases with an increase in the mass of animals, therefore the energy output per unit mass of a large moving land animal is quite great. It is beyond any doubt that this is one of the principal factors restricting the sizes of land animals and in particular, of flying animals, whose greatest mass is far less than that of land forms. The energy consumption in vertebrates also increases when they have an effective mechanism of heat regulation, which will be discussed later.

It should be emphasized that differences in the amounts of consumed energy in representatives of different groups of animals are very great. For example, when the transition takes place from an aquatic to a terrestrial environment, the energy consumption for animals of the same species can be tens and even hundreds of times greater. A considerable increase in used energy also occurs in animals with an effective apparatus of heat regulation. Undoubtedly such significant changes in the energy balance take place only with noticeable improvements of the morphophysiological structure of animals.

It is evident that the level of energy consumption in vertebrates depends on both their structure and the amount of oxygen in the atmosphere. It is known that the oxygen content of the atmosphere at the level of the Earth's surface varies little in different regions of the globe, but diminishes appreciably with altitude. The oxygen content of water bodies depends to a great extent on its concentration in the lower air layer, water temperature and a number of other factors. The oxygen content of water usually decreases with depth. The sensitivity of aerobic organisms to diminishing partial pressure of oxygen differs greatly. Laboratory tests and observations in nature show that higher animals are very sensitive to a drop in the oxygen partial pressure. When this pressure decreases several times, hypoxia develops in many animals, which leads to their death.

It is worthy to note that lower vertebrates are usually less sensitive to oxygen shortages than higher animals, particularly mammals (*Ecological Physiology of Animals*, 1982).

Of considerable interest is the effect of a decrease in the oxygen partial pressure by several tens of per cent against its magnitude at sea level on the vital functions of animals. Land vertebrates can, as a rule, adapt to such decreases, which can be seen, in particular, in mountains several kilometres high where a diverse fauna occurs.

An empirical study of the effects of oxygen shortage on vital activity of animals living in mountains is impeded by a number of other factors affecting their vital functions, among which are decreased air temperatures and dwindling food resources. It is typical, however, that in some cases poikilothermal animals sensitive to low temperatures, can live at considerable altitudes. This is presumably due to the fact that they are less sensitive to the reduced oxygen content of the atmosphere.

In water bodies the content of dissolved oxygen is usually low at considerable depths, although such conditions are also observed in surface waters in some areas. Only poikilotherms (mainly invertebrates and fishes) can live permanently in water bodies, not emerging at the water surface.

It might be thought that fluctuations in atmospheric oxygen in the geological past produced an appreciable influence on the evolution of animals, affecting the appearance of aromorphosis.

The concept of aromorphosis developed by Severtsev (1925, 1939) and Schmalgauzen (1940) is among the significant achievements of 20th century evolutionary biology. Severtsev established that the origin of animals of higher taxonomic groups was generally associated with noticeable morpho-physiological changes in organisms, raising the energy level of their activity. Schmalgauzen slightly modified the notion of aromorphosis, supplementing it with cases when changes in organisms lead to a considerable expansion and complexity of their relations with the environment. It is evident that this definition of aromorphosis does not contradict Severtsev's idea, because large progressive modifications of organisms usually took place on the basis of higher energy of their vital activity. A classic example of aromorphosis examined by Severtsev and Schmalgauzen was the appearance of new classes of vertebrates.

As known, almost all the existing species of vertebrates belong to six classes: cartilaginous fishes, bony fishes, amphibians, reptiles, mammals and birds. According to the palaeontological record, these classes emerged within two relatively short periods of time. The first embraces the Devonian and a part of the Carboniferous (two classes of fish, amphibians and reptiles). Then for a long time no new classes of vertebrates arose, and only at the boundary of the Triassic and Jurassic mammals appeared and in the Jurassic birds emerged. Thus the formation of classes of vertebrates was completed.

It is natural to ask why the appearance of the classes of vertebrates took place within these limited time intervals. One might suppose that to a great extent fluctuations in oxygen content of the atmosphere were responsible for this. Figure 8.2 shows the ratio of atmospheric oxygen in the past to that of the present obtained as average values of the results of two calculations presented in Fig. 7.1.

As can be seen in Fig. 8.2, along with a general tendency to increasing atmospheric oxygen content in the Phanerozoic, two sharp rises in the oxygen content also occurred (in the Devonian—Lower Carboniferous and at the end of the Triassic—Jurassic). These intervals are close to the time at which the different classes of vertebrates first appeared.

Such an agreement can hardly be accidental. An increase in the

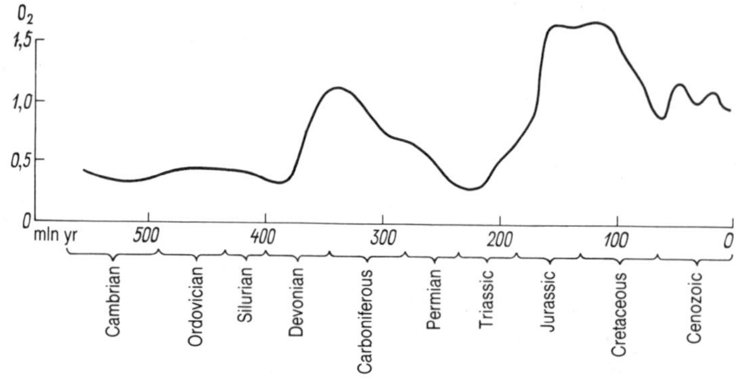


Fig. 8.2.

metabolism of animals was an important feature of their progress in the course of evolution. The phylogenetic development of the cardiovascular system and other organs of vertebrates, which ensures the higher energy consumption by more progressive forms, is well known. However, an increase in the use of energy could have also been achieved with an increase in the oxygen content of the atmosphere, which would create some advantages in the struggle for existence for more complex organisms, whose vital functions, other conditions being equal, required more energy.

Careful study of changes in the metabolic level with the formation of new classes of vertebrates is hampered by the absence of direct data on the physiology of earlier representatives of the relevant classes. It might be supposed, however, that an increase in metabolism was indispensable for the transition of vertebrates from water to land and for the formation of endothermal animals. In other cases an increase in the amount of oxygen might have improved the functioning of individual organs of more progressive animals, without a considerable rise in metabolic level of the organism as a whole.

It may be thought that an increase in atmospheric oxygen should have accelerated the evolution of animals and its decrease slowed it down.

Returning to the origin of the modern classes of vertebrates, it might be suggested that the distribution of early reptiles depended not only on the amount of oxygen in the atmosphere (although oxygen content was rather high in the mid-Carboniferous), but also on the first occurrence of arid conditions on part of the continents which took place after the emergence of life on land and corresponded to a decrease in the mean air temperature in the middle of the Carboniferous. This supposition is in accordance with specific features of reptiles that enable them to exist in the conditions of insufficient moisture, as distinct from amphibians, which appeared earlier and were to a great extent aquatic animals in the Devonian and Carboniferous (Tatarinov, 1972).

When considering causes of the relationship between the epochs of the formation of aromorphosis of vertebrates and variations in atmospheric oxygen, one might be reminded of the conclusion once made by Darwin (1859) that organisms are highly sensitive to even relatively small variations in the environment.

It is probable that of the different components of the external environment, the amount of oxygen in the air for terrestrial animals (or the amount of oxygen in the water for aquatic animals) is one of the most essential factors of their life activity. The level of metabolism of aerobic animals is directly dependent on the amount of oxygen in the environment, increasing, other conditions being equal, with an increase in oxygen.

An additional energy received by an animal in an environment with an increased oxygen content may be used for different purposes, including the development of a more complex structure in the course of evolution. Of especially great importance in this case might be the improvement of organs maintaining metabolism, such as the circulatory and respiratory systems in vertebrates.

According to Gilyarov (1975), in such cases the evolutionary process may take place in the manner of a positive feedback. As known, these feedbacks intensify variations in the system, stipulated by external factors, and as a result, the system may acquire a qualitatively new state. In the example under consideration an initial slight improvement of the circulatory and respiratory organs, induced by a higher influx of oxygen, additionally increases the energy flux to the organism, which creates conditions for their further improvement and increased effectiveness. The influence of this positive feedback, along with the effects of a gradually increasing amount of oxygen in the environment, was presumably the cause of the development of aromorphosis at the formation of new classes of vertebrates.

We shall not dwell on the description of the morphophysiological structure of aromorphosis, which is considered in detail in the aforementioned works of Severtsev and Schmalgauzen. However, it should be mentioned that the greatest changes in the energy level of life activity of vertebrates took place when amphibians emerged on land (where the presence of a gravitational force sharply increased the amount of energy used on movement) and in the process of the formation of endothermy in mammals and birds. It might be stated with certain assurance that these events could take place only with highly favourable changes in the environment, in particular with an increase in the amount of oxygen.

The examples given here do not accidentally refer to vertebrates, as this is the most progressive group of animals and one characterized by a high level of energy use in the course of their life activity. The evolution of vertebrates was evidently determined by the factors limiting the energy consumption to a greater extent than that of invertebrate animals. We think, however, that the evolution of invertebrates was also dependent on the external conditions affecting their energy regime.

Turning to the evolution of plants, it might be suggested that the appearance of aromorphosis associated with an increase in energy of life activity is also typical of this group of organisms. Evidently, such increases occurred with the enhancement of the productivity of photosynthesis, which made possible a more complex structure of plants and greater mass of their non-reproductive organs. The probability of the appearance of aromorphosis was presumably higher with a rise in carbon dioxide concentration, which determined the productivity of photosynthesis. It is also possible that the functions of some plants were appreciably dependent on the oxygen content of the atmosphere. In particular, the development of higher plant flowers and the germination of their seeds are often accompanied with a noticeable increase in temperature, which is the result of intensification of the process of oxidation. It is evident that such intensification is only possible with a sufficient amount of oxygen in the environment. The relationship between the evolution of plants and changes in the chemical composition of the atmosphere is treated below.

It should be noted that great changes in the morphophysiological structure of organisms associated with the appearance of aromorphosis were made up of many small variations, each of them producing a comparatively slight effect on the conditions affecting the organism's existence. One of the reasons for the effective operation of natural selection for organisms that change little compared to unchanged organisms is, evidently, that many ecological systems are highly sensitive to external factors.

As mentioned in Chapter 5, all ecological systems are subject to a constantly changing density of animal and plant populations. These fluctuations may take place both due to the variability of external factors and with steady external conditions (the auto-oscillatory process of variations in the density of populations). As a result, less stable ecological systems will not last long, and their destruction will threaten their populations with extinctions. More stable ecological systems have better chances of preserving their populations for a long time.

Natural selection supports the dissemination of organisms, whose evolution increases the stability of ecological systems. At the same time, natural selection maintains more stable ecological systems and eliminates less resistant systems, which in this case are treated as superorganisms that also take part in the struggle for existence.

Numerical modelling of energy relationships in ecological systems leads to the conclusion that many systems reached a high level of stability at the earlier stages of their existence, after which the stimuli for further development appeared to be greatly limited. A large number of these systems with ancient forms of organisms are found in open sea where external conditions have always been more stable.

A rapid development of evolutionary process leading to great changes in living organisms seems to be less probable in stable ecological systems under relatively fixed conditions of the environment. In such systems, changes in the organisms they contain should be more or less interconnected, which considerably decreases the probability of fixing the mutation for each organism treated separately.

Nevertheless, even the most stable ecological systems evolve, both as a result of changes in natural environment (for instance, owing to changes in atmospheric gas composition) and due to the purely biological factors. The rate of evolution of such systems is evidently the smallest.

Of considerable importance in accelerating the evolutionary process might be changes in the environment that disturb the stability of ecological systems. In some cases, such disruptions might result in the extinction of certain groups of organisms.

It can be argued that the resistance of an individual organism (or population) to changes in the abiotic environment usually differs from its resistance to the same changes when this organism is a part of ecological system. Let us assume, for example, that an animal can endure a change in the environmental temperature of about 5 °C. Such a temperature change would not result in the death of the animal if it inhabits a hypothetical environment provided with all the conditions necessary for its life. In a real ecological system this change in temperature would produce not only a direct influence on the animal but affect it indirectly through changes in its diet, the behaviour of animals preying on it, the activity of parasites, and so on.

If only one of all these diverse changes in the environment surpasses the limits of the conditions that make possible the existence of this animal, it will inevitably be destroyed. The extinction of species does not require a simultaneous death of all its representatives or even of a considerable part of them. As shown in the calculations presented in Chapter 5, an insignificant decrease in birth rate or a slight increase in death rate is sufficient for the extinction of the species within relatively brief interval of time. Therefore, a small variation in the external conditions is often sufficient for wiping out earlier existent animals and their replacement by new forms, which results in the dissemination of new species.

It might be noted that these considerations are directly associated with the idea formulated by Darwin (1859), who wrote that if in a country some physical condition, for instance, climate, changes even insignificantly, the relative number of its inhabitants is also immediately subjected to change, and certain species will probably die out completely. Darwin concludes on the basis of our knowledge about the intimate and complex interactions among the inhabitants of any country, that any change in the relative number of some species will profoundly influence the others irrespective of changes in climate. Thus, the sensitivity of ecological systems to changes in the environment is the most important factor determining the rate of evolution.

### 8.1.3. THE DIVERSITY OF SUCCESSIVE FAUNAS

As mentioned earlier, the higher taxonomic groups of animals, such as phyla or classes, include different numbers of lower taxonomic groups. The change in the number of species and genera belonging to a particular phylum or class reflects successful adaptation of representatives of this phylum or class to the environment in which the organisms live.

It may be supposed that in epochs with an increased atmospheric oxygen content, conditions were conducive for the diversification of forms entering into the higher taxonomic groups, whereas in the epochs with a decreased oxygen content the diversity of forms grew less. This supposition is based on the idea that with a large amount of oxygen, forms with relatively higher metabolic level, whose existence would have been impossibe at smaller oxygen content, might emerge.

Of considerable interest is the evolution of animals through the long time interval from the mid-Carboniferous to the mid-Triassic, when the atmospheric oxygen level was decreasing. As a result of this decrease the oxygen content was only about one third of its present level by the end of the Permian and throughout a part of the Triassic. This level, which is close to the amount of oxygen that existed in the early Palaeozoic, roughly corresponds to its present level in the upper troposphere. The lower (exothermal) vertebrates that used relatively little oxygen could more easily adapt to such conditions. It is very doubtful that the existence of higher (endothermal) vertebrates at such low oxygen concentrations was possible.

It is evident that with a decrease in the atmospheric oxygen content, the life activities of animals and their mutual relations with other components of ecological systems must have been changing. A gradual deterioration in the conditions necessary for the existence of animals must have been reflected in the specific features of the evolutionary process in the time interval in question. These features have been noticed by palaeontolgists for a long time, although their causes were unknown. In particular, Simpson (1961) wrote: “It is interesting that no phylum has expanded steadily from the time of its appearance to the present day . . . The most nearly general feature is that most of the phyla contracted in the Permian, Triassic, or both.”

This contraction which Simpson describes has been confirmed by many other investigators. For example, Raup and Stanley (1971) present a scheme of temporal variations in the number of taxa of fossil animals in the Phanerozoic. This scheme shows that although a gradual increase in the number of taxa occurs over this period, there was an abrupt decline in the number of taxa occurs over this time, there was an abrupt decline in the contraction began in the middle of the Permian. Robinson (1971) states that from the Permian to the Upper Triassic the number of genera of therapsids was reduced by a factor of ten while the number of genera of sauropsids increased considerably. This shows that for the greatest part of the Triassic, conditions were unfavourable for the existence of the mammal-like reptiles. A tendency towards a decrease in the sizes of therapsids, which belonged to the most progressive groups, occurred in the Triassic. This tendency may be due to a growing shortage of oxygen, since maintenance of the metabolism of large animals requires more oxygen than for small animals, other conditions being equal.

It might be supposed that a decrease in oxygen content throughout the Upper Permian, and Lower and Middle Triassic, a time interval of about 100 million years, considerably retarded the evolution of the mammals. This class of vertebrates spread at the end of the Triassic when a new rise in atmospheric oxygen occurred.

Another considerable decrease in atmospheric oxygen concentration occurred in the second half of the Cretaceous. At this time, the oxygen content decreased by about 1.5—2 times and approached its modern concentration. Although this level was rather high compared to the average conditions of the Phanerozoic, it is, however, probable that the oxygen decrease in the Cretaceous was of importance for the fauna, and in particular promoted the extinction of some groups of animals at the end of the Mesozoic.

Following these considerations it might be concluded that the evolution of the modern classes of vertebrates was determined to a large extent by global changes in the environment, caused in part by variations in degassing of the Earth's mantle. This concept of a link between the concentration of atmospheric oxygen and the evolution of higher taxonomic groups of animals is based on the idea that an improvement in the energy supply of an animal was of decisive importance for large adaptive modifications, which in turn through a positive feedback permitted an increase in the energy used by the animal.

The influence of enhanced oxygen concentrations on evolution over many millions of years gives rise to the possibility of definite evolutionary trends occurring for a number of animals under these conditions. The suggestion that such evolutionary trends are associated with long-term variations in the external conditions surrounding animals has been made by Huxley (1962) and other authors. We should also mention the supposition of Sokolov (1975), Mayr (1976), Sochava (1979) and others about the probability of atmospheric oxygen variations influencing the process of macroevolution. In this connection, the substantiation of Dobzhansky (1951) of the idea relating the role of vacated ecological niches to the rate of evolution is important. An increase in the amount of oxygen created new adaptive zones which gave rise to the appearance of new ecological niches. In this way, a high rate of development of adaptive changes is encouraged.

It is possible that similar mechanisms of evolution may also occur when other abiotic components of the environment change, enlarging the adaptive space of the biosphere. Some examples of such changes are given below.

As mentioned earlier, with an increasing amount of oxygen in the atmosphere throughout the Phanerozoic, the possibility arose of evolving new groups of animals, which use more energy in the course of their vital activity. Therefore, it is natural to suppose that in epochs of increased oxygen concentrations the diversity of faunas grew larger while in epochs of decreased oxygen content the diversity diminished. To study this it is possible to make use of data on animals that lived throughout the Phanerozoic, namely the marine invertebrates. Using information on the number of families of these animals for each period of the Phanerozoic (Pitrat, 1970) and the above data on variations in oxygen, the diagram in Fig. 8.3 has been constructed. This figure shows the dependence of the number of families of marine invertebrates (N) on the average amount of atmospheric oxygen in a given period expressed as fractions of the present amount of oxygen. The existence of a definite relation between these two variables confirms the supposition made above.

Typical data on successive changes in the diversity of certain groups of animals throughout the Phanerozoic are given in Fig. 8.4. This figure shows variations in the number of families of marine invertebrates, genera of echinoderms, and orders of fish according to the data from the monograph edited by Hallam (1977). As can be seen, the general principles suggested by Simpson are clearly pronounced for all these groups. The data in Fig. 8.4 show that the diversity of groups was greater in the Devonian—Carboniferous and in post-Triassic time, i.e. in the epochs of increased oxygen content of the atmosphere. At the same time, comparison of the data given in Figs. 8.2 and 8.4 makes it possible to conclude that differences occur between variations in atmospheric oxygen and fluctuations in the diversity of the three animal groups. It can be observed that in addition to the effects of oxygen content, animal diversity also exhibits a gradual increase in the number of taxonomic groups. This tendency probably reflects in part the known loss of palaeontological information back with time. Taking this into account, we

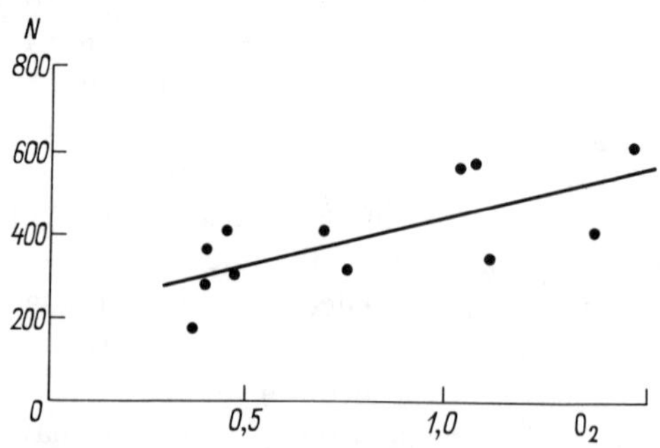


Fig. 8.3.

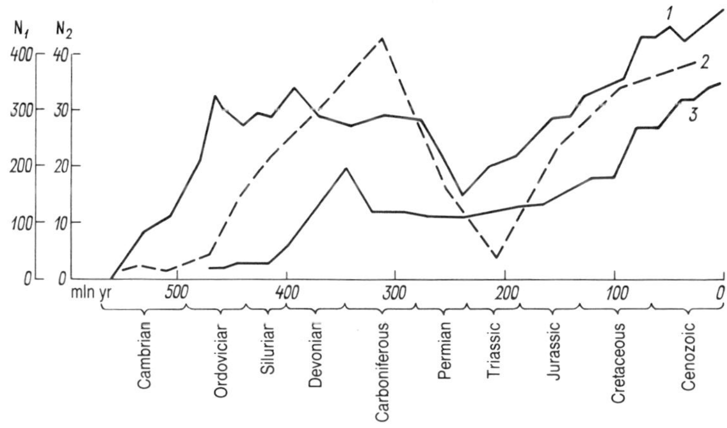


Fig. 8.4. 1, families of marine invertebrates; 2, genera of echinoderms; 3, orders of fish; $N_1$, the number of genera and families; $N_2$, the number of orders.

may considerably improve the agreement between the data on variations in oxygen content and diversity of the indicated groups of animals.

Analysis of data on variations in the diversity of other higher taxonomic groups of animals shows that although 'the rule of Simpson' is fulfilled in most cases, we do not always observe a close correspondence between variations in the diversity of groups and fluctuations in atmospheric oxygen. The lack of such correspondence may be explained by a less strong influence of oxygen content on the evolution of organisms whose phylogenesis was less dependent on their metabolism (for instance, organisms that were small in size).

It is worth noting that animals expending the maximum energy on movement flourished the most during the epochs of greatest oxygen increases. These included, for example, flying animals such as large winged insects of the Carboniferous, pterosaurs and birds of the Jurassic and also the largest land animals such as the giant reptiles of the Jurassic and Cretaceous.

During the time intervals when oxygen concentrations were more or less constant, diversity of the fauna could vary under the influence of other ecological factors. In considering this problem it is possible to make use of Kurtén's data (1971a) on the number of mammalian families in different epochs of the Tertiary (see Table 8.1).

Examination of the data in this table shows that in addition to a well-known rapid increase in the number of families in the Palaeocene—Eocene, there were certain fluctuations in the number of mammalian families in later time. The causes of the 'explosive evolution' of mammals at the beginning of the Tertiary are quite clear: this was the time when mammals occupied

TABLE 8.1

Number of mammalian families and natural conditions of the Tertiary

| Epochs | | Total number of families | $CO_2$ concentration (%) | Difference between mean global air temperature and modern temperature (°C) |
|---|---|---|---|---|
| Palaeocene | Early | 17 | | |
| | Middle | 27 | 0.120 | 7.6 |
| | Late | 41 | | |
| Eocene | Early | 51 | | |
| | Middle | 64 | 0.185 | 9.1 |
| | Late | 91 | | |
| Oligocene | Early | 95 | | |
| | Middle | 82 | 0.050 | 4.5 |
| | Late | 77 | | |
| Miocene | Early | 95 | | |
| | Middle | 88 | 0.110 | 7.0 |
| | Late | 92 | | |
| Pliocene | Early | 93 | 0.055 | 3.6 |
| | Late | 85 | | |

numerous ecological niches that were vacated as a result of the extinction of many reptilian groups at the end of the Mesozoic.

It is interesting to look into the causes of subsequent fluctuations in the number of mammal families. It follows from the data of Table 8.1 that in the Oligocene—Pliocene there was a relationship between the number of mammalian families and ecological conditions. A decrease in the number of families in the Oligocene and Pliocene was presumably the result of declining carbon dioxide concentration, which led to climatic coolings and a decrease in the productivity of autotrophic plants. The number of mammalian families was particularly affected by the Oligocene cooling, which appears to have been a major ecological crisis. It might be thought that as a result of cooling, the 'ecological capacity' of the biosphere for mammals decreased, i.e. the number of ecological niches which could be used by different groups of animals grew smaller. It should be noted that when the Oligocene cooling came to an end, the number of mammalian families in the early Miocene expanded greatly, probably as a result of the increased 'ecological capacity' of the biosphere.

Let us examine more closely the variations in the rate of emergence of new families and the extinction of the earlier existent families throughout the Tertiary. According to Kurtén's summary and the length of the Tertiary epochs adopted in his book,[1] we have constructed Table 8.2, containing the average rate of appearance and extinction of mammalian families per million years.

TABLE 8.2

Appearance of new families and extinction of earlier existent families of mammals (changes in number per million years)

| Epoch and its duration | | Appearance rate | Extinction rate |
|---|---|---|---|
| Palaeocene | Early | 1.8 | 0 |
| ($10 \times 10^6$ yr) | Middlc | 3.0 | 0 |
| | Late | 4.2 | 2.7 |
| Eocene | Early | 3.2 | 2.0 |
| ($18 \times 10^6$ yr) | Middle | 4.2 | 0.8 |
| | Late | 5.3 | 3.3 |
| Oligocene | Early | 6.0 | 5.0 |
| ($12 \times 10^6$ yr) | Middle | 1.8 | 2.5 |
| | Late | 1.2 | 2.0 |
| Miocene | Early | 5.2 | 2.2 |
| ($15 \times 10^6$ yr) | Middle | 0.8 | 0.2 |
| | Late | 1.0 | 0.2 |
| Pliocene | Early | 0.6 | 2.6 |
| ($7 \times 10^6$ yr) | Late | 0.3 | 0.6 |

As can be seen in Table 8.2, the number of mammalian families was growing with an increasing rate from the beginning of the Palaeocene to the early Oligocene. This might be explained by an increase in the number of initial forms, from which new families arose. The rate of extinction of earlier existent families was close to that of the emergence of new families in the early Oligocene only, which presumably corresponded to the filling of most of the ecological niches accessible for mammals. On termination of the early Oligocene the number of families, did not stabilize, which, as already mentioned, was probably the result of changes in natural conditions. However, in the middle and late Oligocene the rate of extinction decreased rather than increased. A decline in the total number of families at this time was the result of a still greater decrease in the rate of appearance of new families. This shows that the deterioration of environmental conditions, which led to a decrease in the 'ecological capacity' of the biosphere, did not develop sufficiently fast to cause mass extinction of the different groups of organisms. In this respect the Oligocene crisis differed from other critical epochs of geological history such as, for example, the end of the Triassic and

[1] Kurtén gives two scales of absolute ages of the Tertiary epochs, which are not fully identical. Our calculation has been based on the second scale, in reference to which Kurtén compared the numbers of mammalian families.

the end of the Cretaceous, when the number of taxonomic groups of different animals contracted greatly as a result of their extinction. This difference is not due entirely to the fact that the external conditions surrounding the organisms deteriorated. It is possible that mammals adapted more easily to the conditions of cooling in the Oligocene than the organisms without heat regulation that became extinct in the Mesozoic.

As soon as the Oligocene crisis was over the rate of appearance of new families returned to a level close to maximum, and as a result, the total number of mammalian families in the early Miocene was exactly the same as in the early Oligocene. It might be supposed that this number corresponded to the 'ecological capacity' of the biosphere for Tertiary mammals that can be maintained under favourable external conditions. Relatively low rates of appearances and extinctions of mammalian families from the middle Miocene to the late Pliocene appear to be typical of the epoch when the number of mammalian families was approaching the maximum possible, with the more archaic groups of animals having already become extinct under the earlier conditions of drastic environmental fluctuations.

In addition to considering the evolution of animals, it is also possible to consider the influence of the evolution of the chemical composition of the atmosphere on changes in the vegetation cover of the continents.

As already mentioned, the most important factor in the energetics of autotrophic plants is carbon dioxide concentration, which determines the rate of photosynthesis. It is known that when other conditions are equal, an increase in carbon dioxide increases the amount of chemical energy obtained by plants. From Fig. 7.1 it can be seen that in the Phanerozoic the greatest increase in carbon dioxide concentration took place in the middle of the Devonian, as a result of which the carbon dioxide concentration reached a high maximum sustained up to the beginning of the Carboniferous. The next highest maximum of carbon dioxide concentration, which occurred in the first half of the Permian, was much less prolonged. The longest, although not very great, increase in carbon dioxide embraced the period from the middle of the Triassic to the end of the Cretaceous, the highest concentrations occurring in the Upper Cretaceous. In the Cenozoic against a background of generally lowering $CO_2$ concentration, two relatively short maxima are pronounced in the Eocene and in the Miocene.

Comparison of variations in carbon dioxide with principal events in the development of vegetation is impeded by the lack of a generally accepted chronology of considerable changes in the vegetation. Krasilov (1977) identifies four major epochs of evolution of higher taxonomic groups of plants: the second half of the Devonian, the Permian—early Triassic, the Cretaceous and the Miocene. The first of these epochs is associated with the formation of progymnospermous forests, the second with the expansion of coniferous forests, the third with the emergence of archaic flowering plants, and the fourth with the appearance of steppe plant communities.

It can be noted that these epochs correspond to four of the five maxima in carbon dioxide concentration described above. A full coincidence is observed in three cases (the second half of the Devonian, the Cretaceous and the Miocene). The difference between the time of carbon dioxide maximum in the first half of the Permian period and the expansion of coniferous forests may correspond to the earlier formation of this type of vegetation cover which is not evident in the available palaeobotanic data.

The widespread expansion of flowering plants in the Cretaceous might have been promoted by a high level of atmospheric oxygen, which, according to the data presented in Fig. 7.2, took place in the first half of the Cretaceous. Takhtadzhyan (1980) considers that the angiosperms emerged at the beginning of the Cretaceous.

The only maximum of carbon dioxide concentration that has no equivalent in the summary of Krasilov occurs in the Eocene. It is possible that this maximum was important in promoting the formation of humid tropical forests at this time.

Our data confirm Krasilov's idea that a relationship exists between the epochs in which the higher plant taxons appeared and tectonic processes. As can be seen in Fig. 7.1, fluctuations in volcanic activity corresponded to variations in carbon dioxide concentration, which in turn were to a certain extent in agreement with the development of the vegetation cover.

Although the evolution of animals was to some extent associated with variations in atmospheric oxygen, the association between fluctuations in oxygen and volcanic activity was of a complex nature. It is possible that, other conditions being equal, an increase in carbon dioxide due to intensive volcanic activity, might have raised the oxygen concentration due to an increase in photosynthesis. However, these 'equal conditions' usually did not exist. In particular, a release of photosynthetic oxygen into the atmosphere was not only a result of the productivity of photosynthesis but also depended on organic carbon burial which sharply deteriorated when the climate became arid.

Although climatic conditions on the continents and in particular the moisture regime depended on the shape of the relief and the position of the continents (i.e. in the long run on tectonic processes) this complicated relationship could not be expressed as a function of a single variable. Because of this it is hardly possible to find simple relations between the global characteristics of tectonic processes and the evolution of animals.

It follows from this in particular, that mutual relations between major events in the evolutionary history of plants and animals are also rather complex. One of the exceptions from this rule is an evident relationship between the wide distribution of plants on the continents and the diversification of continental fauna. The relative complexity of the factors that relate most of the essential stages of the plants' history with that of animals is no reason for denying the existence of such relationships.

In conclusion we shall note that the idea of a dependence of evolutionary process on variations in the atmospheric composition was first expressed by an outstanding French biologist E. Geoffroy St-Hillaire (1772—1844). In one of his works (1833) he says: "... If we suppose that during a slow and gradual course of time, the proportions of different elements of the atmosphere changed, an absolutely indispensable consequence of this fact was that the animal world correspondingly endured these changes." We think that this idea was far ahead of the science of the first half of the 19th century.

## 8.2. Critical Epochs in Geological History

### 8.2.1. GLOBAL ECOLOGICAL CRISES

As already mentioned, the contemporary fauna is divided into fairly non-homogeneous higher taxonomic groups. The number of animal species belonging to different classes varies considerably. While the class of insects includes more than a million species, many other classes consist of only several tens of species and some of them are represented by less than ten species. The extent of distribution and the density of populations of certain species are also very diverse. Some animal species are spread over most of the Earth's surface while others are confined to areas that do not exceed a few square kilometres. Still greater relative differences occur in the number of animals belonging to particular species.

Leaving aside the impact of man on present-day fauna, it may be supposed that the groups of animals with many different species, high population density and vast distributions, may be regarded as the stable units of the modern biosphere. At the same time the groups of animals comprising only a few species, with negligible numbers of animals, and very limited distributions, appear to be the remains of formerly flourishing forms that are now on the verge of extinction.

There is no doubt that throughout Phanerozoic time the faunas were in many respects similar to those of our age. These ancient faunas also included animal groups of different ages, variety and number of individuals. The groups that consisted of only a few species and with a small number of individuals could have vanished with time, leaving behind ecological niches to be filled by the animals who successfully competed with these groups, promoting their contraction.

This mode of extinction of various species of animals undoubtedly took place in the past, but there were probably other modes of extinction.

Based on various studies of the palaeontological record it has repeatedly been suggested that in some cases extinct groups of animals included not only individual species with small numbers of individuals, but also large taxonomic groups, such as orders and families, with large numbers of species, high population densities and vast distributions. In such cases the extinction took place over a relatively short time and the vacated ecological niches often remained empty for a long time period. This latter observation gives the impression that in these cases the extinction was not the result of competition among animals of different species. The causes of such mass extinctions are very important in understanding the mechanisms of change in successive faunas.

As mentioned above, the idea of the existence of critical epochs in the past when massive mortality of animals has taken place was first suggested by the founder of palaeontology, Cuvier. Since Cuvier's concepts were quickly rejected, the idea about critical epochs in geological history attracted comparatively little attention for a number of years. However, it has been established more recently that mass extinctions of animals within relatively short time intervals have occurred in the past. For example, a great extinction of many groups of land and particularly marine animals took place at the transition from the Permian to the Triassic. Other changes in terrestrial and marine faunas, associated with mass extinction, occurred at the boundary of the Triassic and the Jurassic, and in the Cretaceous and Tertiary periods as well as at other times (Newell, 1963; Ostrom, 1969; Valentine, 1968).

The mass extinction of animals occurred at the end of the Cretaceous period. Not only the tremendous scale of this extinction (according to Russel (1979) about three-quarters of animal species presumably vanished at that time), but also the disappearance of the dinosaurs during this epoch make this period particularly interesting. The importance of the latter event is discussed further.

It should be noted that some authors do not agree with the concept of critical epochs and only acknowledge the existence of certain cases of relatively fast extinction of animal groups, because of interactions between these groups and other organisms and not because of abiotic factors. This point of view is particularly pronounced in the works of Davitashvili. For example, in his monograph (1969), which contains a detailed review of studies concerned with extinctions of animals and plants, his objection to the concept of rapid global extinctions in particular epochs is based on his assertion that such events could not be securely established because the palaeontological record is incomplete. However, it should be mentioned that Davitashvili considered the same palaeontological evidence quite reliable if it gave rise to the conclusion that a certain group of animals became extinct as a result of a gradual decrease in diversity and number of animals.

In considering the influence of an incomplete palaeontological record on studying the processes of mass extinction, it is important to remember that conclusions concerning the existence at any given time of particular species (or genera) are usually made on the basis of relevant finds of fossils. If there are no finds for the relevant time period, or later periods, it is assumed that this species (or genus) has become extinct. Conclusions of this kind are inevitably inaccurate. Firstly they refer, not to a specific moment, but to a rather long time interval, whose duration depends on the error in dating the available finds. Secondly, in many cases a number of causes may prevent the preservation of the remains of the existent animals. The simplest of these causes is the contraction of the number of animals and of the area of their distribution, after which the probability of preservation of their remains is negligible. For example, it is very unlikely that in millions of years of time, the palaeontological record would be sufficient to establish the existence of representatives of the rhynchocephalians or crossopterygians living today.

It can be seen thus that there are considerable difficulties in using palaeontological evidence when studying the epochs of mass extinctions.

As already mentioned, some authors doubt the reliability of the evidence for mass extinctions, considering that since the accuracy of dating the last finds of certain fossil types for different regions is limited, the extinction of these animals might not have in fact been simultaneous. At the same time it has been noted that in a number of cases the epochs of mass extinctions were preceded by a gradual decrease in the populations of the animal groups that finally vanished in the critical epochs. The first of these objections, pointing to the limitations in using the palaeontological record for studying changes in successive faunas, does not answer the question of whether short-term massive extinctions really took place or not. The situation described in the second objection may, in some cases, reflect the actual changes in the composition of faunas and in other cases may be the result of a simple statistical principle that has nothing to do with the process of extinction. Because of the limited number of finds of remnants of formerly existent animals, it is quite likely that remains have not been found in the time interval before the mass extinction took place. Thus it is inevitable that a mass extinction will be preceded by the disappearance of a certain number of species (or genera) of animals, which actually have not died out. The less complete the palaeontological record, the higher the number of such 'pseudoextinctions' that precede the actual dying out of the relevant animals.

Here is a simple example illustrating this principle (Budyko, 1982). Let us assume that in a critical epoch a certain number of animal groups died out almost simultaneously and the probability of fossil finds for each group at the time of its existence during the time interval $T$ is 50%. In this case it is probable that 50% of the indicated animal groups will vanish from the palaeontological record not during the 'critical epoch' but $T$ years earlier; 25% will vanish $2T$ years earlier, and so on. Since the time interval $T$ may represent millions of years, it is easy to draw the wrong conclusion that the process of extinction was developing gradually for a long time interval and came to an end in the critical epoch. The possibility of such error is often not taken into account when studying the processes of extinctions of animals.

The reality of mass extinctions may, however, be confirmed by palaeontological data if methods of mathematical statistics are applied for this purpose.

Studying the information on the time of disappearance of certain taxonomic groups (for instance the genera or families of marine invertebrates), it is possible to establish that for consecutive time intervals of several millions of years throughout the Phanerozoic, the ratio of the number of extinct groups to the duration of the time interval usually varies only a little. However, during some intervals this ratio is considerably increased. Employing statistical methods, it is possible to find out if these increases were random events or related to certain causes that led to mass extinctions.

The results of the statistical analysis given below of the extinction of reptiles at the end of the Mesozoic have confirmed the reality of this event (Budyko, 1971). In recent years a number of studies have been carried out along this line, among which is the work of Raup and Sepkosky (1982) on variations in the number of families of marine animals during Phanerozoic time. They have established that according to the available data, it might be concluded that there were undoubtedly four mass extinctions, one of which took place at the end of the Ordovician, the second at the end of the Permian, the third at the end of the Triassic and the fourth at the end of the Cretaceous. Moreover, they found that mass extinction also occurred in the late Devonian, but the statistical significance of this is less evident.

The weak point of such calculations is that they only yield the minimum possible number of reliable mass extinctions, while the number of extinctions that really took place might be far greater. Taking this into consideration, the conclusion should be made that it is difficult to study critical epochs of geological history confining oneself only to the analysis of palaeontological data. In this connection, investigations of large-scale changes in the environment in the geological past are of great importance in studying massive mortality of organisms. If such changes took place, they could have easily led to mass extinction of organisms.

There are many hypotheses of possible causes for mass extinctions of living creatures, but leaving aside the entirely groundless suppositions, the total number of hypotheses on this subject is relatively small. There are still fewer causes of a drastic global change in the environment, which not only could have taken place but undoubtedly did take place in the past.

Environmental changes that were unfavourable for many organisms can be divided into two groups, the first occurs over long time intervals comparable to geologic periods or epochs. In Phanerozoic time the most significant long-term change in natural conditions unfavourable for many animals took place during the Permian–Triassic periods, when the diversity of the majority of phyla decreased noticeably. This change, which was called 'the rule of Simpson' in the previous section, is probably due to a decrease in the amount of oxygen in the atmosphere. Undoubtedly such changes, although on a smaller scale, have taken place repeatedly in the past. In some cases they might have been associated with relatively gradual climatic changes. It is possible that an example of such a climatic change was the Oligocene cooling which, as previously shown, resulted in a reduction of the number of mammalian families.

It should be pointed out that the statistical studies described above do not show that such unfavourable changes in the environment cause mass extinctions. In such cases the reduction in the diversity of successive faunas can be explained not only by a higher rate of disappearance of the earlier existent groups, but also by lower rate of appearance of new groups of animals. If with slow changes in natural conditions there is an increase in the number of extinct groups, this increase over a long time interval is not connected with an abrupt rise in the rate of extinction, which can be determined by statistical analysis.

The second type of adverse changes in the environment occurs over relatively short time intervals and corresponds to the mass extinctions considered above. Among the most evident causes that might have led to mass mortality of organisms are fluctuations in volcanic activity and impacts of comparatively large celestial bodies on the Earth's surface.

In Chapter 7 the results of investigations which established that after explosive volcanic eruptions the temperature of the lower air layer decreases all over the world (or in one of the hemispheres) due to the reduction in atmospheric transparency, were presented. After a single eruption the air temperature decrease can be up to several tenths of a degree but a series of successive eruptions might result in a mean air temperature decrease by 5 °C or even more. Such events could take place at intervals of about 10 million to 100 million years, which corresponds to the time intervals between the critical epochs.

It has been repeatedly suggested that the critical epochs are associated with the collisions of comets, large meteorites or asteroids with the Earth. As mentioned at the beginning of this chapter, the idea that the impact of comets on the Earth's surface caused the extinction of organisms was proposed by Laplace at the end of the 18th century. More recently, this hypothesis has been revived by a number of authors, including Urey (1973) a well-known chemist, who suggested that mass extinctions could be explained by the collisions of comets with the Earth.

It is quite possible that the impact of celestial bodies on the Earth was an important factor leading to mass extinctions. Observations of the Moon have long shown that there are numerous craters on its surface formed by the impact of more or less large celestial bodies. It has been found through space exploration that the surfaces of Mars and Mercury have a similar structure. As mentioned in Chapter 6, the Earth's surface has experienced the impact of a similar number of celestial bodies to these planets, which has resulted in the formation of many craters. It is probable that impact craters (astroblems) with a radius of 100 km have appeared on average once every 14 million years and those of 500 km radius, once every 600 million years. Although in most cases their traces on our planet have been smoothed as a result of atmospheric and hydrospheric processes, some astroblems are still preserved today (*Geology of Astroblems*, 1980).

It is beyond any doubt that such grandiose phenomena as the collisions of comparatively large celestial bodies with the Earth must have produced a great influence on animate nature. Such collisions were accompanied with explosions, whose energy was extremely great. A short-term rise in air temperature following the explosion would have changed to a longer period of cooling because of a considerable increase in aerosol particles in the stratosphere. The physical mechanism of such a cooling is probably, to a great extent, similar to that which leads to a cooling after a single major volcanic eruption, although the temperature drop in the former case might be much more pronounced due to the tremendously powerful explosion. It might be thought that after the impact of a relatively large celestial body great quantities of aerosol particles would be released into the atmosphere as happens after explosive volcanic eruptions. In recent years attempts have been made to find dissipated meteorite matter in layers corresponding to the epochs of major animal extinctions. These attempts have yielded interesting results.

For instance, excessive amounts of iridium, compatible with the matter of extraterrestrial bodies, have been discovered in the layer corresponding to the end of the Cretaceous. Similar anomalies have also been detected in the concentrations of some metals of the platinum group as well as of nickel and cobalt. This phenomenon has been observed in different continents and islands, including Europe, Africa, North America and New Zealand, as well as at the bottom of the Central Pacific, which proves that it is global in nature. According to the amount of iridium in sedimentary layers, the size of the extraterrestrial body that arrived at the Earth has been calculated. This body appears to have been between 5 and 16 km in diameter with a mass of an order of $n \times 10^7$ g. Since these magnitudes are characteristic of asteroids, it has been assumed that the celestial body which collided with the Earth was an asteroid. Kinetic energy generated by the impact of such an asteroid is about $10^{30}$ erg, which roughly corresponds to an explosion of $10^{14}$ tonnes of TNT (Alvarez *et al.*, 1980; Ganapathy, 1980; Hsü, 1980; Newell, 1980; Smit and Hertogen, 1980; Hsü *et al.*, 1982; O'Keefe and Ahrens, 1982; and others).

In the different studies of this problem, several hypotheses have been advanced to explain the likely mechanism by which the asteroid impact caused extinctions at the end of the Cretaceous. In particular, it has been suggested that this impact resulted in the injection of dust into the atmosphere. The quantity of dust probably exceeded by three orders of magnitude the amount of aerosol released into the stratosphere after the greatest explosive eruption of the last 150 years (Krakatoa in 1883). It has been suggested that for several years after the impact the stratospheric dust screen did not permit solar radiation to reach the Earth's surface, which resulted in the disruption of photosynthesis and caused mass extinction (Alvarez *et al.*, 1980). It has been argued by several authors that the extinction of animals was associated with the heating of the atmosphere that took place either immediately after the impact of the asteroid or thousands of years later, as a result of the accumulation of carbon dioxide in the atmosphere, whose absorption in the ocean decreased because of the destruction of phytoplankton due to the cessation of photosynthesis (Hsü *et al.*, 1982).

These suggested mechanisms for mass extinction in the late Cretaceous seem to be improbable. A cessation of photosynthesis over the entire planet for a few years should have led to the extinction of almost all living organisms. Since a large number of organisms survived the mass extinction, this supposition is evidently based on a considerable overestimation of the amount of dust that was released into the stratosphere following the asteroid's impact. It might also be thought that this impact was not accompanied by an appreciable heating of the whole atmosphere. If the asteroid impact was in the ocean (the evidence for which is the absence of traces of a crater of appropriate size and age on the continents), then the conversion of the entire energy of the impact into heat would not have led to any considerable increase in the temperature of the ocean. Moreover, the major portion of the generated energy would have probably been absorbed by the Earth's crust which the asteroid entered deep. The influence of heating of the atmosphere on the extinction also seems improbable because of the selective nature of vertebrate extinction. Many groups of vertebrates that had no heat regulation vanished, while the warm-blooded animals (mammals and birds) were almost unaffected. This observation makes it very probable that the extinction at the end of the Cretaceous was caused by cooling.

The mechanism of such a cooling has been discussed in the author's earlier books (Budyko, 1971, 1980, 1982). In the second of these an estimate was given of the climatic change caused by the impact of a celestial body, whose volume was hundreds of millions of cubic metres. The rough calculation shows that in this case, the total radiation reaching the Earth's surface is reduced by more than 50%, which results in a 5—10 °C decrease in mean temperature near the Earth's surface over a period of one or two years.[2]

A more detailed estimate of the effect on the Earth's climate of an impact of a large celestial body is presented in the work of Toon *et al.* (1982). This study concludes that because of the coagulation and sedimentation of aerosol particles produced by the explosion of the asteroid, high aerosol concentrations are maintained only for a time interval up to several months after the arrival of the asteroid. During this time, incoming solar radiation drops to a level that is insufficient to support photosynthesis. The calculation shows that

[2] The results of this calculation have not appeared in the English translation of this book (*The Earth's Climate: Past and Future*, 1982) which is slightly abridged as compared to the original.

the reduction of solar radiation will lead to a decrease in the mean temperature of the lower air layer above the oceans by about 2—3 °C for more than two years, and above the continents by several tens of degrees for about half a year. It is possible to estimate from this work that mean global temperature is likely to decrease on average by 9 °C within the first ten months following the asteroid impact, and by 6 ° C during the first 20 months. These magnitudes are rather close to the results obtained in our calculation, although the contents of the comparable calculations do not coincide completely.

The work of Toon and his collaborators presents more accurate estimates of variations in the persistence of high atmospheric aerosol concentration after the asteroid impact and establishes that high aerosol concentrations are maintained during a relatively short time period. At the same time it seems that in this work the difference between the temperature decreases above the oceans and the continents is overestimated. It is probable that these authors did not take into account the likelihood of a considerable intensification of heat transfer by air currents from the warmer oceans into the continents following the asteroid impact, which would have made the difference between the temperature decreases on the continents and the oceans appreciably smaller. A temperature decrease of 5—10 °C was undoubtedly sufficient for the extinction of stenothermal life forms, which at the end of the Mesozoic were abundant since these life forms evolved during an era characterized by a warm or hot climate at all latitudes.

It should be noted that an abrupt decrease in the temperature of the Earth's surface as a result of an increase in the amount of dust in the atmosphere has also been observed to occur on Mars (Golitsyn, 1974).

Besides the volcanic and meteoritic hypotheses, there are some other hypotheses concerning the causes of the sharp changes in natural conditions during the geological past, although these hypotheses are often insufficiently substantiated and have provoked several objections (see reviews — Newell, 1967; Herman, 1981; and others). In particular, a well-known idea about the possible effect on terrestrial organisms of the explosions of supernovae at a relatively close distance to the solar system has been tested by investigating data on the isotopic content of a number of elements for the epochs of mass extinctions. However, such analyses have yielded negative results ('The Quest . . .', 1980).

Although the problem of mechanisms for causing abrupt changes in natural conditions is still far from exhausted and requires further investigations, taking into account the fact that the 'climatic zone of life' (the range of atmospheric conditions in which organisms can exist) is narrow, it should be considered highly probable that for the long time period of the biosphere's existence, this zone has disappeared at different times for individual groups of organisms. This has occurred as a result of the effects on the environment of various external factors, including those discussed above. The most important example of such changes in natural conditions is the ecological crisis that took place at the end of the Mesozoic.

### 8.2.2. EXTINCTION AT THE END OF THE MESOZOIC

The boundary between the Cretaceous and Tertiary periods is one of the most important landmarks in geological history. This boundary corresponds to the mass extinction of many organisms, including the dinosaurs.

In a similar way to other epochs of mass extinction, the end of the Upper Cretaceous is marked by the extinction of organisms belonging to different groups. Thus, at this time a considerable (perhaps, the greatest) number of the marine plankton species disappeared. Three of the four families and 17 of the 23 genera of planktonic foraminifera became extinct at the end of the Upper Cretaceous. At the close of the Cretaceous the majority of families and genera of the then existent bivalves vanished. A lot of brachiopods, ostracodes, ammonites, belemnites and other invertebrate animals became extinct at this time. Considerable changes took place in the group of bony fish. In particular only 8 of the 38 genera and 11 of the 50 species of bony fish survived the transition from the last stage of the Cretaceous period (the Maestrichtian) to the first stage of the Tertiary period (the Danian).

These data are derived from the monograph *The Development and Change* ... (1978), where some conclusions are offered concerning the extinction of various animals at the end of the Cretaceous. One of these conclusions suggests that the extinction of individual groups was a gradual process, and the diversity of these groups, in some cases, started to decrease long before the end of the Cretaceous. Generally speaking, this conclusion does not contradict the hypothesis that the extinction at the end of the Mesozoic took place over a short time interval. As is known, with more or less stable environmental conditions for sufficiently long time intervals there are usually noticeable fluctuations in the number of species and genera of the higher taxonomic groups. However, other conditions being equal, short-term mass extinctions might annihilate less diverse groups more quickly because of the more limited adaptation of representatives of these groups to environmental conditions.

It is possible that in certain cases, the reduction in the diversity of the groups, might not be real and, rather, reflect the lack of sufficient palaeontological evidence for the time interval preceding the extinction (see earlier discussion). The data on the most abundant fossils, in particular the foraminifera, clearly show that the extinction in question was a sudden event. Examination of the planktonic changes at the transition from the Cretaceous to the Tertiary permits the conclusion that the extinction took place within a time period of no more than 200 years (Smit and Hertogen, 1980). Since the authors of this study obtained information confirming the increased abundance of iridium and some other metals in the layer corresponding to the epoch of extinction, it is natural that they supported those who explained the mass extinction as a result of asteroid impact.

Of particular interest in the late Cretaceous extinction is the fate of the representatives of the class of reptiles. This group of animals occupied numerous ecological niches in the Mesozoic era and included various land, marine and flying animals. Among the Mesozoic reptiles were many large and gigantic forms, some of which were the largest land animals that have ever occurred in the history of the Earth.

According to Colbert (1965), ten orders of reptiles existed at the close of the Cretaceous, five of which were extinct by the beginning of the Tertiary. It can be seen in Table 8.3, based on the table constructed by Colbert, that by the beginning of the Tertiary only 30% of the reptilian families that existed in the Upper Cretaceous survived.

TABLE 8.3

Extinction of reptilian families at the close of the Cretaceous

| Order | Number of Upper Cretaceous families | Number of families at the beginning of the Tertiary |
|---|---|---|
| Turtles | 15 | 8 |
| Eosuchians | 1 | 1 |
| Rhynchocephalians | 1 | 1 |
| Lizards and snakes | 6 | 3 |
| Crocodilians | 6 | 2 |
| Saurischians | 6 | 0 |
| Ornithischians | 9 | 0 |
| Pterosaurs | 1 | 0 |
| Ichthyosaurs | 1 | 0 |
| Plesiosaurs | 4 | 0 |
| Total | 50 | 15 |

The data in this table probably overestimate the number of reptilian families that became extinct at this time. In particular the ichthyosaurs probably died out before the end of the Cretaceous. The data on the extinction of many families of turtles, lizards and snakes at the end of this period are evidently out-of-date now. However, it is probable that at the close of the Cretaceous such important groups of reptiles as the dinosaurs, pterosaurs, plesiosaurs and giant marine lizards, the mosasaurs, became extinct. It is worth noting that in the transition from the Cretaceous to the Tertiary the extinction of more ancient forms of crocodiles, belonging to the Mesosuchia, accelerated.

The data on mammals are more limited compared to those on reptiles and do not reveal a marked change in the groups of animals which existed at the transition from the Cretaceous to the Tertiary. A similar conclusion might be drawn for birds, for which palaeontological record is even more incomplete.

Examination of the data on changes in faunas at the end of the Cretaceous leads to the conclusion that these faunas were subject to mass extinction that took place within a limited time interval. Even the most conservative estimate allows one to conclude that many of the above-mentioned large groups of organisms vanished within the time interval of no more than several million years.

The limited accuracy of dating the available fossils and inadequacy of the palaeontological record do not permit a fully reliable determination of the duration of the epoch of mass extinction, and of the synchronism of these epochs for each group, using palaeontological evidence. Palaeontological evidence allows us to evaluate only the upper limit of the duration of extinction.

As a result of the events that took place at the end of the Mesozoic, the fauna of the early Cenozoic turned out to be very different from that of the Upper Cretaceous.

The extinction of the dinosaurs, which flourished since the second half of the Triassic, when their various forms rapidly spread across the entire land surface is a particularly interesting feature of the late Cretaceous mass extinction. At the transition from the Triassic to the Jurassic, the first extinction of dinosaurs occurred, resulting in the disappearance of some of the Triassic groups of these animals. Little information is available on the dinosaurs of the Lower and Middle Jurassic. The second expansion of the dinosaurs occurred in the Upper Jurassic. For the time of this expansion up to the end of the Cretaceous, numerous forms of herbivorous and carnivorous dinosaurs existed, among which large species prevailed and giant forms were often found. Of particularly large sizes were the sauropods which could weigh up to 50—80 tonnes. This is approximately ten times greater than the weight of the largest land animals of today.

From the end of the Triassic to the end of the Cretaceous (i.e. for more than 100 million years), the dinosaurs existed simultaneously with mammals, who were unable to remove the dinosaurs from their ecological niches of large carnivorous and herbivorous land animals. The dinosaur's success in the struggle for existence was a part of the general success of the archosaurs (including the pterosaurs, which in Mesozoic time were more widespread than birds) and reptiles on the whole. These animals occupied a vast number of different ecological niches.

The fate of the dinosaurs has often attracted the attention of investigators, frequently drawing interest from scientists of non-biological specialities and also from a wide circle of laymen. Sometimes this interest is attributed to the extraordinary appearance and gigantic sizes of many dinosaurs. However, it should be hoped that there are deeper reasons for this attention to the history of dinosaurs. In particular, this history is of considerable importance in understanding the origin of the noosphere, which will be more closely discussed below.

While studying the influence of changes in the environment on the evolution of dinosaurs, it is possible to use the data from Fig. 8.2, which describes fluctuations in the amount of atmospheric oxygen in the Phanerozoic. It follows from this figure that the dinosaurs first appeared and spread in the epoch of a comparatively low atmospheric oxygen content, which, however, was rapidly increasing throughout the second half of the Triassic. The dinosaurs appeared approximately 20 million years earlier than the mammals; this is possibly explained by a lower level of energy expenditure by reptiles, which could evolve with smaller amounts of oxygen compared to mammals. The earlier emergence of dinosaurs was probably of great importance for their success in subsequent competition with mammals which appeared incapable of taking the ecological niches occupied by these more primitive animals.

Although not considering in detail the first extinction of dinosaurs at the end of the Triassic, it should be mentioned that this extinction was evidently analogous to that of the late Cretaceous, but smaller in scale and did not lead to the total annihilation of dinosaurs. A rapid radiation of dinosaurs in the Jurassic was probably associated with increasing concentrations of atmospheric oxygen, which made it possible for giant forms to occur.

The existence of the largest dinosaurs for a considerable part of the Mesozoic was possible as a result of the high level of atmospheric oxygen, which was characteristic of the second half of the Jurassic and the first half of the Cretaceous. The effect on the evolution of dinosaurs due to the decrease in atmospheric oxygen in the second half of the Cretaceous is of considerable interest. Although, as seen in Fig. 8.2, the oxygen amount still remained rather high at the end of this period (higher than at the beginning of the Jurassic), the conditions for the existence of the largest dinosaurs (as well as large flying reptiles) deteriorated appreciably at this time. Since the oxygen decrease took place slowly, the dinosaurs, like other animals of that time, could gradually adapt to the changing environmental conditions. This is confirmed by the existence of gigantic species of dinosaurs up to the very end of the Cretaceous. One might have thought, however, that the resistance of these species to the unfavourable changes in the environment at the close of the Cretaceous, in a similar way to all other large animals, was less than in earlier epochs.

In considering the causes of the dinosaurs' extinction, it should be emphasized that it is illogical to consider their extinction separately from the causes of the extinction of many other groups of organisms in the late Cretaceous. Taking this into account, it is immediately possible to reject the hypotheses about the local changes in ecological conditions of the life zone of dinosaurs, since they are, as a rule, inappropriate for explaining the causes of mass extinction of marine plankton and of many other groups of marine animals. It seems probable that, as in all critical epochs of geological history, the late-Cretaceous extinction of organisms resulted from global changes in the natural environment and was selective in its effect, i.e. its influence on various groups of organisms was different. This notion is important in clarifying the mechanism of the extinction.

Considering the fate of vertebrates at the close of the Cretaceous, it should be reminded that mammals and birds, as far as this can be seen in palaeontological record, suffered only slightly or not at all. At the same time the extinction led to the disappearance of the majority of the reptilian families. The opinion has long been advanced that this difference is explained by exothermal properties of reptiles, which are more susceptible to decreases in environmental temperature than endothermal animals (mammals and birds). This hypothesis has been questioned in recent years when it has been claimed that dinosaurs, as distinct from other reptiles, were endothermal animals. In view of the importance of this problem, it is now discussed in more detail.

### 8.2.3. THE THERMAL REGIME OF THE DINOSAURS

Although the hypothesis concerning endothermy of dinosaurs was made in the middle of the 1960s, this hypothesis has attracted most attention since the publication of Bakker's paper (1971), in which the relationship between the thermal regulation of dinosaurs and the structure of their limbs was considered. Bakker noted that almost all primitive land vertebrates moved about on broadly spaced sprawling legs, whereas land mammals (except for monotremes) and birds are supported by straight legs, which lightens the load on the muscles of their limbs, by shifting it partly onto the bones and joints. According to Bakker, birds and mammals are 'walking erect'. Considering that the upright posture corresponds to a higher activity of animals, Bakker came to the conclusion that the erect dinosaurs were homeothermal animals like mammals and birds. In this paper he suggested that the first representatives of archosaurs, which moved on sprawling legs, might possibly have been poikilotherms, but their mid-Triassic forms with straighter and elongated limbs possessed a higher level of metabolism, and the dinosaurs that appeared later attained homeothermy.

The work of Bakker was discussed by Feduccia (1973) who pointed out that since among lower vertebrates there are examples of partly developed endothermy (warm-bloodedness), the same possibility could also apply to dinosaurs. However, Feduccia considered that the relationship between the ability to walk erect and endothermy was not sufficiently grounded, because in modern animals these properties do not necessarily coincide. Feduccia considered the upright pose of dinosaurs an adaptation necessary for large animals. He referred to Jerison's conclusion (1973) that the brain size of dinosaurs did not exceed that of reptiles and was much less than the brain size of birds and mammals. Although the centre of thermoregulation occupies a relatively small part of the brain, the existence of this centre in modern vertebrates is associated with comparatively large brain sizes. Although in the warm climate of Mesozoic time exothermy could have been in many cases an advantage in the struggle for existence, Feduccia concluded that there are no grounds to suggest that dinosaurs differed from typical reptiles.

Another paper by Bakker (1974) presented additional arguments in favour of endothermy in dinosaurs. One of these is the dinosaurian carnivore to herbivore ratio calculated from data on the number of fossils found. According to Bakker, this ratio appears to be relatively low, close to that in existent mammals entering the predator—prey system and considerably lower than a similar ratio for Palaeozoic reptiles. Since exothermal (cold blooded) predators use much less food than endothermal animals, their relative number might be greater. Because of this, Bakker thought that such a comparison was enough to prove the occurrence of endothermy in dinosaurs. Another argument put forward by Bakker for endothermy is the more complex bone structure of dinosaurs compared to that of amphibians and reptiles. The dinosaur bone structure is similar to the bone structure of higher vertebrates. Bakker's third argument (partly advanced in his first paper) states that large dinosaurs with long limbs must have moved quickly, therefore they needed expend an enormous amount of energy which, in his opinion, is difficult to provide with a low level of metabolism.

Several objections to this concept of the dinosaurs' endothermy have been raised. As mentioned earlier, Bakker considered that one of the arguments in favour of the concept of endothermy in dinosaurs is the small value of predator/prey ratio, which is similar to that of mammals. This ratio has been studied by Farlow (1976), who through analysis of empirical evidence came to an opposite conclusion to that of Bakker. At the same time Farlow emphasized that it is difficult to solve this question since the available information on the density of dinosaur populations is limited. It is evident that to satisfactorily resolve this question it is necessary to have mass data on the fossil dinosaurs that entered the predator—prey systems. Unfortunately, palaeontology does not possess such information. For example, the data on Tyrannosaurus used by Bakker refer to a small number of animals and therefore have no statistical value. Taking into account that the conditions for burial of the remains of carnivorous and herbivorous dinosaurs might not coincide, it is likely that we are a long way from obtaining realistic estimates of the density of dinosaur populations for specific ecological systems.

Another of Bakker's arguments (i.e. the peculiarity of the bone structure of dinosaurs compared to that of other reptiles, established in the work of Riqlet (1969)) does not solve the question of endothermy of dinosaurs. Tatarinov (1976) notes that there is no direct relationship between the bone structure of animals and their thermal regulation.

The third argument put forward by supporters of the concept of endothermy of dinosaurs is based on the suggestion that dinosaurs were highly active organisms, while such activity is not possible in reptiles with low or average levels of metabolism. Bakker as well as Desmond (1976) have suggested in particular that the giant herbivorous dinosaurs (the sauropods) were not, as generally thought, semi-aquatic animals but fed on land like giraffes, using their long neck for browsing the leaves from tree-tops. We shall consider this hypothesis below. Expanding this point of view, Desmond wrote that if a small lizard weighing only a few ounces could endure only short periods of activity, how could an 8 tonne predator like Tyrannosaurus be successful in the struggle against its prey and where did it obtain its energy from. However, questions such as these are not logical. For example, the crocodiles of the Nile which are much more closely related to dinosaurs than the lizard mentioned by Desmond, have enough energy to attack not only buffaloes but even rhinoceroses.

The arguments of Bakker and Desmond about the brain size of dinosaurs are difficult to understand. The small size of the brains of the large dinosaurs compared to the weight of their bodies is well known. This is also characteristic of many of the smaller forms of dinosaurs (Jerison, 1973). Suggesting that the relative brain size of some small dinosaurs was comparable with that of birds, Bakker and Desmond thought that this is sufficient grounds for concluding that dinosaurs were endothermal. However, it would seem that if significance is attached to brain size as a factor in heat regulation (in other cases these authors deny such significance), the indicated comparison gives rise to the conclusion that the overwhelming majority of dinosaurs were exothermic.

To study the thermal regime of dinosaurs we have used the physical theory of the thermal regime of animals presented in Chapter 5 and some considerations concerning their energy balance (Budyko, 1978). Below are given the main conclusions of this study. More recently, similar views have been put forward by McGowen (1979).

Since dinosaurs had no hairy coat and evaporation from the surface of their bodies was evidently insignificant, the equation for the heat balance at the body surface of a dinosaur can be written in the form:

$$R_0 + M = (\rho c_p D + 4\sigma T^3)(T_s - T), \tag{8.1}$$

where $R_0$ is the radiation balance of the animal's body surface when determining the heat loss due to longwave radiation at the air temperature $T$, $M$ the heat production of the animal ($R_0$ and $M$ refer to unit surface of the animal's body); $\rho$ the air density; $c_p$ the specific heat capacity of the air under constant pressure; $D$ the coefficient describing the intensity of turbulent heat exchange between the animal's body surface and the atmosphere; $\sigma$ the Stefan constant; $T_s$ the mean temperature of the body surface of the animal.

This equation characterizes the steady-state of the thermal regime of the animal. Under unstable conditions, the rate of variations in the body temperature can be calculated by the equation:

$$\rho_1 c_1 W \frac{\partial T_s}{\partial t} = [R_0 + M - (\rho c_p D + 4\sigma T^3)(T_s - T)]S, \qquad (8.2)$$

where $\rho_1$ and $c_1$ are the density and heat capacity of the animal's body; $W$ is its mass; $S$ is the body surface. It follows from Equation (8.1) that the difference between the temperature of the animal's body and of the air under stable conditions is equal to

$$T_s - T = \frac{R_0 + M}{\rho c_p D + 4\sigma T^3}. \qquad (8.3)$$

This formula can be used for evaluating the thermal state of animals whose temperature varies little in time. It also characterizes the thermal state of small animals both under stable and under variable conditions of their thermal balance, since in this case the left-hand side of Equation (8.2), proportional to the weight of the animal, is usually much smaller than the greatest values of the terms on the right-hand side of the equation, proportional to the body surface.

Let us assume in the first approximation that the radiation balance of the animal is close to the radiation balance of the Earth's surface. As the observational data show, the magnitude of $R_0$ for the Earth's surface varies approximately from +400 to −150 W m$^{-2}$. Since the coefficient $D$ is a function of wind velocity, the sum of $\rho c_p D + 4\sigma T^3$ depends on this velocity, increasing with its increase. For average conditions, the indicated sum is equal to 20 W m$^{-2}$ °C$^{-1}$. Taking this into account, it is possible to find from Equation (8.3) that the body temperature of exothermal animals (for which $M = 0$) in the open space may rise to approximately 20 °C above the air temperature and drop to 7 °C below this temperature. Observations of present-day small reptiles have shown that in the sun their body temperature may indeed be up to 20 °C higher than the air temperature and can reach even greater values in windless weather.

Using Equation (8.3), it may be concluded that endothermy could not have exerted any essential influence on the temperature of small dinosaurs, whose mass did not exceed the body mass of man. Let us suppose that the heat production of these animals was equal to the heat production of man, which, as indicated in Chapter 5, is about 55 W m$^{-2}$ in the state of rest. This would have raised their body temperature by approximately 3 °C, which is insufficient for making up for the regularly occurring decreases in the air temperature. Thus, the relatively small dinosaurs, irrespective of the presence or absence of endothermy, must have been able to maintain the optimum temperature of their body in the same way as other reptiles.

The calculations based on the formulas for the thermal regime of animals given in Chapter 5 show that endothermy is an effective method of maintaining the body temperature of small animals provided they have heat insulating covers such as the hairy coats of mammals.

Now let us consider the heat regime of large dinosaurs. The heat loss per unit mass diminishes with increasing animal size due to a decrease in the ratio of surface area to mass (the law of Max Rubner); in endothermal animals this decrease is to a great extent compensated for by a reduction in the heat production per unit mass of the body.

From the results of experiments on mammals and birds, the following formula was obtained (Kleiber, 1932, 1961):

$$M = 3\,W^{3/4}, \tag{8.4}$$

where $M$ is the energy release at the expense of metabolism of the animal in kilocalories per hour, $W$ is the mass in kilograms. Considering that the surface area of the animal's body is proportional to its mass to the power of 2/3, we find that with increasing animal size, other conditions being equal, the metabolism per unit surface of the body increases in proportion to its mass to the power of 1/12. According to zoogeographical comparisons, on which the rule of Bergman is based (larger specimens of endothermal animals are usually found under the colder climatic conditions), this dependence is of some ecological importance.

Let us estimate how much the body temperature of the largest dinosaurs, the sauropods, could differ from the air temperature, provided they were endothermal. Considering that the mass of these dinosaurs could be almost 1000 times greater than the mass of man, we find by the above formulas that the corresponding difference is 4—5 °C.

It is worth mentioning that the largest dinosaurs could be homeothermal animals to some extent even at low levels of metabolism. It follows from Equation (8.2) that fluctuations in the body temperature of the animal depends considerably on the value of $W/S$ (i.e. on the animal's body size) decreasing with an increase in size. By solving Equation (8.2), it is possible to find that the body sizes of the dinosaurs did not produce an appreciable influence on their thermal regime during relatively long-term temperature fluctuations of the order of weeks and months. However, for short-term variations in air temperature, which take place particularly during the course of the day, an increase in the animal's body size can noticeably enhance the stability of the temperature regime of the animal.

The daily temperature fluctuations of small reptiles in the open roughly correspond to fluctuations in the temperature of the Earth's surface, whereas the amplitude of variations in the body temperature of the largest dinosaurs could have been more than ten times smaller than the amplitude of daily variations in the temperature of the Earth's surface. According to the earlier opinion of Colbert *et al.* (1946), this resulted in the large dinosaurs being practically homeothermal animals, even if they did not attain endothermy, provided that they lived under climatic conditions with an insignificant annual cycle of temperature. As mentioned above, such a climate was widespread in Mesozoic time.

Let us evaluate the thermal regime of the large dinosaurs, with both exothermy and endothermy. In the former case the mean body temperature of a dinosaur can be calculated by formula (8.3) with $M = 0$. The characteristic average daily magnitude of the radiation balance of the Earth's surface is 100—150 W m$^{-2}$. Consequently, if the dinosaurs inhabited open land, their mean body temperature would have been higher than the air temperature by 5—7 °C. Taking into account that the climate of the Mesozoic era was much warmer than it is today, it may be inferred that the mean body temperature of large dinosaurs in the open, even with their exothermy, would have been higher than the air temperature under contemporary climatic conditions of the tropics by approximately 10 °C. With endothermy of the giant dinosaurs, this difference could have reached 14—15 °C.

As there are no data on the range of body temperatures within which the existence of dinosaurs was possible, we shall use data from crocodiles for evaluating the upper lethal temperature for dinosaurs since crocodiles are the closest living relatives of the dinosaurs. As established by Colbert, Cowels and Bogert in the afore-mentioned work, the upper lethal body temperature of an American alligator is 38 °C and only slightly exceeds their optimum temperature of 32—35 °C. By exposing the alligators to the open sun, with not very high air temperatures, the experimenters observed that alligator death from overheating took place in a time interval of between one hour (small animals) to three hours (a large alligator). Since under the present climatic conditions of the tropics, air temperature often rises above 38 °C, it is unlikely that the large dinosaurs could have existed with their body temperatures above the lethal temperature of alligators. In view of this, the conventional concept that the sauropods, in a similar way to modern crocodiles, were exothermal animals, seems highly probable. Like crocodiles, they spent considerable time in water, which prevented their overheating by the solar rays.

It is relevant to this discussion that the largest modern land mammals in the tropics, such as elephants, rhinoceros and hippopotamuses, are almost hairless. As mentioned above, this greatly reduces the influence of endothermy on the heat regime of animals. The presence of such an adaptation, as well as the tendency of these animals to avoid sunlit places in the hottest weather, clearly shows that overheating is dangerous for large land endotherms under the conditions of a warm climate. For giant endothermal dinosaurs this threat would have been even more acute because of their large sizes and the higher Mesozoic air temperatures. Thus, this examination of the heat balance of the largest dinosaurs allows us to conclude that they were exothermal animals, refraining from coming out of water in the heat of the day. The lesser forms of dinosaurs could probably protect themselves against the solar rays more easily than the giant forms while remaining on land.

It is clear that in the absence of a heat-insulating coat, endothermy added no essential advantages for protection against cold in small dinosaurs, while it increased the risk of overheating. The principal difference between the heat regime of large and small dinosaurs was that while large dinosaurs could only avoid overheating by staying in water bodies, small dinosaurs, like many modern reptiles, could find various sites on land to escape the adverse effects of temperature fluctuations.

The data presented here show the low probability of the development of endothermy in the giant dinosaurs. Since in present-day members of different classes of vertebrates, the mechanism of temperature control within each class is more or less the same, it might be thought that the above conclusion also applies to other, not so large forms of the dinosaurs.

The study of the thermal regime of the dinosaurs can also be carried out by examining their feeding habits. In the discussion of the dinosaur's endothermy the opinion was offered that erect walking was an adaptation necessary to provide the dinosaurs with ample food, the amount of which corresponded to the energy expenditure at a high metabolic level. However, the possibility cannot be excluded that walking on upright legs was also advantageous for exothermal animals under conditions of the desperate struggle for existence among various Mesozoic reptiles.

Let us consider the energy balance of the largest plant-eating dinosaurs, the sauropods. It has long been noted that there is a contradiction between the requirement of these animals for enormous amounts of food and their very small skulls with tiny and weak jaws. With these jaws it was impossible to seize a large mass of the plants these dinosaurs fed on. This contradiction has been mentioned in the discussion of the thermal regime of the dinosaurs, although it has not been paid due attention.

To study this contradiction we will use the afore-mentioned analogy between some of the sauropods and the giraffes and assume, according to the idea of supporters of the concept about the dinosaur's endothermy, that these dinosaurs, like modern giraffes, subsisted on leaves which they browsed off the top branches of trees. Because the mass of the largest dinosaurs was approximately 100 times greater than the mass of a giraffe and because energy consumption increases in proportion to the animal's mass to the power of 3/4, we find that for the endothermal dinosaur the required amount of food must have been thirty times greater than is necessary for the giraffe. According to field observations, giraffes spend most of the day eating up the available food. It is evident that to increase the time they spend on their daily nutrition by thirty times is impossible. Consequently, the existence of an endothermal animal, whose body structure and way of feeding is similar to that of a giraffe but whose mass is far greater, is very unlikely.

It is now logical to consider how the energy balance of the largest living herbivorous mammals (elephants, rhinoceros, hippopotamuses), whose mass can exceed that of the giraffe by up to ten times, is maintained. It should be noted that all these animals have massive skulls and jaws, which are in much better proportion to the total size of the body, compared to the giraffe. Therefore each of these indicated large animals can take hold of a much greater amount of vegetable mass per unit time than the giraffe. It is typical that the skulls of the largest extinct land herbivorous mammals (some species of hornless rhinoceros) were also relatively large.

A similar relationship between body size and jaw size occurs in the whalebone whales, the group which includes the largest animals that have ever existed. Since the whalebone whales have to trap for their food a great number of small organisms occurring in low concentrations in the water, the relative size of their skulls is considerable and can attain one-third of the total length in the largest Greenland whale. Since a further increase in the relative size of the whale's skull is hardly possible, it may be thought that the maximum sizes of the whalebone whales (as well as the largest herbivorous mammals) are to a great extent limited by their energy balance.

It should be mentioned that this ratio between the amount of food necessary for a giraffe and for a giant endothermal dinosaur coincides with the ratio between the metabolic level of exothermal and endothermal animals (after Bakker). This coincidence would seem to permit the conclusion that the sauropods' way of life could have been similar to that of the giraffe if they were exothermal animals. However, such a conclusion is erroneous. Exothermal animals would not have been able to maintain the level of activity attainable by endothermal giraffes throughout the day. This activity is necessary to allow the browsing of numerous portions of vegetable material, each being very small in mass. Furthermore, the jaws and teeth of sauropods were much less fit (compared to the giraffes) for browsing off and ruminating the individual parts of plants.

It seems probable that the feeding method of sauropods with small skulls differed considerably from that of modern land herbivorous mammals. It has been suggested that these dinosaurs, as with some contemporary lizards (e.g. marine iguanas), subsisted on algae, which often occurred in high concentrations in shallow water.

Assuming that other groups of dinosaurs are also exothermal, it should be emphasized that the way of life of smaller herbivorous dinosaurs differed from that of the sauropods. For instance, the herbivorous dinosaurs of the group *Iguanodon* evidently consumed land plants. This is indicated in particular by relatively large size of their skulls.

Therefore, the conclusion that the sauropods were exothermal and to a great extent aquatic can be tested by several independent methods. The first of these is the evaluation of the heat balance of the dinosaurs, which shows the likelihood of their overheating when remaining for a long period in the sun, provided they were endotherms. The second is the comparison of the food budgets of dinosaurs and other large plant-eating mammals, from which it follows that it was impossible to provide the energy necessary for giant dinosaurs if they dwelled constantly on land. The third method of clarifying the life style of the large dinosaurs relates to the great difference in the energy used by animals moving in water and on land. Taking into account that the data on the thermal and energy balances of the giant dinosaurs implies that they were exothermal animals, it is easy to understand how extremely difficult it was for these animals to maintain the necessary energy consumption while moving, if they stayed long out of water. A similar conclusion has been made in numerous studies concerned with the morphology of dinosaurs by outstanding palaeontologists of the past. These considerations show that it is highly improbable that the dinosaurs were endothermal animals.

This question of endothermy of dinosaurs has been discussed in recent years at a number of scientific meetings, including a Conference of the American Association for the Advancement of Science held in 1978. In the report of this Conference, the supporters of the idea that dinosaurs were endothermic could not satisfy the supporters of the traditional opinion of their exothermy (Miller, 1978).

### 8.2.4. THE CAUSES OF THE LATE MESOZOIC EXTINCTION

A simple statistical analysis shows that the extinctions of many groups of reptiles at the close of the Cretaceous period were not a chance coincidence, but the result of a general change in environmental conditions. We shall now give an example of such an analysis (Budyko, 1971). According to Colbert, five of the ten reptilian orders vanished at the end of the Cretaceous. Let us assume that the average length of life of an order of reptiles that became extinct in Cretaceous time was approximately 100 million years and the time of disappearance of individual orders can be determined with an accuracy of up to five million years. In this case, the probability that the extinction of five out of ten orders during five million years was a chance coincidence, calculated by the Poisson formula, is 1/10 000 (i.e. an extremely small probability). Consequently, the coincidence in the extinctions of these groups of reptiles suggests that they were caused by certain changes in the natural environment, rather than happening by chance. A similar conclusion on the nature of the extinction of animals at the end of the Cretaceous has been drawn by Raup and Sepkosky (1982) on the basis of statistical analysis of data on marine animals.

Hypotheses about the nature of the environmental changes leading to mass extinction can be based on the survival of endothermal animals (mammals and birds) while many exothermal reptiles became extinct. Because of this, the hypothesis that a global cooling was the main cause of the late Mesozoic extinction is highly probable. This hypothesis can be confirmed by three additional considerations. One of them is concerned with the disappearance of a large number of marine organisms at this time. As shown, a considerable number of these organisms are stenotherms, which cannot tolerate even relatively small fluctuations in temperature. The second additional consideration in favour of this hypothesis is the observation that mainly comparatively small land reptiles, that could burrow or move away from the extremes of cold, were truly successful after the extinction, while all large reptiles failed to survive the end of the Cretaceous. The third consideration is that short-term decreases in temperature, which in our age often take place in the middle latitudes and sometimes even in the tropics, occurred very rarely or not at all in the Mesozoic era. This occurred because of the prevalence of a warm or hot climate all over the world and the absence of a zone of cold climate, where in the modern epoch, air masses of low temperature are formed which in some cases can penetrate into the zones of warm climate. Because of this Mesozoic organisms must have been less resistant, compared to modern forms, to decreases in the environmental temperature.

The influence of sharp temperature decreases on the vital activity of different organisms is quite varied. As mentioned earlier (Budyko, 1971), for each organism the magnitude of a temperature lowering can be established: (a) leading immediately to its death; (b) reducing its activity to the limits where it perishes in the course of the struggle for existence; (c) reducing the organism's resistance against infectious diseases, as a result of which it falls their inevitable victim; (d) disturbing the process of reproduction. Moreover, the destruction of the organism can be evoked by the extinction of plants or animals, on which its existence depends (for example, the sources of its nutrition).

For stenothermal organisms the temperature decreases need only be very small. It is known that many contemporary tropical animals perish when the temperature of their environment decreases by only a few degrees. It follows from these considerations that Mesozoic animals could be even more sensitive to temperature decreases.

In addition, the data on the decreasing content of atmospheric oxygen in the second half of the Mesozoic are relevant. This could lessen the resistance of many animals (particularly of large size and also flying forms) to the effects of decreases in environmental temperature. Furthermore, palaeotemperature data indicate that a cooling trend occurred at the end of the Cretaceous. This cooling could have led to worsening of conditions necessary for the survival of the groups of organisms that were particularly sensitive to thermal conditions. However, it is unlikely that these two factors, which produced an unfavourable influence on the vital activity of many animals at the close of the Mesozoic, could have been the principal reason for the mass extinction at the end of the Mesozoic. Both the decrease in the oxygen concentration and the cooling revealed by palaeotemperature data did not occur suddenly but developed over the course of millions of years. This long time period would allow many organisms to adapt to such environmental changes in the process of evolution.

Very different consequences would have arisen from environmental fluctuations of the same scale if these took place over very short time periods and therefore did not permit either evolutionary changes in organisms or their acclimatization to the less favourable environment. In view of this, the volcanic and meteoritic hypotheses of the causes of mass extinctions mentioned earlier have more advantages compared to suggestions that these extinctions were the result of gradual changes in the environment. If these hypotheses are true, the extinction of organisms might have occurred within very brief time intervals.

It is possible that both of these factors (volcanoes and meteorites) caused the mass extinction at the end of the Mesozoic. The impact of a heavy asteroid might have led to the destruction of the Earth's crust over a considerable area and to the intensification of subcrustal processes, affecting the level of volcanic activity. As a result of this, a series of explosive volcanic eruptions might have taken place, causing great coolings. The duration of such a series of eruptions was probably hundreds or even thousands of years. Even in this case the mass extinction could have occurred, geologically speaking, within a relatively short time interval.

Returning to the extinction of the dinosaurs, it should be mentioned that according to some authors (Russel, 1979, and others), the dinosaurs became extinct simultaneously in all the areas where they lived. However, some other scientists think that the palaeontological data indicate a non-simultaneous disappearance of dinosaurs from different regions of the Earth. In particular, it has been suggested that in the west of North America, dinosaurs still existed at the end of the Maestrichtian stage, when they were already extinct on the other continents (Colbert, 1965). At the same time it has been noted in some works that the number of dinosaur species was declining in the last stages of their existence. As already mentioned, this suggestion may not been correct because of the incompleteness of the palaeontological record. It is difficult, however, to draw a final conclusion on this matter.

We think that the most plausible cause of the mass extinction at the end of the Cretaceous is the suggestion that a short-term global cooling occurred, probably associated with an asteroid impact which led to an intensification of volcanic activity. However, it is unlikely that the magnitude of such a cooling was very large. In particular, the conclusion that after the impact of the asteroid, mean air temperatures above the continents decreased by up to 40 °C, and were below the freezing point for half a year, seems doubtful. Such a cooling would have led to the extinction of a far greater number of species of the continental flora and fauna, compared to what actually took place at the end of the Cretaceous.

It seems likely that the result of another study (Alvarez *et al.*, 1980) also overestimates the reduction of radiation beyond the limit at which photosynthesis is possible, for several years. In the later work of the same authors (1982) they estimated that the time of such a reduction in radiation is several months. A similar conclusion was also made by Toon and his colleagues (1982). It seems to us that such an estimate is possibly too high and the main cause of the extinction of marine organisms was a decrease in the water temperature by several degrees. As already mentioned, the late Mesozoic animals would be very sensitive even to small fluctuations in the thermal regime.

The mass extinction at the close of the Cretaceous was one of the greatest extinctions that took place during Phanerozoic time. It is possible that the extinction at the end of the Permian was of even greater scale. According to Raup (1979), 96% of the earlier existent species of marine animals vanished as a result of this extinction. Although it has been suggested that this estimate is somewhat high (Gould, 1981), the revised estimate of the relative number of extinct species (80—85%) is still a large value. Undoubtedly there were a great number of smaller extinctions in the Phanerozoic as well as several great extinctions, but it is difficult to find evidence for these because of the inadequate palaeontological record. There is evidence that some of the smaller extinctions were also caused by the impact of celestial bodies on the Earth (Asaro *et al.*, 1982; Ganapathy, 1982; McGhee, 1982; Palmer, 1982).

In recent years mass extinctions have been discussed at some scientific meetings. One of these was held in October, 1981 in the U.S.A. and was devoted to considering the way in which the impact of large celestial bodies would cause massive extinctions of organisms. It is noted in the Report of this Meeting (Simon, 1981) that as a result of the debates, the possibility that the impact of large celestial bodies caused the extinctions has been acknowledged. It is characteristic that this recognition was treated as a 'scientific revolution'. Broader issues were discussed at a meeting in West Berlin in May, 1983, where the role of 'sudden events' in the history of the Earth was discussed. In Report of this Meeting (Fifield, 1983), a number of papers note the impossibility of understanding a great many events of the Earth's history on the basis of the uniformitarian concept. The meeting discussed the role of explosive volcanic eruptions as a factor leading to the extinctions of organisms and considered the significance of the impact of extraterrestrial bodies on mass extinctions. The reports delivered at the meeting presented calculations of the probable frequency of collisions of asteroids and comets of different sizes with the Earth and gave new estimates of the occurrence of mass extinctions throughout Phanerozoic. The frequencies of these events agree well with each other.

Many works have appeared recently which confirm the reality of mass extinctions in the history of the Earth and establish the relationship between these extinctions and abrupt changes in the abiotic environment. Although the actual mechanism of mass extinctions is far from being fully understood, there is a high probability that at least some of these extinctions are associated with relatively short-term changes in natural conditions due to the impact of celestial bodies of the Earth's surface and major volcanic eruptions.

It may be of interest to briefly consider the widely discussed question about the correspondence of the foregoing conclusion to the concept of actualism. Although this conclusion shows that to universally use the principle of uniformitarianism in studying the Earth's history is impossible, it does not contradict the actualistic approach for clarifying the general principles of the biosphere's evolution.

Since the impact of large celestial bodies on the Earth and the sudden intensification of volcanic activity take place comparatively rarely, these events have not been detected in the course of historical time (i.e. the last few thousand years). Therefore it is easy to come to the erroneous conclusion that such phenomena do not occur in our epoch. However, fairly reliable information shows that these phenomena which have been characteristic of the entire history of the Earth can also recur in our age, but with a relatively low frequency which prevents their direct observation. It is evident that such events should be included in the actualistic concept of the biosphere's evolution. Because of this it is important to more precisely formulate the actualistic approach to the study of the Earth's history. This approach must be based, not only on the principles governing the processes that are directly observed by man, but also on the general principles of the contemporary processes that undoubtedly take place but for some reason cannot be directly observed.

## 8.3. Macroevolution

### 8.3.1. MICROEVOLUTION AND MACROEVOLUTION

The views about the influence of abiotic factors on evolutionary process discussed in the last section may be useful in understanding differences between microevolution (evolution at the levels preceding the formation of species) and macroevolution (evolution at the levels above species, i.e. at the levels of genera, families and so on). The suggestion that differences existed between the two forms of evolutionary process was made in the 19th century by K. E. Baer and other scientists. The hypothesis which postulated that higher taxonomic groups could evolve when particularly pronounced mutations appeared, precipitated a long debate which has continued until the present day.

The present widespread view on the correlation between microevolution and macroevolution is discussed in the book by Timofeev-Resovsky *et al.* (1977), which states that there are no differences of principle in the way that both processes take place. At the same time it is noted in this book that "the reduction of macrophylogenesis to microphylogenesis does not mean that at the macrolevel we cannot come across some different, specific features". Accepting this point of view, it may be supposed that these specific features are associated with differences in the spatial and temporal scales of the factors influencing macroevolution and microevolution. It appears from the available data that the time of existence of a species of animals is on average of the order of magnitude of several million years. For higher taxonomic groups this time period is noticeably longer, for instance, it is on average of the order of 10 million years for a genus and about 70 million years for a family.

In the previous sections of this chapter, information was presented on global changes in the abiotic environment which must have been of importance in the evolution of organisms. The first type change is associated with fluctuations in the chemical composition of the atmosphere, which could be significant over time intervals of the order of 10 million years or more. The second type of global change in the abiotic environment is due to abrupt short-term climatic changes caused by the intensification of volcanic activity or by the impact of comparatively large celestial bodies on the Earth's surface. Such events take place very rarely and they are evidently separated from each other by tens of millions of years.

The influence of these two types of change in abiotic environment on evolutionary processes is in some respects similar, while in others different. It is evident that organisms can, to a certain extent, adapt to gradual changes in the environment of the first type, and therefore in some cases these changes cause the appearance and dissemination of new higher taxonomic groups. The role of environmental changes of the second type in the evolutionary process is mainly in the elimination of groups of organisms that are less resistant to a sudden deterioration of natural conditions.

A feature which is common to the impact of both types of environmental change on evolution is that the probability of these influencing the formation of species is small, because of the relatively short time characteristic of the existence of species. However, the probability of these factors influencing the formation of genera and particularly of higher taxonomic groups is much greater, because the time period characteristic of macroevolution is comparable to the time over which changes in abiotic environment occur. Not only do the two types of environmental change influence microevolution and macroevolution on different time scales but differences also occur in the space scale of these changes in the environment which affect the evolutionary process.

While in many cases the formation of species is affected by local changes in the environment, the two types of environmental change influencing evolution (as discussed above) are global in nature and affect either the whole biosphere or a significant part. Because of this it is necessary to examine the interaction of organisms and the environment at the level of the global ecological system when studying macroevolution.

Living nature can be studied at all levels of its organization, including the molecular and cellular levels, as well as the levels of tissues, organs, organisms, populations, ecological systems and the biosphere as a whole. This list (which is not always identical in different studies) should also be supplemented with the level of life in the cosmos (this is discussed in Chapter 10). It should be noted that the study of living nature at the level of the bisophere has attracted the attention of only a few scientists, and therefore the significance of this in understanding the general principles of the evolution of organisms is still obscure. It might be thought that the study of the biosphere's evolution would be helpful in understanding the role of global processes in the formation of higher taxonomic groups of organisms.

Returning to the impact of changes in abiotic components of the environment on evolution, it should be remembered that the evolutionary process of organisms entering into stable ecological systems is usually retarded. Therefore this process is evidently microevolution, i.e. it is limited to the rearrangement of mainly lower taxonomic groups of organisms.

Changes in the environment that disrupted the stability of ecological systems were of considerable importance in accelerating the evolutionary process. If these disturbances were sufficiently great, they could have led to the extinction of many groups of living organisms, whose ecological niches remained vacant, sometimes for a long time. In such cases an increase in the rate of evolution was the greatest, when environment changes were strongest.

To illustrate the influence of large-scale changes in the abiotic environment on changes in successive faunas the following example is given. South America was isolated from the other continents soon after the beginning of the Tertiary period, as a result of which its fauna developed for a long time under conditions of isolation. An important place in this fauna was occupied by representatives of two orders of herbivorous animals, the Litopterna and Notoungulata. At the end of the Pliocene, South America was connected to North America and numerous mammals, which differed greatly from the animals already existing in South America, expanded southwards. As a result of this invasion, many species of the South American fauna became extinct, among which were the overwhelming majority of litopterns and notoungulates. These species were replaced by tapirs, horses, deer, mastodons and carnivorous animals such as, pumas, jaguars and bears. These animals formed new stable ecological systems. For almost the entire Tertiary period the herbivorous animals of South America occurred in ecological systems that included the marsupial carnivores, which, evidently, were less active than the placental flesh-eaters of North America. It is quite natural that the original carnivorous animals of South America disappeared at the same time as the animals they preyed on.

A widespread explanation for the extinction of many species of endemic South American animals at the end of Pliocene is that these animals were replaced by the more advanced animals from North America (Colbert,

1958). Such an explanation seems to be simplified and it would be more correct to consider that the reason for this phenomenon was the destruction of South American ecological systems after the disruption of their isolation.

By using the equation for the energy balance of ecological systems, it is possible in principle to calculate quantitatively the rate of extinction of various species and learn which of the factors caused the extinction of the herbivorous animals of South America (extermination by the new species of carnivorous animals, trophic competition with the northern Herbivora, infection from parasites brought in by the new groups of animals, or combination of all these factors).

In addition to disturbance of the stability of ecological systems due to the appearance of new organisms, there might be a reverse process, when the distribution of new organisms is the result of the disturbance of ecological systems. This process was examined by Gilyarov (1966).

The first section of this chapter has been devoted to the effects of changes in the chemical composition of the atmosphere on evolutionary process. In some cases these changes, occurring over millions of years, have accelerated the process of evolution and promoted the appearance of new higher taxonomic groups. In other cases these changes have retarded evolution and restricted the progressive develoment of organisms.

In the second section of this chapter the problem of the dependence of evolution on relatively fast changes in abiotic environment, that took place in critical epochs of geological history, has been discussed. These changes resulted in mass extinctions of organisms, which sometimes promoted an acceleration of evolutionary process. This occurred because the disappearance of less progressive forms allowed more progressive forms to occupy the vacated ecological niches and encouraged a rapid development of these forms. The best-known example of such acceleration of evolutionary process is the expansion of mammals and birds into a variety of ecological niches after the extinction of many groups of reptiles at the end of the Mesozoic.

In connection with the above considerations the question arises about how evolutionary process developed in the absence of large-scale changes in the environment. The existence of many 'living fossils' (i.e. organisms that have changed relatively little over tens and sometimes even hundreds of millions of years), gives the impression that in this case the progressive development of living nature would have been very insignificant. Although ancient organisms that have changed comparatively little are usually found in regions with more stable environmental conditions, in the past they were more susceptible to the consequences of large-scale changes in natural conditions. Because of this, it is impossible to exclude the suggestion that without considerable fluctuations in the abiotic factors influencing the evolutionary process, this process would have been extremely slow for all organisms, particularly at the level of macroevolution.

### 8.3.2. THE SIZE OF ORGANISMS

Macroevolution is associated with various morphophysiological modifications of organisms; one of the most common modifications is an increase or decrease in the size of organisms. It is possible that changes in the size of organisms usually depend on the energy interrelations of the organism and its environment.

The underlying component of the energy system combining all living organisms, is the energy flow from autotrophic plants to various heterotrophic organisms. Because of a great energy loss at each link of the trophic chain, the energy flow gradually decreases as it moves along the chain. The flow is finally extinguished with the mineralization of the organic matter from living things, thus closing the food-chain. The structure of the energy flow is to a great extent non-homogeneous, in some links of trophic chains the energy of living matter is concentrated in organisms that are large in size while in other links the energy is distributed among numerous small living organisms. The energy transfer may also proceed in both directions (for instance, the feeding of herbivorous whales on plankton and the consumption of their biomass by various microorganisms).

The size of living organisms is an essential factor in evolution because it determines the amount of energy that organisms incorporate and the amount of energy necessary to maintain their vital activities. The role of size in the ecology of land plants is determined by the advantages of large plants in obtaining the greatest amount of solar energy compared to small plants that are often shaded to a great extent. In the thin vegetation cover of arid regions, plants with developed roots usually have the opportunity of obtaining water from deeper soil layers, large succulent plants contain a great volume of water accumulated in their tissues. Large plants generally live longer than small ones. In animals an increase in body size enhances their resistance to changes in the thermal regime, reduces the danger of encounters with other animals and, in the majority of cases, appreciably prolongs their life. At the same time, for both plants and animals, an increase in size augments the energy uptake necessary for their vital activities, which limits the greatest possible size of living organisms.

It should be noted that for many plants and animals the amount of energy obtained from external medium is in the first approximation proportional to the area of their surface, whereas the amount of energy used by their vital functions is proportional to their mass. Since under average conditions both of these flows of energy are equal to each other, it is clear that the intensity of the energy uptake from the external environment per unit surface area of the organism has to increase proportionally to its size. This dependence is partly compensated for in warm-blooded organisms, whose specific loss of energy in maintaining a constant body temperature somewhat decreases with an increase in their size. But in this case the compensation is not full and therefore an increase in the size of organisms should be, as a rule, accompanied with an increase in the energy uptake per unit surface area. Energy uptake necessary for living organisms not only limits the greatest possible size of organisms but determines the optimum size of each organism in a particular ecological niche (i.e. at the point where the energy balance disadvantage is compensated for by the advantages of increasing size).

It should be emphasized that populations of the same biomass per unit area but consisting of organisms of different sizes are characterized by a different resistance to changes in natural conditions. For example, we will consider two populations of similar animals having the same energy source (nutrition) but differing considerably in size and life expectancy. In this case the productivity of populations represented by smaller animals will be higher than the productivity of population consisting of larger animals. This is associated with, in particular, a higher rate of metabolism in the population of smaller animals. Based on equations describing animal population dynamics it can be shown that an increase in the death rate, corresponding to a loss in biomass for small and large animals by the same magnitude, will lead to a faster extinction of the large animals compared to small animals (see Chapter 9). Thus, large animals with a low coefficient of biomass turnover, although having a number of advantages against smaller animals under stable environmental conditions, are in a more precarious situation when their death rate increases because of an unfavourable change in the environment. Because of this, many large species of organisms are likely to become extinct when the stability of ecological systems is disrupted by a change in environmental conditions. This is why many ancient organisms of smaller size have been able to preserve their ecological niches, even during unfavourable changes in environmental conditions, while large organisms have become extinct and more progressive species now occupy their niches.

### 8.3.3. GENERAL FEATURES OF MACROEVOLUTION

The ideas presented above make it possible to clarify some principles governing the formation of taxonomic groups above the species level (i.e. macroevolution). The following conclusions concerning the macroevolutionary process can be drawn.

(1) The process of macroevolution, more than that of microevolution, depends on large-scale changes in abiotic factors of the environment. It is probable that many of the principal events of macroevolution could have taken place only as a result of large-scale changes in the environment.

(2) In the course of macroevolution, directional changes in organisms may be produced by long-term regular changes in abiotic factors influencing the process of evolution (e.g. atmospheric chemical composition, climate, etc.).

(3) A general tendency towards higher oxygen content of the atmosphere and natural waters which appeared long before the beginning of the

Phanerozoic era must have been advantageous for progressive changes in animals.

(4) There were epochs that were particularly favourable for the development of organisms when large progressive modifications of successive floras and faunas took place. These epochs were followed by time intervals when the evolutionary process was retarded.

(5) Along with relatively slow modifications in the diversity of organisms as a result of gradual changes in the natural environment, mass extinctions of organisms took place under the impact of external factors within brief time intervals.

(6) The influence of global changes in abiotic factors on evolution could lead to the formation of similar aromorphosis in groups of animals that differed in origin. The possibility of a polyphyletic origin of certain higher taxonomic groups is a result of this.

It is possible that the process and history of macroevolution will be more fully clarified as a result of studies concerned with the relationships between the evolution of organisms and the development of the biosphere. Such studies have only been undertaken recently. The results of these studies must be included into the modern theory of evolution, whose principal ideas were formulated several decades ago. The main part of the modern evolutionary theory is the synthetic theory of evolution, the foundations of which were laid in the 1930s and 1940s as a result of a synthesis of the traditional Darwinism with the achievements of genetic studies. In the second quarter of the 20th century the works of Severtsev and Schmalgauzen appeared and substantially supplemented the ideas of the creators of the modern evolutionary synthesis. Subsequently, considerable progress has been made in investigations of evolutionary processes at the molecular and cellular levels as well as at the level of populations. These investigations have appreciably added to our knowledge of microevolution but have been of less value in understanding the mechanism of the macroevolutionary process.

One of the objectives of further development in evolutionary studies should be the examination of the principles of macroevolution on the basis of synthesizing the results of microevolution studies and studies of the impact of large-scale changes in the abiotic environment on evolution. As mentioned above, the significance of such an approach can be seen in a remark of Darwin, who wrote that in the absence of changes in the environment natural selection is inefficacious. The ectogenic view of the evolutionary process was most consistently presented in the works of Severtsev. In the monograph *The Principal Trends of Evolutionary Process* (1925) he wrote: "All evolutionary modifications of animals depend, directly or indirectly, on changes in the environment." The mechanism of progressive development of animals was considered closely in the works of Severtsev and Schmalgauzen, Severtsev indicating the unevenness of this process. In his monograph *Morphological Features of Evolution* (1939) he pointed out: "The process of progressive evolution has as if a rugged pattern, the periods of rise alternating with the periods that are marked by no morphological progressive evolution at all." One of the first specialists in the field of evolutionary biology was Schmalgauzen. He supported the notion of critical epochs and in his monograph (1940) notes: "At the boundary of many geological periods quite sudden changes in the whole faunas actually took place, and it is difficult to make responsible for this something else but the changes in physical factors of the environment (which gave rise to drastic perturbations of the biological conditions for life)." This view is now shared by a great number of scientists. In particular, Mayr (1963) writes that there is no doubt that mass extinctions always correlate with highly changed external conditions.

In conclusion we quote Simpson, one of the founders of the synthetic theory of evolution: "I do not doubt that the inorganic features of the various environments have influenced evolution as have also the organic features. If a group of organisms is well adapted to its total environment and that environment remains more or less constant, those particular organisms will evolve very slowly or not at all. If an environment, inorganic and/or organic, does change, organisms in it will also change. If they maintain change at a rate near that of change in the environment they will survive with derived adaptations. If the environmental change is markedly more rapid than the biological possibility of evolutionary change in a given taxon (at any taxonomic level from species to classes or even phyla), that taxon becomes extinct" (Simpson, 1982).

Accepting these ideas, we would like to emphasize that according to the general contents of our book, the problem of evolution of organisms as stated in this chapter has been treated only in regard to the relationship between evolutionary process and global changes in the biosphere. All other aspects of this problem are approached only in passing.

CHAPTER 9

# MAN AND BIOSPHERE

## 9.1. Formation of the Noosphere

### 9.1.1. THE NOOSPHERE

The idea that the biosphere, in which intelligent human activity develops, is being converted into the noosphere (a sphere of intelligence) arises from the notion presented by P. Teilhard de Chardin and E. Le Roy. The concept of the noosphere was developed by Vernadsky, who considered it an important part of the science of the biosphere. The work of Yanshin (1981) describes the principal concepts of Vernadsky about the conditions, which when fulfilled, lead to the formation of the noosphere. Yanshin formulates these conditions as follows:

"(1) Mankind has become a completely integrated unit. The history of mankind has embraced the entire Earth as a single unit, thus completely replacing the indigenous and historically isolated cultural regions of the past . . . ."

"(2) Transformation of the means of communication and exchange of information. The noosphere is an orderly functioning unit, whose parts are harmoniously linked and act together at all levels . . . ."

"(3) The discovery of new sources of energy. The formation of the noosphere requires a radical transformation of the natural environment by man and cannot be performed without immense quantities of energy . . . ."

"(4) The growth of the welfare of the working masses. The noosphere is being created by the intelligence and labour of people, therefore the special significance of this premise is beyond any doubt . . . ."

"(5) The equality of all people. Embracing the entire planet as a whole, the noosphere by its very essence cannot be the privilege of a single nation or race. It is produced by the skill and mind of all nations . . ."

"(6) The exclusion of wars from the life of human society. In the twentieth century war, threatening the very existence of mankind, has arisen as the greatest obstacle to the development of the noosphere. It follows that without the elimination of this obstacle the attainment of the noosphere is practically impossible. On the contrary, with the elimination of war mankind will have made a great step towards the noosphere."

Yanshin notes that the "noosphere is the result of the two greatest revolutionary processes of our time that have merged into a single flow, namely the progress in the field of scientific ideas on the one hand and of social relations on the other. Therefore the formation of the noosphere is possible only as a result of a stable alliance of the forces that make up the basis of these processes. . . ." It might be thought that the premise of the formation of the noosphere constituted the process of anthropogenesis, including the evolution of primates leading to the appearance of modern man. The development of the noosphere has been determined by growth in the scale of human impact on the biosphere and by an increasing level of intelligent guidance of man's activities.

Since the modern biosphere is at present only partly embraced by purposeful and expedient human activities, it may be considered that the formation of the noosphere is still far from complete. With the development of the noosphere it becomes possible to control the state of the biosphere in conformity with the needs of human society. The likelihood of such a control occurring can already be seen, in particular with the possibility of influencing global climate (which is discussed in the next chapter). It is probable that in the near future, deliberate change of the biosphere's state will become increasingly possible. At the same time, following the achievements in conquering cosmic space, the noosphere will expand far beyond the limits of our planet. The appearance of the noosphere has been the result of the process of anthropogenesis, in the course of which there appeared an intelligent creature — *Homo sapiens.*

### 9.1.2. THE BRAIN

Evolution of modern man resulted from the increase in the size and the complexity of the mammal brain and, in particular, of the brain of primates, who were the ancestors of man.

Information on the evolution of the brain in vertebrates is presented in the monograph by Jerison (1973), who found that in the majority of animals the volume of the brain is proportional to the volume of their body to the power of 2/3. This indicates a close connection between brain volume and the surface area of the body, where, in particular, the sense organs are located. The coefficients of proportionality of the indicated relationship differ greatly among various groups of animals. For modern mammals this coefficient is on average somewhat greater than for birds and much greater than for fish, amphibians and reptiles.

As Jerison's data show, the volume of the mammalian brain is on average ten times greater than the brain of similar sized reptiles. The effectiveness of the brain's activity depends on both the relative size of the brain and also its absolute size, because with increasing brain size the elements it consists of may attain a more complex structure. Because of this only more or less large animals can have a highly developed brain. This conclusion is confirmed in particular by the fact that the size of the human body exceeds the body size of the overwhelming majority of animals. At the same time, not all large animals have today, and particularly had in the past, highly developed brains.

In the Mesozoic era when reptiles were flourishing, there were many large and gigantic animals. However, their brains were insignificant in volume and practically did not increase in volume during the long time period of reptilian dominance. Since the mammals of the Mesozoic were competitively confined by reptiles to ecological niches suitable only for small animals, the possibility of mammals developing large brains were excluded. As a result, this epoch was an evolutionary 'blind-alley' when the formation of the noosphere was not possible. It was only with the disappearance of the majority of reptilian groups at the end of the Mesozoic and, especially, the extinction of dinosaurs, that a rapid progressive evolution of mammals became possible. In the course of this evolutionary process many representatives of the mammal class attained great sizes and their brains increased in volume and structural complexity.

The size of the brain is often described by the coefficient of encephalization, which is equal to the ratio of the brain volume to a conditional volume which is proportional to the body volume raised to the power of 2/3. The coefficient of encephalization for mammals varies from about 0.1—0.2 for the most primitive animals to about 6 for modern man. Although the coefficient of encephalization, as well as other indices characterizing the volume of the brain, are only approximate indicators of the level of higher nervous activity of an animal, it is possible to make use of these data for obtaining important information on the development of the brain in fossil animals.

Table 9.1, constructed according to Jerison's data, presents average values of relative brain size in different forms of mammals. This table shows that the evolution of the mammalian brain was a long process that proceeded throughout the entire Tertiary period. The tabulated data can be supplemented with information on the evolution of the brain size in Mesozoic mammals.

TABLE 9.1

Relative brain volume in mammals (the average magnitude for modern animals is assumed as 1.0)

| Forms | Herbivores | Carnivores |
|---|---|---|
| Early Tertiary, primitive | 0.18 | 0.47 |
| Early Tertiary, progressive | 0.38 | 0.61 |
| Middle and late Tertiary | 0.63 | 0.76 |
| Modern | 0.95 | 1.10 |

Although it is difficult to obtain such information, because of limited data for the comparatively rare mammals of the Mesozoic, Jerison concluded that even at this stage representatives of the oldest groups of mammals had brains whose relative size was greater than that of reptiles. However, for more than

100 million years the relative brain size of mammals did not alter by much and it was only in the Tertiary that brain size started to increase. Jerison concluded that if we assume the average coefficient of encephalization for modern animals to be 1.00, then it is 0.25 for archaic mammals of the Eocene and 0.50 for the Oligocene animals. At the same time, Jerison noted that the evolution of the brain during the Tertiary was accompanied with an increasing diversification (i.e. the range of variations in relative brain sizes in various groups of animals).

It might be thought that the evolution of the mammalian brain was very dependent on the conditions of their environment. Pretertiary mammals were small night animals, whose activity was enhanced under the conditions of lower temperatures at night. They evidently competed little with the then dominant reptiles. The fast evolution of mammals at the beginning of the Tertiary, following the late Cretaceous extinction of the majority of reptilian groups, was not accompanied with a noticeable increase in the relative brain size of mammals. This was because mammals were able to fill the large variety of ecological niches vacated by the extinct reptiles without acute mutual competition.

As Jerison mentions, considerable changes in the brain structure of mammals took place in the late Eocene; in many mammals the brain became increasingly complicated. As seen in the previous Chapter, the number of mammalian families in the late Eocene approached, for the first time, the maximum corresponding to the 'ecological capacity' of the biosphere. Under such conditions new families could only appear if they successfully competed with earlier existent groups that had similar ecological requirements. This could only be achieved if the new organisms were more advanced. In this situation the appearance of new families had to be accompanied with the extinction of the older groups that occupied the same ecological niches. This is confirmed by the data in Table 8.2, which shows that in the late Eocene the rate of appearance of new families and the rate of extinction of older families were high and almost equal in value. The increase in the average volume of the brain in the Oligocene, as established by Jerison, is presumably a result of changes in natural conditions, which in particular promoted the extinction of archaic forms with smaller brain sizes. Although climatic changes in the Miocene and Pliocene were not as great as during the Oligocene cooling, these changes made it more difficult for animals to adapt to varying natural conditions, which led to the survival of animals with an enhanced level of higher nervous activity.

It follows from the above discussion that the evolution of the brain accelerates, firstly, when progressive groups of animals occupy all or nearly all ecological niches (i.e. the 'ecological capacity' of the biosphere is reached) and, secondly, when considerable changes in natural conditions occur. In addition to the data given above, this conclusion can be confirmed by findings concerning the evolution of Tertiary mammals on two isolated continents, namely South America and Australia. During the Tertiary period both these continents were largely located in low latitudinal belts, where climatic fluctuations were less pronounced. At that time both South America and Australia were more or less isolated from other continents.

In South America the major groups of placental mammals were various ungulates. Jerison remarks that the brain size of these animals changed little for 50 million years. A similar situation applied to Australia, which was inhabited by marsupial animals. It is possible that the slow development of the mammal brain on these continents can be explained in part by the relatively stable natural conditions and in part by an inadequate usage of the available 'ecological space' for mammals on these continents. In South America in particular, big flesh-eaters were represented by carnivorous marsupials up to the end of the Tertiary. These animals were less effective in pursuing ungulates in comparison with placental carnivores. Although the Australian marsupials are similar in appearance to ecologically analogous placental animals of other continents, this does not mean that the marsupials could maintain a high level of interspecies competition, which is characteristic of placental animals and which contributes to a fast rate of evolution. Furthermore, the land areas of South America and Australia were small compared to the vast system of interconnected continents, which included Africa, Eurasia and North America. Since the appearance of new groups of animals is associated with random processes, the area on which the relevant 'evolutionary experiment' is being carried out often turns out to be a decisive factor in its successful completion.

In conclusion let us consider the evolution of the primate brain. The earlier lower primates already possessed a fairly large relative brain size. The growth of brain size accelerated with the appearance of higher primates and particularly with the appearance of anthropoid apes, which branched from the main genealogical tree of primates in the Oligocene. Somewhat later the ape-man ancestors of human beings became a separate branch. The relative brain size of all higher primates is noticeably greater than the average brain size in other groups of mammals. The brain size was even greater in *Australopithecus* and other immediate ancestors of modern man (Jerison, 1973).

It can be suggested that the rate of evolution of primates was essentially dependent on the extent of variability of their environment. Although primates existed almost throughout the Tertiary period, during the first part of the Tertiary when environmental conditions were more stable the progress of this group of mammals was relatively slow. The rate of evolution of primates (including the growth rate of the coefficient of encephalization) increased in the second part of the Tertiary, starting in the Oligocene when considerable changes in natural conditions occurred. It has been repeatedly suggested that the descent of modern man was promoted by abrupt changes in natural conditions during the Quaternary (Gerasimov, 1970; Budyko, 1980).

The relationship between the evolution of primates and changes in environmental conditions can help in locating the initial centre of anthropogenesis. There are two points of view with regard to this location. Accepting the fact that the first stages of this process took place in low latitudes, some scientists (among them Darwin and Wallace) considered the centre of anthropogenesis to be the African continent which is still inhabited by the species of apes most closely related to human beings. Other scientists consider that some palaeontological findings show that the first direct ancestors of man appeared in South Asia.

Ecological inferences are, however, in favour of the former view point. As indicated above, even with drastic changes in global climate, temperature in the tropics varied relatively slightly. The principal result of climatic fluctuations in the tropics was a change in the precipitation regime, which in many cases was considerable. Since in South Asia conditions of excessive moisture prevail, the influence of precipitation changes on vegetation cover and other components of the natural environment was comparatively insignificant for most of this area. However, the consequences of fluctuations in global climate in Africa, much of which has always been under varying conditions of moisture stress, are different. In Africa, fluctuations in the precipitation regime inevitably led to pronounced changes in the boundaries of geographical zones. This was accompanied by the destruction of the former ecological systems and created conditions conducive to the emergence of new forms of ecological interrelations among organisms. Under such conditions the rate of evolution of many groups of living things, including the primates, was higher.

It can be inferred from the considerations presented in this section and in the previous chapter, that the descent of man (and consequently the prospects for the formation of the noosphere) became possible as a result of the following changes in the conditions surrounding organisms:

(1) Relatively slow fluctuations in the gaseous composition of the atmosphere, in the course of which the amount of atmospheric oxygen increased appreciably. During the epoch of increased oxygen content many advanced groups of animals appeared, including the principal classes of vertebrates.

(2) The short-term drastic changes in the thermal regime that repeatedly resulted in the extinction of numerous groups of animals, but promoted the wide expansion of the more progressive forms that survived the epochs of extinctions.

(3) Climatic fluctuations throughout the past few million years that influenced the evolution of the ancestors of modern man.

It is possible that with stable environmental conditions the rate of evolution throughout the entire period of the biosphere's existence would have been insufficient not only for the descent of man but also for the appearance of any complex organisms.

## 9.2. The Origin of Man

### 9.2.1. THE DIFFERENCE BETWEEN MAN AND ANIMALS

The hypothesis that man originated from ape-like ancestors was advanced as early as the 18th century. In particular, Kant supported this hypothesis and in the early 19th century so did Lamarck. A major role in clarifying man's origin was played by the works of Darwin and Wallace, which were published at the end of the 1850s. They contained a theory of the origin of species by means of natural selection of the fittest organisms. Although these studies did not consider the problem of human evolution (Darwin made only the observation that his conception may shed light on man's origin), the possibility of applying the theory of Darwin and Wallace to explain human evolution became immediately apparent to many scientists.

In 1860 the famous debate between Huxley and Wilberforce was held, in which Huxley defended the hypothesis that man originated from ape-like animals by means of natural selection. The 1860s saw the publication of the studies by Huxley, Haeckel, Vogt and other authors, who developed this point of view.

The position advocated by Wallace was somewhat different. Although in his first work concerned with the origin of man, Wallace (1864) presented views similar to those of Huxley, a few years later he came to the conclusion that natural selection could not have been the only cause of man's origin (Wallace, 1869). In his opinion, natural selection could bring about only a slight improvement in man's mental facilities by comparison with those of higher apes. He noted that the evolution of man's brain was far ahead of the requirements of his existence at the early stages of social development of human society and explained this fact by the influence of a 'higher rational will' on the process of human development. It is worth mentioning that the assumption about the influence of non-material factors on man's origin was the only one of this kind in Wallace's numerous biological studies (George, 1964).

Darwin's own view on the evolution of man was presented in his study *The Descent of Man and Selection in Relation to Sex* published in 1871. This study contained comprehensive data in support of man's origin from ape-like ancestors, which were similar in a number of respects to contemporary higher apes. Darwin pointed out that the probable place of man's origin was Africa, where one finds today the two apes that are closest to humans, namely chimpanzees and gorillas (Wallace was of the same opinion).

In his study Darwin only briefly mentioned Wallace's arguments concerning the difficulties in explaining the development of the human brain and expressed his disagreement with these considerations. Apparently, this disagreement was based on his comparison of the mental activities of humans and animals, which according to him, had many common features. As a result of this comparison, Darwin concluded that there is no qualitative difference between the mentality of animals and of humans.

In the light of contemporary views on this subject it is evident that such a position cannot be accepted. Nevertheless Darwin's conclusion in his book (1871) deserves attention: "We must, however, acknowledge, as it seems to me, that man, with all his noble qualities . . . with his godlike intellect which penetrated into the movements and constitution of the Solar System — with all those exalted powers — man still bears in bodily frame the indelible stamp of his lowly origin". It is clear from this sentence that in a certain measure Darwin did recognize the fundamental difference between man and the most advanced forms of animals.

Although Wallace survived Darwin by three decades and witnessed many new achievements in the study of man's origin, he maintained his views on this problem to the end of his days. Because of Wallace's authoritative position and the fact that he greatly valued the studies of Darwin and disagreed with his conclusions rarely and reluctantly, the differences between his view and that of Darwin in respect to such important question should have attracted attention. Yet this did not happen, probably, because Wallace's assumption concerning the influence of non-material factors on the evolutionary process did not seem acceptable.

In the middle of the 19th century there were very few palaeontological data shedding light on the origin of man. During the last century many outstanding discoveries were made in this field, among which the discovery of remains of *Australopithecus* in south and east Africa and the discovery of remains of *Pithecanthropus* and of related creatures (*Homo erectus*) in Java, China and Africa, were particularly notable.

Data provided by palaeontological studies confirmed Darwin's and Wallace's views on the African origin of man. However, *Australopithecus*, which was closely related to the ancient ancestors of man, appeared to differ appreciably from contemporary higher apes. *Australopithecus* was largely a predator and hunted various animals with the help of the simplest weapons. Thus, the use of animal food that is characteristic of modern man has very ancient roots.

*Australopithecus* existed at the end of the Pliocene and throughout a part of the Quaternary period, i.e. during a period from several million to one million years ago. At the end of the Pliocene (more than two million years ago) the primitive representatives of the genus *Homo* first appeared. *Homo erectus*, which in many respects is intermediate between *Australopithecus* and modern man, occurred for several hundred thousand years in the Pleistocene.

Modern man (*Homo sapiens sapiens*) descended approximately forty thousand years ago, and is supposed to have evolved from Neanderthal man (*Homo sapiens neanderthalensis*), who had lived a longer period of time and vanished at approximately the same time as modern man appeared. Since later forms of Neanderthal man differed even more from modern man than earlier ones, the origination of modern man from late Neanderthal man is often questioned.

The greater part of modern man's existence corresponds to the first of the cultures he created, namely that of the Upper Palaeolithic. During that time, the hunting of large animals constituted the economic basis of human life. The findings of the studies presented in the next section have led to the conclusion that the Upper Palaeolithic culture ended with an ecological crisis that embraced vast territories and was associated with the destruction by hunters of some of the animals of the Upper Palaeolithic.

It should be noted that the appearance of such a crisis was rather unusual in the history of the biosphere, since ecological systems are, as a rule, subject to self-regulation in regard to the number of organisms that enter into relevant food chains, which excludes the possibility of a rapid extinction of certain species of plants and animals. The disturbance of this regulation indicates that the emergence of modern man constituted a turning point after which man ceased to be subject to biological laws governing the density of animal populations.

The annihilation of large animals presumably led to a reduction in the human population of certain areas until a time when hunting and gathering of useful plants came to be replaced by cattle-breeding and tillage. This happened in the Neolithic, approximately 10 thousand years ago. Thus, the period, during which modern man has been producing food to satisfy his needs represents only about one-fourth of the total duration of his existence.

Although available palaeontological and archaeological data shed light on many stages of man's evolution, the nature of the mechanism of this evolution requires further research. One of the difficulties in explaining this mechanism was noted by Darwin. Having concluded that man originated from ape-like ancestors he added: "The high standard of our intellectual powers and moral disposition is the greatest difficulty which presents itself, after we have been driven to this conclusion on the origin of man" (Darwin, 1871). In order to overcome this difficulty it is important to clarify the question formulated by Wallace more than 100 years ago.

Natural selection usually preserves the changes in organisms that are useful for their vital activities and are employed in the course of their existence. Yet one has the impression that throughout almost the entire history of modern man the enormous capacities of his brain have only been used to a very small extent. There are grounds to believe that man's central nervous system has not changed much since Upper Palaeolithic times. Evidence in support of this is provided by the images of animals in the caves of Western Europe, which are often regarded as the greatest achievement of animalistic painting throughout the human history. The impression that an enormous gap exists between modern man's capacity for mental activity and the actual usage of this capacity in the early stages of cultural development is in particular derived from observations of how the achievements of modern civilization can be used by representatives of the most backward tribes living at the level of Stone Age cultures. Today, when representatives of such tribes receive relevant education from early years they can easily enter into a modern technological civilization. Therefore, it would seem that under earlier conditions the mental capacities of such persons were used only to a small extent.

Attempts to explain this gap in terms of the conventional conception of evolution by natural selection (survival of the fittest) are met with difficulties. Although many animals acquire, as a result of domestication, certain forms of behaviour that are not typical of their wild existence, there is no analogy for such a gap in the animal world. The supposition that an organ as complex and perfect as the brain of modern man could have developed as a result of preadaptation (i.e. to a certain extent accidentally), seems highly improbable. Therefore, in order to explain the origin of man, it is necessary to take into consideration some special factors that have led to a qualitative distinction between man and all animals. The problem of defining the nature of these factors deserves great attention.

### 9.2.2. THE ROLE OF CHANGES IN THE ENVIRONMENT

According to the views of a number of scientists (Gerasimov, 1970; and others), it may be thought that the evolution of man was influenced to a certain extent by changes in climate during the Pliocene and Pleistocene epochs. It was noted in the previous chapter that at that time, a cooling occurred at middle and high latitudes, which caused considerable changes in the climate of all continents. It is quite probable that these climatic fluctuations served to intensify the evolution of both man and many other living things.

It is difficult to estimate quantitatively the rate of human evolution. According to Kurtén's data (1965), in the Pleistocene European mammal species existed on the average for one million years. This magnitude is comparable to the length of life of the species that were allied to man during the Pleistocene. It would thus appear that the rate at which the ancestral species of man replaced each other was not very high. On the other hand, such an estimate of the rate of evolution appears to be rather conventional.

The existing systematics of living organisms is largely based on morphological and phylogenetic principles and does not generally reflect their ecological characteristics. Yet there are many examples of organisms that are similar in an ecological sense but very distant from each other in terms of their origin. At the same time closely allied organisms are often related to their environment in very different ways.

In the course of early studies of man's origin, based on the theory of natural selection, one of the arguments in favour of this concept was the assertion that higher apes are morphologically closer to man than to lower apes. But it is also possible to state that in terms of ecology the differences between modern man and Neanderthal man are in fact greater than the differences between Neanderthal man and many mammals that do not belong to the order of primates.

Such a view rests on modern man's enormous influence on the environment, which was already evident during the Upper Palaeolithic (when man destroyed dozens of species of large animals), and has reached tremendous proportions in the present epoch. Even though Neanderthal man existed for a period of time which was several times greater than modern man, he did not disturb the biosphere's balanced ecology, and in this respect was much more similar to other mammals than to modern man. Because of this, it is evident that the available systematics does not correspond to the actual differences in the ecological properties of man and of his ancestors.

In view of the scope of this difference it must be concluded that changes in man's ecology have generally occurred much more rapidly compared to the average rate of change in the ecology of different animals. It may be supposed that this process was associated with the development of man's central nervous system. This attribute is not taken into account sufficiently in the system of classification pertaining to man and organisms that are related to him. Available data show that during a period of two million years the volume of the brain of man's ancestors increased by nearly three times (Kurtén, 1971b). At the same time the brain structure became more complex.

One of the causes of such a rapid development of the central nervous system could have been the critical situation in which man's ancestors found themselves during the Pliocene and Pleistocene epochs, when noticeable changes in the climate of the tropics took place. These changes were the consequence of temperature decline at high latitudes, which led to an increase in the average meridional gradient of air temperature. As a result the atmospheric circulation system underwent alteration and the high-pressure belt shifted to lower latitudes. Since there was little atmospheric precipitation within this belt, moisture conditions changed in many tropical regions, which resulted in the replacement of tropical forests in some areas by savannas and semi-deserts.

There is every reason to believe that before this change in climate occurred, man's ancestors lived in forests and their way of life differed little from that of other higher primates, except for their more developed ability to walk erect. It is quite probable that like almost all primates they were largely plant-eating animals. The immediate ancestors of *Australopithecus* (which can be called pre-*Australopithecus*) were morphologically similar to *Australopithecus* and were relatively small in size. The rate of locomotion of the erect pre-*Australopithecus* was much lower than that of most four-legged animals of the same size, and they also lacked the means of attacking others and defending themselves such as large fangs and claws. In open areas, where it was impossible to seek refuge in trees, pre-*Australopithecus* was nearly defenceless even from relatively small representatives of the numerous predators of African savannas. At the same time, the disappearance of many edible plants, on which they fed in tropical forests, made their reliance on plant food much more difficult.

Under conditions of an increased death rate, their low birth rate (characteristic of higher primates), and the long period of growth of their progeny to adulthood made them particularly vulnerable to predators and represented a great obstacle in enabling them to maintain their population level. The principal factors that enabled pre-*Australopithecus* to survive were their relatively developed brain and ability to walk upright, which fully freed their hands. It is likely that pre-*Australopithecus* lived under extremely harsh conditions of natural selection, which led to a high rate of evolution with low population density. Such an evolutionary process, which Simpson called quantum evolution, does not usually leave palaeontological evidence because of the small number of relevant animals (Simpson, 1944).

Palaeontological studies have shown that *Australopithecus* mastered the use of the simplest tools in defending himself and attacking others. With their help, *Australopithecus* was able to successfully attack such large apes as baboons and various hoofed animals and as a result, obtain substantial amounts of animal food. The discovered remains of australopithecines indicate that the number of individuals in their populations was not very small. They must, therefore, have overcome the crisis situation described earlier.

One should note that even in using weapons such as stones, sticks and bones of hoofed animals, the groups of relatively small australopithecines would still be in a difficult situation when they were attacked by large predators. It was equally difficult for slow moving *Australopithecus* to pursue any rapidly moving animal. Their success in struggling for survival could be achieved only through their substantial superiority in intellectual capacities in comparison with all animals that attacked them or could be their prey. Therefore, in the course of their evolution, natural selection greatly favoured the development of their brain, and this, in turn, contributed to their gradually increasing success in defending themselves against predators and in hunting various animals.

The increasing mental capacities of australopithecines enabled them to make wider use of various objects as weapons and start making different tools. Because of this, collective activities such as defending themselves and attacking other animals became important. It is on the basis of collective activities that elements of social organization first emerged, which further led to the development of ethical conceptions determining the behaviour of individuals within communities. The role of the labour process, the development of language and of social relations in man's evolution has been described in the well-known study of Engels, *The Origin of the Family,*

*Private Property and the State.*[1] Engels has also drawn attention to the importance of starting to use animal food by man's ancestors, which greatly intensified man's physical development.

Palaeontological data provide evidence that some lines of *Australopithecus* developed a tendency towards large sizes. This considerably reduced the pressure of natural selection on the evolution of the nervous system and led to a reduction in the need to use weapons and tools. The corresponding lines of development led to evolutionary 'dead ends' and to the extinction of their representatives.

Besides such regressive lines, there also existed the lines of man's ancestors of which progressive development of the central nervous system was characteristic. It may be presumed that this development was uneven; it intensified during periods in which the pressure of natural selection on populations of man's ancestors was stronger. It is quite probable that such periods coincided with the environmental changes that occurred in connection with Quaternary glaciations.

As glaciers advanced and receded, climatic fluctuations took place over vast areas. These fluctuations could greatly disturb the conditions under which man's ancestors existed. This confronted them with new problems and intensified their evolution. It seems that the last major stage in man's development coincided with the Würm glaciation, when *Homo sapiens sapiens* developed.

While these ideas shed light on the role of climatic changes in man's origin, they are nevertheless insufficient to explain the causes of the development of the human brain, which according to Wallace, did not correspond to the needs of primitive society, exceeding its requirements. In our opinion, in order to answer this question, attention should be paid to the lack of correspondence between the technological means employed by man's ancestors and the problems they confronted.

Until recently the hunting of large animals and defence against predators was accompanied with great difficulties and risks, even for representatives of civilized societies possessing firearms. These difficulties were much greater for man's ancestors, who relied on primitive weapons for hunting, and especially so for his distant ancestors, who were apparently much smaller in size and weaker than modern man. Encounters by man's ancestors with large predators were nevertheless inevitable. In addition, dangerous situations arose in the course of searching for animal food when only large animals could provide a sufficient quantity of meat for numerous tribes. There is no doubt that man's ancestors were usually victorious in such encounters. In particular, this is confirmed by the remains of very large animals such as elephants, mammoths and cave bears discovered at sites where ancient men established their camps. Because man's ancestors possessed only primitive

[1] K. Marx and F. Engels, *Selected Works*, vol. 3, Moscow, 1983, pp. 211—370.

weapons they could only fight large animals successfully by developing special hunting techniques that far exceeded those needed by a well-equipped hunter of today.

We will make a paradoxical assertion: the more imperfect were the hunting weapons available to man's ancestors, the greater was the extent to which they had to be supplemented with a higher level of mental activity to make it possible to overcome large animals. This is why the need for man's mental development was very intense in the remote past, and this led to a consolidation of progressive developments in man's brain by natural selection. Such a concept makes it possible to assume that man's evolution was more intensive in regions where the hunting of large animals constituted the primary food source. Such conditions existed over large forestless areas close to the ice sheets of the glacial periods. It is possible that it was in those regions, which were located at middle latitudes, that modern man developed. Modern man could have evolved under ecological conditions which required considerable intellectual capacity, both for the successful struggle against large and dangerous animals and possibly for competition with related primates.

Following the appearance of the subspecies *Homo sapiens sapiens*, further progress in man's cultural development proceeded most rapidly in regions where natural conditions did not provide sufficient edible plants and required the continued hunting of large animals.

Representatives of this subspecies spread over all continents and sometimes found themselves in zones in which food could be provided through easily accessible edible plants (in the humid tropical and subtropical areas) or else from animals that were unable to offer serious resistance to hunters. This latter situation may have occurred when modern man spread into Australia, where marsupial animals represented easy prey for the experienced hunters of the Upper Palaeolithic. It is possible that a similar situation occurred when man first spread into America, where the existing animals had never encountered man and did not possess instincts favouring their protection against hunters (Martin, 1973). Under such conditions cultural development was less intense and perhaps occasionally led to the loss of earlier cultural achievements.

These considerations suggest that with the development of a material culture that facilitated man's struggle for existence, his enormous mental reserves became free. In the presence of stimuli produced by complex conditions of the environment, these reserves could be employed for a further development in material culture. In this way, cultural development became more intense, and at a particular stage this made possible the appearance of the first civilizations. The development of social relations in human societies played a leading role in this process. However, progress was highly uneven in both space and time. In the cases when environmental conditions placed large stresses on human society this resulted in a high level of utilization of man's potential mental capacities. In the presence of relatively favourable conditions of the environment that level was substantially reduced. These considerations appear to contribute to overcoming the difficulties described earlier in explaining man's origin.

### 9.2.3. THE STAGES OF ANTHROPOGENESIS

Recent investigations of the origin of man have yielded many important results making it possible to more clearly present the foregoing information on the basic stages of anthropogenesis.

Using palaeontological data it is difficult to reliably determine, when the first primates appeared because morphological differences between the first primates and their ancestral forms are not clear. Data on the remains of ancient primates are scanty. Recently, Bakker has examined some relatively well-preserved fragments of an animal's skeleton that resembles some contemporary prosimians. The age of these fragments is about 50 million years (Herbert, 1982a). This animal, called *Cantius Trigonodus*, was undoubtedly able to move freely in trees, hooking its fingers over branches. According to Bakker, such an adaptive trait is the most distinctive feature of primates, so the spread of the primates presumably corresponds to the age of *Cantius Trigonodus*. Bakker indicated that an adaptation for life in trees was important as it served to protect the earlier primates from predators. This prolonged the time that adult animals spent with juveniles, increasing the possibility for a continued development of the brain through ontogenesis and increasing the time spent in teaching young animals.

In studying anthropogenesis it is important to determine the time at which the evolutionary lines leading to the appearance of the apes most closely related to man (gibbons, orang-outans, gorillas and chimpanzees) separated from the man's genealogical line. However, progress has been slow because of contradicting evidence from biomolecular and palaeontological studies. Comparison of the structure of serum proteins and genetic data from man and from the four great apes has shown a close relation among these species. This relationship is similar to that between such comparatively close species as horses and zebras. According to calculations based on the biomolecular differences between man and the apes, the time of departure of gibbons from the hominid genealogical tree was about 12 million years ago, orang-outans about 9 million years ago, and chimpanzees and gorillas about 6 million years ago. The probable error of these estimates is $\pm 3$ million years (Andrews, 1982). It has been assumed that the immediate ancestor of man was *Ramapithecus*, the remains of which have been dated at 8—14 million years ago. This implies that the time of separation of the higher apes from man's ancestors was at least 14 million years ago.

The widespread belief that *Ramapithecus* played an essential role in anthropogenesis has been brought into question by the work of Pilbeam

(1982), who examined recently discovered remains of *Sivapithecus*, the animal who either belonged to the *Ramapithecus* group or, according to some views, was identical to *Ramapithecus*. Pilbeam established that the animal he examined is similar in many respects to the orang-outan. This conclusion, which is supported by a number of other scientists, excludes *Ramapithecus* as a direct ancestor of man and makes it probable that the lines of man and the higher apes separated comparatively recently, possibly, as indicated by Pilbeam, some 7—9 million years ago (Greenberg, 1982).

The recently discovered remains of an ancient form of *Australopithecus*, called *Australopithecus afarensis*, appear to be much more closely related to the ancestral forms of man. The first discovery of this form, aged at 3.6—3.7 million years, was examined by Johansen (*New Species of Man . . .* , 1979) and consists of a fairly complete skeleton (called Lucy) belonging to a relatively small individual of *Australopithecus*. Lucy had the ability to walk erect combined with a very small skull volume not exceeding that of a chimpanzee. Further discoveries have made it possible to establish that the age of this species was up to four million years (Herbert, 1982b).

At present there are no generally accepted views about interrelations among the ancestral species of the hominid group. An obvious reason for this is the incompleteness of the palaeontological record, which only permits subjective interpretations of the available information. However, a thorough search for new discoveries could well help to clear up many of the unanswered questions.

Following the review of Cronin *et al.* (1981), we shall assume that the genus *Homo* has existed for about 1.8—2.2 million years. The majority of investigators think that this genus was first represented by the species *Homo habilis* which is closely related to *Australopithecus*. More than one million years ago a more progressive species, *Homo erectus*, first appeared and was followed several hundred thousand years ago by the early representatives of the species *Homo sapiens*. It is generally assumed that the predecessors of the genus *Homo* were *Australopithecus*; some scientists believe that the ancestor of man was the species *Australopithecus africanus* while other scientists favour the more ancient form, *Australopithecus afarensis*.

Cronin *et al.* (1981) discuss the suggestion of Gould and Eldredge that hominid evolution took place during separate and relatively short time intervals and can therefore be regarded as an example of punctuational evolution. Acknowledging the uneven rate of hominid evolution, Cronin *et al.* (1981) do not, however, consider that the evolution of the hominid group corresponds to the concept that brief periods of evolutionary development were replaced by long intervals, during which organisms did not change. At the same time they note the possibility that there could have been a particularly fast evolutionary development during the time period not yet covered by palaeontological evidence and during which the higher ape groups separated from the immediate ancestors of man.

## 9.3. The Impact of Ancient Man on the Biosphere

Until recently it was assumed that man's influence on the environment did not begin to acquire large proportions until the last few decades, when technological development became more intense and the population on our planet started to grow rapidly. However, recent studies have shown that several thousand years ago the spontaneous activities of man produced large changes in his natural environment and in some cases threatened the further existence of human society. The first case of this type appears to be associated with the development of the Upper Palaeolithic culture in Europe, Asia and America.

The Upper Palaeolithic was the first culture created by modern man. Its economic basis was provided by hunting large animals using weapons which enabled hunters to kill animals such as mammoths and the wool-bearing rhinoceros. The Upper Palaeolithic culture lasted a very long time, at least two-thirds of the period of modern man's existence. During the Upper Palaeolithic, which coincided with the last (Würm) glaciation, many of the large animals inhabiting the middle latitudes of the Northern Hemisphere became extinct. In Europe, for example, many large, and previously numerous, herbivorous animals, such as the mammoth, the wool-bearing rhinoceros, the steppe bison and the giant deer, as well as some large predators, including the cave lion and the cave bear, disappeared. At the same time major changes took place in the distribution of several animal species that had survived the extinction. For instance, the large herds of reindeer that had inhabited Western and Central Europe moved to other areas. Similar changes occurred in the animal world of Asia and North America. It is also possible that a number of large animals vanished on other continents (e.g. in Africa and South America).

In the 19th century, some authors were suggesting that primitive man was able to destroy mammoths and some other animals. Abundant data supporting the view that man played a major role in annihilating animals at the end of the Pleistocene are contained in the works of Martin (1958, 1966, 1967). In his opinion, primitive man played a decisive role in the disappearance of many animals not only in middle but also in tropical latitudes. It should nevertheless be noted that some researchers are skeptical about exactly how substantial the influence of man's activities at this early time was on the disappearance of large and widely distributed animal species.

In a monograph by Butzer (1964) it is noted that the animals that became extinct in the late Pleistocene were highly specialized, and could not adapt to changes in natural conditions in the late Würm period. Butzer also noted that, in addition to large herbivorous animals, some of the predators, which became extinct at the end of the Pleistocene were not the objects of the large-scale hunting activities of primitive men. But he also noted that not all hoofed animals in Europe became extinct at this time, and that until recently, many herds of large herbivorous animals have inhabited tropical countries. These animals continued to live and multiply in spite of intensive hunting activities of the local human population. In Butzer's opinion, these considerations contradict the view that it is because of man's hunting activities that a number of animal species became extinct at the end of the Upper Palaeolithic. Butzer nevertheless recognizes that traces of the mass destruction of large animals found at many Upper Palaeolithic sites, suggest that man exerted a definite influence on the overall balance of the biomass of animals that have now became extinct.

In considering the reasons for the extinction of mammoths Colbert (1958) writes:

"Why did the mammoths became extinct? This question like so many of the questions having to do with problems of extinction, is extremely difficult to answer. In fact it is probable that we shall never know the real reason for the disappearance of the mammoths a few thousand years ago, after their successful reign through Pleistocene time. Very likely the extinction of the mammoths was the result of complex causes. Man may have had something to do with it, but we can hardly believe that primitive man was of prime importance in bringing an end to those numerous, widely spread, and gigantic mammoths."

In considering the causes of substantial changes in the animal world of Europe at the close of the Upper Palaeolithic, studies so far have been confined to general considerations that are difficult either to prove or to disprove without quantitative interpretations of the problem. For example, it may be pointed out that the high degree of specialization of animals that became extinct at the end of the Upper Palaeolithic did not in fact prevent their survival under the sharp changes in the environment that took place repeatedly during the Pleistocene.

The disappearance of a number of herbivorous animals at the same time as the extinction of large predators does not prove that man's activity did not influence the animal world; the disappearance of these predators may be explained by disruptions in their food chains in connection with the extinction of corresponding herbivorous animals either as a result of natural causes or as a result of man's activity. Similarly, the argument that until recently the existence of herds of large herbivorous animals in tropical areas precluded the possibility of the destruction of similar animals at middle latitudes by hunters in the Upper Palaeolithic is not convincing.

In an interesting survey by Van Valen (1969), qualitative arguments in favour of each of the two alternative hypotheses concerning the causes of animal extinctions at the end of the Pleistocene, namely anthropogenic and climatic, are put forward. Van Valen enumerates 85 arguments, 16 of which he considers the most convincing. These include twelve arguments in favour of the anthropogenic hypothesis and four in favour of the climatic one. However, it is quite clear that such a calculation does not explain the cause of these extinctions. In his conclusion, Van Valen underlines the point that the available data are insufficient but he believes that the evidence necessary to elucidate the causes of the extinction at the end of the Pleistocene will subsequently be obtained.

The possible influence of man's activities on the extinction of animals at the end of the Pleistocene can be examined, with due consideration to the major factors that affect the number of animals, in a quantitative manner. If such calculations confirm that man played a major role in the process of animal extinction, then the significance of other factors, including climatic changes, may be viewed as secondary.

About 40 000 years ago modern man first appeared in Europe replacing earlier Neanderthal man. Modern man developed a more advanced technique for producing stone and bone weapons, and created an effective system for the mass hunting of large herbivorous animals. During the Upper Palaeolithic a substantial growth in Europe's population occurred, the level of material culture increased appreciably and the first achievements were made in the field of art (for instance, the famous animal drawings in the caves of France and other European countries). The Upper Palaeolithic came to an end about 10 000—13 000 years ago, shortly before the termination of the Würmian glaciation.

The end of the Upper Palaeolithic was associated with major changes in the way of life of man in Europe. This was marked by a dissemination of the culture of the Middle Stone Age — the Mesolithic. At this time some of the achievements of the Upper Palaeolithic culture (e.g. the art of cave painting) was forgotten, although new techniques for producing stone tools were developed.

There are grounds for believing that the end of the Palaeolithic was a turning point in the early history of mankind and is associated with the cessation of mass hunting of large herbivorous animals. As a result, man was deprived of the traditional means of supplying himself with food and many other indispensable materials, and instead had to search for new sources of livelihood.

A calculation procedure, using palaeogeographical and biological data, that makes it possible to clarify the way in which hunting exerted an influence on animal numbers during the Upper Palaeolithic has been suggested by Budyko (1967). This calculation procedure is based on the equation presented in Chapter 5 which describes changes in the number of animals of a particular species over time:

$$\frac{\mathrm{d}n}{\mathrm{d}t} = \alpha n - \beta n, \tag{9.1}$$

where $n$ represents the number of animals per unit area; $\mathrm{d}n/\mathrm{d}t$ the rate of change in the number of animals over time; $\alpha$ the relative increment in the number of animals resulting from births; $\beta$ the relative decline in population produced by the number of deaths per unit time. Parameters $\alpha$ and $\beta$ may depend on the magnitude $n$; such relationships differ for different animal species.

Calculations based on Equation (9.1), as well as data from field studies, show that the number of animals of a particular species under natural conditions usually fluctuates within a relatively wide range. The periods of such fluctuations are comparable with the length of life of one generation of animals. These fluctuations are usually explained by changes in feeding conditions produced by differences in meteorological regimes in particular years, epidemics, and complex interactions between the number of plant-eating and flesh-eating animals that in some cases have the character of autofluctuations (Severtsev, 1941; Lack, 1954). However, when considered over a time period substantially longer than the mean life span of the animals, the average number of animals of a particular species should be more or less constant under stable external conditions. Changes in the average population may occur either as a result of changes in natural conditions or as a result of the evolutionary development of the animal species under consideration or of other species with which it interacts (Severtsev, 1941; Schmalgauzen, 1946).

The principles that govern changes in the animal numbers vary substantially when the animals are being systematically hunted. In such cases Equation (9.1) may be represented as

$$\frac{dn}{dt} = \alpha n - \beta_1 n - g, \tag{9.2}$$

where $g$ is the number of animals annihilated over a given period of time by hunters. It is evident that, for animals that were pursued by hunters during the Palaeolithic, the coefficient $\beta_1$ characterizing mortality from natural causes, other conditions being equal, was smaller than the coefficient $\beta$ since the hunters were more successful in destroying sick and weakened animals than healthy and strong ones.

Since the products of hunting activities represented the principal source of human food during the Upper Palaeolithic in Europe, it should be assumed that $g = \gamma m$, where $m$ is the number of persons per unit area and $\gamma$ is the relative quantity of biomass of the animal being pursued that is expended on one person over a given period of time (the ratio of the expended biomass to the average weight of one animal). As a result of this, Equation (9.2) acquires the following form:

$$\frac{dn}{dt} = \alpha n - \beta_1 n - \gamma m. \tag{9.3}$$

It should be noted that in calculating losses of animals as a result of systematic hunting activities, the 'principle of encounter' cannot be used in studying mutual relations between the number of carnivorous and herbivorous animals. According to this principle the number of animals killed is assumed to be proportional to their total number. This is a natural assumption in the case of small animals that may hide from hunters pursuing them. It may be considered that man, having mastered mass hunting techniques, could always find his prey within the areas that he himself inhabited, particularly if he hunted large animals such as mammoths that lived in vast open tundras and steppes. Because of this, the value $\gamma m$ in Equation (9.3) must be regarded as being independent in relation to $n$ until the animals being hunted survive on a territory inhabited by a particular human tribe in numbers that meet the tribe's current needs.

In calculating the population dynamics of large herbivorous animals during the Upper Palaeolithic it is necessary to estimate changes in the human population at that time. It is apparent that such changes are determined by an equation that is similar to Equation (9.1), i.e.

$$\frac{\mathrm{d}m}{\mathrm{d}t} = am - bm, \tag{9.4}$$

where $a$ and $b$ are the birth and death rates respectively. While these coefficients depend on a number of factors, in the absence of appropriate data we must limit ourselves to a rough estimate of the average magnitude of the difference between these two coefficients: $c = a - b$, assuming that it is independent of time and of population density. In such a case we obtain the following formula from Equation (9.4):

$$m = m_0 \, \mathrm{e}^{ct}, \tag{9.5}$$

where $m$ represents the human population at time $t$ and $m_0$ is the population at the initial time.

Data based on available ethnographic studies of human tribes that had reached various levels of historical development provide some information concerning the population density in Europe at the beginning and at the end of the Upper Palaeolithic. With the help of such an approach a number of authors have concluded that the population density in Europe at the end of the Upper Palaeolithic ranged from 5 to 50 persons per 100 km$^2$. These values should be compared to the much lower population density of human tribes that had not mastered mass hunting methods, and for which the average value of one person per 100 km$^2$ is generally accepted (Braidwood and Reed, 1957; Butzer, 1964). Assuming that the corresponding increase in population occurred throughout the Upper Palaeolithic, which lasted approximately 25 000 years, we find from Equation (9.5) that the value of $c$ ranges from $0.64 \times 10^{-4}$ to $1.56 \times 10^{-4}$ per year. Such coefficients of population growth, which appear to be very low today, are characteristic of the early periods of human history. These values agree well with a rather short average longevity of the Upper Palaeolithic man that has been established on the basis of studying skeletal remains.

Let us now consider changes in the number of animals of some species of large herbivores inhabiting European tundra in the Upper Palaeolithic period. According to the available estimates (see Chapter 5), the food resources of tundras were sufficient to sustain a population of animals corresponding to approximately 800 kg of their biomass per 1 $km^2$. If in a given area the particular species prevailed over other herbivorous animals it could utilize the major part of such food resources. The figure of 800 kg of biomass per 1 $km^2$ determined the upper limit of the possible number of herbivorous animals. It is likely that for animals such as mammoths, whose populations little depended on predators, this limit was in fact attained under favourable conditions. Let us examine how the number of mammoths should change following their subjection to systematic hunting activities. This requires an application of the following equation, which is derived from Equations (9.3) and (9.5):

$$\frac{\mathrm{d}n}{\mathrm{d}t} = \alpha n - \beta_1 n - \gamma m_0\, \mathrm{e}^{ct}. \tag{9.6}$$

By multiplying all the terms in this equation by the average weight of an animal we have

$$\frac{\mathrm{d}N}{\mathrm{d}t} = \alpha N - \beta_1 N - \Gamma m_0\, \mathrm{e}^{ct}, \tag{9.7}$$

where $N$ is the biomass of the animals being studied per unit of area that they occupy; $\mathrm{d}N/\mathrm{d}t$ is the rate of change of that biomass over time; and $\Gamma$ is the biomass of hunted animals consumed by each person per year.

This equation must be solved for an initial condition $N = N_0$, when $t = 0$. This corresponds to a situation in which the biomass of animals at the beginning of the Upper Palaeolithic is equal to the limit determined by food conditions.

In order to avoid the difficulty of estimating the coefficient $\beta_1$, we shall assume that for $\Gamma m_0\, \mathrm{e}^{ct} < \alpha N_0$ the right-hand term of Equation (9.7) is equal to zero, i.e. the average number of animals does not change when losses from hunting are less than the increase in biomass. Subsequently, as the term $\Gamma m_0\, \mathrm{e}^{ct}$ is increased to values that are equal to or greater than $\alpha N_0$, we will neglect the term $\beta_1 N$ and will employ Equation (9.7) in the form

$$\frac{\mathrm{d}N}{\mathrm{d}t} = \alpha N - \Gamma m_0\, \mathrm{e}^{ct}. \tag{9.8}$$

It is evident that both of these simplifications will lead to a certain reduction in the rate of decline in animal numbers by comparison with reality.

Denoting by $\tau$ the length of time during which hunters do not greatly influence the number of the animals being hunted, we are able to determine this magnitude from the relation

$$\Gamma m_0 \, e^{c\tau} = \alpha N_0, \tag{9.9}$$

and we find that

$$\tau = \frac{1}{c} \ln \frac{\alpha N_0}{\Gamma m_0}. \tag{9.10}$$

In order to determine the length of time during which the number of mammoths declines up to their full extinction, it is necessary to solve Equation (9.8). Assuming that $N = N_0$ with $t = \tau$, we obtain through such a solution the dependence of the biomass of the animals being studied on time

$$N = \frac{\Gamma m_0}{\alpha - c} \, e^{ct} + e^{c(t-\tau)} \left[ N_0 - \frac{\Gamma m_0}{\alpha - c} \, e^{c\tau} \right]. \tag{9.11}$$

By this formula it is possible to determine the period of time that is required before $N$ declines to zero, i.e. the given species of animals becomes extinct.

As well as the parameters whose values have already been mentioned, it is necessary to know the magnitudes of the coefficients $\alpha$ and $\Gamma$, in order to carry out calculations of changes in the population of mammoths during the Upper Palaeolithic. The first of these coefficients describes the ratio of the number of mammoths born in a given year to the overall number of animals in the herd. On the basis of data relating to contemporary elephants the magnitude of this coefficient is probably 0.05. The coefficient $\Gamma$ may be roughly estimated on the basis of data derived from ethnographic observations of northern tribes, whose food is largely obtained by hunting large animals. Noting that the magnitude of $\Gamma$ describes, in addition to the direct consumption of the biomass as food, its economic use and non-productive losses, the value $\Gamma$ should be estimated as approximately 500—1000 kg yr$^{-1}$ per person. In a study by Bibikov (1969), it is suggested that such an estimate of biomass expenditure is too large. While we agree that the actual consumption of meat and animal fat was less than this estimate, we nevertheless consider the estimate to be reasonable in view of the need to expend a large amount of biomass for economic activities (fat, hides, animal bones) and of large losses, for instance, associated with difficulties in preserving animal products.

Assuming that the average value of $\Gamma$ is 750 kg yr$^{-1}$ and that in accordance with the estimates that have been mentioned above, $N_0$ = 800 kg km$^{-2}$ and $m_0$ = 0.01 km$^{-2}$, we have used the equations described above to calculate the characteristics of changes in mammoth populations for two values of the coefficient $c$ equal to $0.64 \times 10^{-4}$ and $1.56 \times 10^{-4}$ yr$^{-1}$. The calculation of $\tau$ according to Equation (9.10) shows that the number of mammoths declined only modestly over a time period from 10000 to 25000 years (depending on the assumed values of $c$) after the start of hunting activities. Then, as can be seen from calculations based on Equation (9.11), the number of mammoths began to decline rapidly and in the course of several centuries following the termination of a period of stable population they became completely extinct.

In evaluating the results of these calculations one should keep in mind that their precision is limited by the approximate values of the parameters used. At the same time, analysis of the equations indicates that appreciable changes in these parameters do not alter the general principles governing the process of population change and that the principal conclusion about the possibility of extinction of large animals such as mammoths over a period of time comparable to the duration of the Upper Palaeolithic is valid.

It may be noted that this conclusion is also valid when the animal species in question (for example, mammoths) represents only a part of the herbivorous animals feeding on the available source of food. For this situation it is sufficient to assume that the ratio of the biomass of the given species to the overall biomass of the animals being hunted approximates a similar ratio for the biomass of the animals being killed by hunters.

Let us also note that such a conclusion may be reached without any complex calculations, through a simple comparison of the available information concerning the balance of the biomass of large herbivorous animals in Europe during the Upper Palaeolithic. In such a context particular importance is attached to the comparison of the annual increase in this biomass with the density of human population occupying the corresponding territory. If, as in the case of mammoths, the corresponding increase did not exceed 4000 kg per 100 km$^2$, it is apparent that, even if this biomass was fully utilized by hunting tribes, this increase would have been sufficient to sustain a human population of no more than a few persons per 100 km$^2$. Yet it is also clear that having achieved such a limiting value, the human population in fact continued growing, for it was able to obtain sufficient quantities of food through the destruction of the main part of the animal herd as well as through utilizing the natural increase in biomass. As a result, the increase in biomass gradually declined and the number of animals in the herd also declined until the animals were finally destroyed. This accords with the results of the above calculation.

It is possible that after mammoths had been annihilated by man over the greater part of the territory they occupied, their herds still survived in some regions of northern Asia where the human population was small and in other areas of a severe climate. The existence of large herbivorous animals, however, in the regions of such unfavourable climatic conditions could hardly be long and may be expected to come to an end with subsequent climatic fluctuations.

By using the foregoing equations it is possible to explain why primitive man was able to annihilate mammoths but was unable to destroy elephants, which were still numerous until recently in a number of tropical regions.

It is seen from Equation (9.10) that the period of time during which species of animals being hunted continued to exist depended on the possible reserve of their biomass $N_0$ and on the relative magnitude of the consumption of that biomass by hunting tribes $\Gamma$. The available data indicate that the magnitude of the biomass of large herbivorous animals in tropical savannas exceeds the corresponding value in tundra by at least ten times. At the same time the food of primitive man in tropical latitudes could include a much greater proportion of plant products than that of Palaeolithic hunters in Europe, who, because of the effeects of glaciation on climate, relied on the products of hunting activities to survive.

Assuming that the value of $\Gamma$ for hunters in tropical regions is at least one-half of the value assumed for the conditions of the Upper Palaeolithic in Europe, and employing the value of $N_0$ that is characteristic of savannas, it may be found from Equation (9.10) that the value of $\tau$ for the tropics must have been approximately three times greater than that for Europe during the Palaeolithic. Thus the number of elephants under conditions of constant hunting activities could be relatively high for several tens of thousands of year following the full destruction of mammoths. This means that until recently elephants have not been threatened with extinction.

This theory makes it possible to explain why during the Upper Palaeolithic in Europe only the largest herbivorous animals became extinct. The magnitude for the increase in the biomass of this type of animal was primarily determined by the birth rate $\alpha$ because for any of these species the upper growth of their maximum number was limited by food resources of the given geographical zone. The birth rate is minimum for the largest animals and increases as the size of the animal declines. Therefore, the largest herbivorous animals whose increase in the biomass per unit area was the smallest were least protected from hunters of the Upper Palaeolithic. The cause of the biological catastrophe that these species encountered at the end of the Upper Palaeolithic was very simple.

The size increase which large herbivorous animals acquired in the course of evolution protected them to some extent against attacks of predators. As the influence of predators on their numbers was weak an appreciable decrease in birth rate became possible and expedient. When Upper Palaeolithic man started hunting these animals he assumed a role similar to that of wolves and other predators in relation to small herbivorous animals. But while small herbivores could compensate for the losses resulting from predators by a high birth rate, large animals were unable to do this and their extinction was only a matter of time.

It should be noted that such an explanation for the causes of the more rapid extinction of large animals towards the end of the Pleistocene differs from that proposed by Simpson (1944), who believed that it was more difficult for large animals with a less frequent replacement of generations to adapt to changing natural conditions of that time.

It is possible that in individual regions of Europe the actual period of time from the start of mass hunting activities to the extinction of large herbivorous animals was much shorter than the calculated period based on the above reasoning. This may be due to the assumptions that have been made, but also to the fact that the calculations did not consider fluctuations in the number of animals resulting from epizootic diseases and especially as a result of changes in the food base, which could be quite significant in the conditions of the late Würmian period. The existence of such fluctuations could intensify the decline of particular large herbivores in some areas for many thousands of years, while the restoration of the species numbers through migration from other regions was impeded by the continuing pursuit of large herbivores by hunters of the Upper Palaeolithic.

From such a view point, intensive hunting activities should have first of all led to the disappearance of the largest animals with the lowest birth rate. Their extinction could have contributed to a temporary increase in the number of smaller herbivorous animals, whose ability to utilize the food base increased. However, more intensive hunting activities could have led to the extinction of herbivores with higher birth rates, as was the case with reindeer in some regions of Western Europe. It should be noted that for such animals, losses from hunting activities were supplemented with substantial losses from predators, which weakened the stabilizing effect of a high birth rate in maintaining the population of the species.

Certain other implications of the above-mentioned calculations are also noteworthy. The process of decline in the number of animals subject to hunting was initially slow, then it gradually increased and the remaining relatively large herds of animals were annihilated rather quickly. Beyond this, long before the final destruction of a particular animal species, there should have been more or less long periods when these animals became extinct in certain regions as a result of natural fluctuations in their number or because of the migration of their herds.

Thus, a threat arose with increasing frequency to the growing human population in Europe during the Palaeolithic that was primarily due to mass hunting activities. This explains why the Upper Palaeolithic men were so anxious to draw animals, to which they attributed magic properties and which were hoped would help them to restore the vanishing herds. Yet continued intensive hunting activities led to the destruction of large herbivores at an increasingly rapid rate, and since its last stage proceeded particularly fast, the hunting tribes faced a difficult situation in which there was not enough time for a gradual transition to other forms of food acquisition. It may therefore be thought that the transition to the Mesolithic was particularly troublesome for primitive societies, and was accompanied with temporary reductions in human population.

It is known that the termination of the Palaeolithic in the middle latitudes of North America was also associated with the extinction of large animals, which confirms the causal relations of these phenomena. At the same time, it is quite probable that perceptive changes in natural conditions at the end of the Palaeolithic produced additional difficulties for the existence of these animals and served to intensify the process of extinction which was already developing.

Thus, the termination of the Palaeolithic culture in Europe may have been partly the outcome of an irreconcilable contradiction between the mass hunting activities developed by man in relation to large animals in the Upper Palaeolithic, which produced a temporary abundance of food and made possible an increase in human population, on the one hand, and the limited natural resources for these activities, which appeared to be exausted after a certain time period, on the other hand.

The following question then arises: What brought about a disruption to the balanced ecology of the pre-Holocene biosphere when the species of animals pursued by predators could exist for hundreds of thousands and even millions of years?

It is possible that the principal reason for such a disruption was an extremely high rate of evolution in man in comparison with the rate of evolution of the animals he pursued. As already mentioned, large animals possessing a low birth rate are less able to maintain a stable population when hunted by predators than are small animals possessing a higher birth rate. Accordingly, the large body size gives such animals advantages in struggling for existence only if it makes them relatively unattainable for predators. If, on the other hand, their size does not protect them from attacks (as was the case even with mammoths in the Upper Palaeolithic), the only way in which populations could be maintained is by an increase in the birth rate, which is usually achieved in mammals through a reduction in their body size. Such a change in the animals that were hunted in the Pleistocene, required a much longer time period than the period during which their extermination occurred.

A similar approach was applied by Martin (1973) in clarifying the reasons for the disappearance of many species of large animal in North and South America at the end of the Pleistocene. It is known that man appeared on these continents relatively late (towards the end of the Würmian glaciation) and that the principles governing the process of his settlement in America are still poorly understood. Martin has assumed that the first small groups of people crossed the ice-bound Bering Strait and entered the northwestern regions of North America where they found enormous herds of large animals that had never encountered man and did not possess instincts favouring their protection when attacked by man. This produced more favourable conditions for hunting in America than in Europe and Asia, where wild animals were in contact with man at various stages of his evolution for hundreds of thousands of years and developed various methods to protect themselves against hunters.

Under such conditions the first inhabitants of America could supply themselves with an unlimited amount of food, and, having rapidly increased in numbers, they could settle all over America, destroying on their way the majority of the large animals they encountered. Martin's calculations show that over a period of a thousand years the initial wave of settlers could have reached Tierra del Fuego at the very end of South America, having destroyed on their way approximately thirty species of large animals that existed in America at the end of the Pleistocene. These included mammoths, mastodons, great sloths, horses, camels, and many others. Following their annihilation, hunting tribes were forced to shift to other sources of food.

The conclusion that the first culture of modern man — the Upper Palaeolithic — led to the emergence of an ecological crisis on large territories that destroyed this culture may be compared to the well-known statement by Karl Marx: "Culture when it develops spontaneously and is not consciously controlled leaves deserts behind itself."[2] Man's subsequent history has produced much data that confirm the difficulty of maintaining an ecologically balanced biosphere under the conditions of spontaneously developing economic activities.

## 9.4. Contemporary Man's Impact on the Biosphere

### 9.4.1. ANTHROPOGENIC ECOLOGICAL CRISES

Since the Upper Palaeolithic man's economic activities have often led to a deterioration of natural conditions, often making it difficult for human society to exist. Many ecological crises have developed in this way and subsequently affected areas of different size.

From earliest times fire was an important way in which man influenced the environment; the application of fire made it possible to destroy vegetation over large areas. Forest fires and steppe fires have long been used in hunting large animals. This method is still used by Australian aborigines, who destroyed vegetation over tens of square kilometres (Dorst, 1965). Similar hunting methods were probably used by the hunters of the Upper Palaeolithic. It is clear that widespread fires led to a great loss of plants, animals and the ruin of natural ecological systems. At the same time the destruction of forests allowed further hunting of large animals, which perhaps explains the rapid destruction of forest vegetation in many regions following the appearance of modern man.

During the Neolithic, when the principal economic activity was the raising of cattle and agriculture, the burning of vegetation covers assumed very large proportions. Fire was used to expand pastures by destroying forests, especially in the slash-and-burn agricultural practices. The resulting fertiliza-

[2] K. Marx and F. Engels, *Selected Works*, vol. 32, Moscow, 1964, p. 45.

tion of the soil by ash produced abundant harvests, even when the plowing was very shallow. However, under such a system of agriculture the fertility of soils declines rapidly and new forest areas must be cut in a few years (sometimes even in one or two years) with agricultural activities being moved to these new areas. This type of agriculture can be used when there are large sparsely populated forest areas. In the past this method was widely used in many middle latitude countries, and is used today in a number of developing countries in tropical areas.

The wide use of such methods of burning vegetation over large areas has caused pronounced changes in natural conditions, including flora, fauna, soil, and, to lesser extent, the climate and hydrological regime. Since the systematic burning of vegetation, both at middle latitudes and in tropical regions, was initiated a very long time ago, it is difficult to estimate the full scope of the environmental changes produced by this activity. Observations show that in many cases the vegetation cover destroyed by man does not restore itself even after the cessation of systematic burning.

As well as the impact of slash-and-burn agriculture on forests, large areas of forest have also been destroyed in order to produce timber. In other areas, the cattle grazing has had a major impact on the natural vegetation cover of many regions. Such grazing activities were often carried out without considering the need to restore vegetation cover. In forest regions with a dry climate the feeding of goats and other animals on the leaves of young trees eventually led to the destruction of the forests. Excessive grazing by cattle has destroyed the vegetation of dry steppes and savannas turning these into semi-deserts and deserts.

The impact of economic activities on the vegetation cover appears to have frequently produced severe losses to human society. One of the first civilizations in human history appeared in the northwest of India in the third to second millennium B.C. The centres of this civilization (Harappa, Mohenjo-Daro and others) were in regions that have since become deserts. Some authors believe that in the past these regions were dry steppes with favourable conditions for animal husbandry and certain types of agriculture. At this time, excessive grazing of cattle could have resulted in the destruction of plant cover, causing an increase in temperature and a decline in the relative humidity of the lower air layer. As a result, the amount of precipitation decreased, which made the renewal of vegetation cover impossible. Therefore, it is likely that anthropogenic changes in climate were one of the factors that led to the demise of this ancient civilization in India.

Another example relates to changes in natural conditions in the Mediterranean basin during ancient times. Originally large forests existed in Greece and some other Mediterranean countries and subsequently were partly cut for fuel and other uses and partly destroyed as a result of excessive cattle grazing. This promoted intensive soil erosion and led to the complete destruction of the soil layer on many mountain slopes. This, in turn, intensified the aridity of the climate and produced a strong deterioration in the conditions for agricultural production. Although in this particular case changes in natural conditions did not cause substantial losses to ancient civilizations, they did influence many aspects of human life.

It has been suggested that the depletion of soils in areas occupied by the Maya State in Central America through slash-and-burn agricultural activities was one of the reasons for the destruction of this highly developed civilization. Europeans who settled in Central America found numerous cities in the tropical jungles that had long been abandoned.

There are many other examples of the creation of adverse environmental conditions by man that were irreversible in the light of the then technical possibilities. Such cases may be described as anthropogenic ecological crises.

In studying anthropogenic ecological crises of the past it can be suggested that their impact on the early stages of human society was considerable. This is especially true of the Upper Palaeolithic crisis that has already been described. Subsequent ecological crises did not embrace such large areas and appear to have produced shorter-term effects on human society.

It should be noted that the influence of ecological crises on the development of human society was ambivalent. Besides the damage that they caused to man's traditional economic activities, crises stimulated the development of new ways of meeting economic requirements of the society. In particular during the crisis of the Upper Palaeolithic the extinction of large animals eventually led to the emergence of new and much more effective types of economic activities, namely the raising of cattle and agriculture.

Substantial deterioration of the conditions for agriculture in a number of ancient Mediterranean countries, as discussed above, exerted a definite influence on the development of sea-faring, different crafts and other types of non-agricultural production. It must be supposed, however, that this stimulating influence of ecological crises on economic progress was not always pronounced and in any event it played a smaller role than did social development.

The history of man's influence on the biosphere shows that technological progress constantly increases the possibility of influencing the environment and creating conditions under which major ecological crises emerge. On the other hand, this progress increases the possibility of overcoming deteriorations in the natural environment produced by man. These two opposing tendencies have been particularly pronounced in the second half of the 20th century.

### 9.4.2. MAN'S INFLUENCE ON THE BIOSPHERE

In our age, human activities exert an enormous influence on the natural conditions of the entire planet. Changes in the flora and fauna on land are particularly pronounced. Many species of animals have been completely destroyed by man and a still greater number of species face extinction. It is believed that more than 120 species and subspecies of mammals as well as about 150 species of birds have vanished (Dorst, 1965). It is likely that this is a considerable underestimate and the total number of mammals and bird species annihilated by man during the last millennium is much greater. An example from the bird fauna of the Hawaiian islands illustrates this. At about 600 A.D. these islands were inhabited by Polynesians and since the end of the 18th century Europeans began to frequent them. It has been established in palaeontological studies that from the time when people first settled on the islands to the time when Europeans appeared, the number of bird species living on the Hawaii was reduced to more than a half of their original number. During the last two centuries, the period of European colonization, about one-third of the remaining bird species have vanished. It has been suggested that the majority of the extinct species disappeared as a result of changes in the birds' habitat under the impact of economic activities and only a minority were directly annihilated by man (Olson and James, 1982).

During the last millennium many large birds (Dinornithes from New Zealand and Aepyornithes from Madagascar mentioned earlier) as well as mammals (*Bos primigenius*, Steller's sea-cow and others) have been annihilated. A comparatively recent annihilation by man are certain large primates such as the giant lemur on Madagascar and gigantic forms of orang-outan on the islands of Indonesia. Many large animals are now on the verge of extinction, including whales, which include the largest animals that have ever lived on the Earth. It has already been explained why man's hunting activities are particularly dangerous for large animals.

Vegetation cover has experienced enormous changes over large areas of the continents. Wild plants have been destroyed and replaced by agricultural fields, while many of the remaining forests are largely secondary, i.e. they have been greatly altered as a result of human activities. Similarly, great changes have occurred in the vegetation cover of many steppe and savanna regions as a result of intensive grazing of domestic animals.

Man's impact on the natural vegetation cover has also influenced the process of soil formation and has led to changes in the physical and chemical properties of soil. Even greater changes have been experienced by soils in agricultural areas as a result of their systematic cultivation, the application of fertilizers, and the removal of a substantial part of the plant biomass. In many regions the cultivation of soil has led to an increased rate of erosion and as a result, soils have been destroyed over large areas.

The influence of human activities on the continental hydrological regime has been increasing rapidly. The run-off of many large rivers as well as of small ones has been appreciably altered as a result of the creation of hydro-engineering facilities. A substantial part of the water of river run-off is diverted in order to meet the needs of industry and urban populations as well as to irrigate agricultural land. The creation of large reservoirs whose area is often comparable to that of great natural lakes greatly alters the evaporation regime and run-off over large areas.

Pollution of the atmosphere and of the continental waters and oceans by man is proceeding on an increasing scale.

Even though the activities of modern man have greatly altered the natural environment of our planet, these changes, as a rule, represented only the sum of local influences on natural processes. They have acquired a global character not as a result of changes by man of the global natural processes but rather because local (or regional) influences have spread over large areas. It seems that human impact on climate is of a particular character, and this will be discussed in the next chapter.

Although many of the changes in natural conditions that have been mentioned exert an unfavourable influence on human life and activities, whether such changes can be viewed as manifestations of an ecological crisis is not easy to ascertain. An ecological crisis for human society has been defined as an irreversible deterioration of man's environment, that is, as a change in the environment that cannot be overcome by man in the course of a reasonably short time period. The example of past ecological crises given above are of this type.

The extent to which current unfavourable changes in natural conditions are irreversible requires close attention. Leaving aside anthropogenic changes in climate for the moment, let us note that the experience of the last few decades confirms that in principle it is possible by relying on up-to-date scientific and technological achievements to overcome a large number of recent changes in the natural environment that have been unfavourable to man. The restoration of forests and other types of natural vegetation cover is widely carried out in many countries. Sometimes this problem is solved by simple means, for example, through limitations on cattle grazing. In other cases more expensive measures are needed that are nevertheless usually accessible, even for countries whose level of economic development is relatively low.

Much experience has been accumulated in preserving and restoring soils and in limiting the destructive effects of erosion. The protection of air and water bodies from pollution and the purification of polluted water bodies often requires large investment. But available data show that it is possible to restore the natural conditions of even very large water bodies, such as some of the Great Lakes in the United States, through investments that represent a relatively modest share of the national budget.

Significant results have been achieved in the U.S.S.R. and some other countries in preserving animal species threatened with extinction and in increasing the number of rare species. Such animals include sea otters, fur-seals, bisons, beavers, sables, saigaks and many others. At the same time measures intended to preserve wild animals do not always produce positive results. This deserves close attention since of all the types of damage man has caused to the environment, the destruction of the genetic bank of currently existing organisms is evidently the only irreparable one.

Man has already destroyed hundreds of animal species and subspecies. This may not appear very important, since the number of lost species represents a negligible part of the total number and the annihilated species are not as a rule, very significant, from an economic view point. But such considerations are completely misleading. The significance of a particular animal species for mankind is not limited to current economic estimates, and the loss of any part of the present genetic bank may deprive man in the future of important means of regulating ecological systems and of possibilities for a comprehensive utilization of living nature for various practical purposes.

It should be concluded that all forms of local changes in the environment produced by man (with the only exception of the destruction of certain animal species) should not be considered irreparable. The possibility of nearly completely overcoming the damage that has been done to nature on our planet allows us to believe that no global ecological crisis is currently occurring. However, there are many local ecological problems which can be settled with the aid of modern scientific and technological achievements, but the probability of finding their solution in the near future is not always obvious. Thus, for example, it is often impossible for capitalist countries deprived of the systems of centralized planning to find the means required to restore favourable natural conditions. This leads to an intensification of local processes of environmental disruption and attaches to them the features of regional ecological crises. Thus, at present the problem of ecological crises is primarily a social problem that is associated with the nature of social development in individual countries.

### 9.4.3. FORECASTS OF A GLOBAL ECOLOGICAL CRISIS

Aside from any particular evaluation of the current ecological situation the view has often been repeated in recent years that a global ecological crisis may develop in the relatively near future, as a result of which a general economic crisis will take place and the world's population will decline sharply. This point of view has first been presented in the study by Forrester (1971) and in the collective monograph *Limits to Growth* (Meadows *et al.*, 1972). As in the studies of the economy of the Upper Palaeolithic mentioned above (Budyko, 1967), Forrester and the authors of *Limits to Growth* solved equations determining increments in human population in the context of other relations characterizing changes in the resources used by man. In a similar way to the calculations for Upper Palaeolithic society the analyses carried out by the American scientists on human society today has led to the conclusion that a further growth in population will produce a crisis that will lead to a sharp reduction in population. The estimated time for the emergence of such a crisis appears to be very short, merely about 100 years.

It should be noted that the numerical models employed in these studies are much more complex than the relatively simple model used in our own study. The series of equations employed can only be solved using computers. Since in the model for the Upper Palaeolithic it was possible to consider food resources alone (mineral resources did not influence human population at that time and there was no environmental pollution), the calculations by this model can be carried out analytically. In the models referring to the present age, the influence of many factors on changes in the human population was taken into account, including food and mineral resources and environmental pollution.

Let us consider more closely the results of the forecast of changes in economic conditions presented in the book *Limits to Growth* that were obtained on the basis of Forrester's model. The relevant calculations were carried out for the period 1900—2100. They took into account the dependence of birth and death rates on a variety of factors reflected in modern empirical data. When such data were insufficient (for instance, in estimating the influence of environmental pollution on the death rate), the authors of the study relied on hypothetically formulated relationships.

The results of their calculations have shown that the industrial production and the production of foodstuff per capita will increase until the beginning of the 21st century and after that both will fall rapidly to values that are far below their levels at the beginning of the 20th century. The reserves of mineral resources at the beginning of the 21st century will also decline sharply and by the middle of the century will be almost depleted. Environmental pollution will increase rapidly during the first half of the 21st century but will then decrease again by the middle of the century as a result of declining industrial production.

Human population will increase until the middle of the 21st century and then decrease and towards the end of that century will be reduced by nearly one-half. In the case the principal cause for the crisis appears to be dwindling mineral resources. This produces a sharp increase in mortality because of a reduction in the output of food products and a worsening of medical services to the population.

The results of these calculations are supplemented in *Limits to Growth* with a number of other calculations relying on various initial assumptions. In most cases, however, these assumptions do not alter the major conclusion of the first estimation. For example, the assumption that the known reserves of mineral resources will be doubled produces a still sharper decrease in human population by the middle of the 21st century, since this results in catastrophic increase in the level of environmental pollution. The assumption that effective control of environmental pollution is introduced under conditions of doubled reserves of mineral resources does not prevent a decline in population attributable to food shortages.

The results of further calculations show that in order to prevent a reduction in the population and to maintain a high level of industrial production a combination of measures is required that includes a cessation in population growth, a sharp reduction in mineral resources use, a cessation of growth in industrial production and a shift of efforts to overcoming the effects of environmental pollution as well as to increasing the output of food products and so on. It also follows from these calculations that in order to stabilize economic conditions in the 21st century such measures must be introduced immediately; their introduction towards the end of the 20th century will be too late and will not forestall the crisis. The authors observe that the probability that such measures will in fact be introduced in time is not very great and expect that the ecological crisis will be followed by 'social disintegration' (i.e. disintegration of the existing organization of society).

A critical evaluaton of Forrester's studies and of the book *Limits to Growth* is given by Fyodorov (1972, 1977), who lists a number of serious shortcomings. These include a failure to consider fundamental changes that are introduced by technological progress in the economic development of society and also the manner in which man's needs are met. At the same time Fyodorov notes that the conclusions of Forrester and of the authors of *Limits to Growth* concerning the forthcoming crisis in society's interaction with nature correctly reflect the principles that govern the spontaneous development of bourgeois societies. The basic socio-economic characteristics of the capitalist system hinder a rational utilization of natural resources, inhibit the supply of food and make difficult a reliance on planned measures in struggling against environmental pollution. Under such conditions the possibility of a crisis in man's interactions with nature appears. Many traits of that crisis may already be clearly seen in our age. Fyodorov also notes that it is only possible to overcome the obstacles that inhibit man's further technological and economic development through a socialist planned economy. The conclusion that long-term economic planning is needed can also be found in *Limits to Growth*, but planning measures that are proposed in order to forestall a crisis are clearly of a utopian character.

The book *Limits to Growth* was the first report of the Rome Club, the international scientific organization founded in 1968, which unites scientists from a number of Western countries. The second report of the Rome Club was the work of Mesarovic and Pestel (1975) and contained calculations of the future economic situation of the main ten areas into which the authors divided the continental surface. Although the conclusions of this study are much more detailed compared to the conclusions of Forrester and of the authors of *Limits to Growth* (who treated economics in regard to the entire mankind), the principal inferences of Mesarovic and Pestel are in many respects similar to those of the earlier studies. Mesarovic and Pestel considered the economic crisis that is now developing in many countries to be the beginning of much more violent economic and ecological crises rather than a temporary phenomenon. These crises could materialize in the course of the next few decades and initially affect the population of developing countries.

Unlike Forrester, Mesarovic and Pestel do not exclude the possibility of averting the ecological crisis by continuing technological progress. However, the probability of implementing the measures they propose to prevent such a crisis is also very low.

In the second half of the 1970s and at the beginning of the 1980s new studies of future economic and ecological conditions have been carried out within the framework of the Rome Club and of some other international and national organizations. A general shortcoming of these studies is that in their calculations they have used the same relationships among the elements of the ecological and economic system that exist at present. Such an approach would have been valid if this system changed only quantitatively and not qualitatively. However, the history of mankind shows that the development of economy in the past was subject to profound qualitative changes. If these changes are not taken into account, the calculations of economic and ecological conditions of the future based on system analysis acquire a conditional character and do not describe the real future of mankind but the future that would result if a number of existing conditions, which will undoubtedly change, were preserved.

An example of the difficulties that are met in such calculations is that they must include forecasts of scientific and technological achievements that are essential for economic development. It is known that the progress of science and technology is characterized (particularly in our age) by swiftness and unevenness, which is expressed in some cases in the fast solution of problems that until recently seemed to be insoluble (the usage of nuclear power, the creation of spaceships). In other cases it is found that it is not possible to obtain the results which appeared to be attainable to the thinkers of many generations. Under such conditions the prediction of scientific discoveries, including the terms of their implementation, is unrealistic.

This is not the only difficulty of realistically forecasting future economic development. Among other reasons are the extreme difficulty in developing accurate future demographic calculations (predictions of human population growth carried out in the past and intended for a relatively short time period have already been shown to be incorrect) and the lack of knowledge concerning the exact amount of mineral resources available.

A general evaluation of modern futuristic studies of Western scientists has been given in the works of Frolov (1979), Zagladin and Frolov (1981) and in the collective monograph *The Global Problems of Our Age* (1981). It is mentioned in the work by Zagladin and Frolov that in these studies current scientific and technological progress and development are extrapolated into the future without adequate reasoning. The concept that the international character of the indicated processes will lead to the entire world integrity of the historical process independent of the socio-economic system within which it proceeds is erroneous. At the same time Zagladin and Frolov indicate: "... in many concrete treatments of specific (first of all scientific and technological) aspects of global problems that are suggested in the West there are a lot of interesting positive moments ... it is important to consider critically these results in terms of our methodology and to make use of them in search of the solution of global problems."

There is no doubt that it is vitally important to undertake a detailed study of the global problems which mankind will encounter in the 21st century. With reference to constructing rational ecological and economic models, it has been stated in one of our previous studies (Budyko, 1977a) that, in view of the grave defects in the methods for numerical modelling of global development, it is advisable to study how these methods may be improved in order that they may be used to identify specific problems concerning the long-term planning of technological progress. Applying numerical models, which make it possible to calculate probable changes in the dynamics of human population in terms of the basic economic indices and characteristics of changes in natural conditions, it is possible to formulate beforehand the technical problems whose solution is indispensable to overcoming different obstacles in the way of economic progress.

The means of constructing such models have been considered in the works of Gvishiani (1978, 1979), in which a number of methodological problems of global modelling have been discussed. At the same time Gvishiani has underlined the exceptional importance of taking into account the necessity of guaranteeing a lasting peace and putting an end to the arms race in global studies.

In recent years works in the field of global modelling have been started in the U.S.S.R. by the Institute of System Studies (*Modelling of the Processes of Global Development*, 1979; *System Modelling of the Processes of Global Development*, 1980) and by the Computer Centre of the U.S.S.R. Academy of Sciences (Moiseev, 1979; Krapivin *et al.*, 1982). In addition to this, the Soviet scientists pay great attention to studying individual problems of global ecology. The most important of these include the possibility of changes in the global climate of our planet under the impact of man's economic activity. The significance of this problem is explained by the fact that, in view of the available data, such changes can have an appreciable influence on economic activities prior to all other global consequences of economic development that are considered in the studies of natural conditions of the future.

CHAPTER 10

# THE FUTURE OF THE BIOSPHERE

## 10.1. Modern Climatic Change

### 10.1.1. THE 20TH-CENTURY CLIMATE

In recent studies it has been established that the near future of the biosphere depends to a large degree on present-day changes of global climate caused by man's rapidly increasing impact on the chemical composition of the atmosphere.

Since the problem of modern climatic change is of great practical importance, in the U.S.S.R. and in many other countries investigations have been organized in which the distinctive features of contemporary natural and anthropogenic climatic change have been studied. In discussing the main results obtained in these investigations, let us dwell first on the study of contemporary natural climatic changes.

Changes of thermal regime during the last hundred years have been studied by Rubinshtein and Polozova (1966), Mitchell (1963), Lamb (1966, 1977), Van Loon and Williams (1976, 1977) and many other authors. In these studies it has been found that climatic changes have been uneven in space and time. These spatially and temporally complicated temperature changes are determined to some extent by the fluctuations in the atmospheric circulation which are random in relation to climatic changes. Time averaging of temperature anomalies for individual stations or limited regions does not completely eliminate the effects of these fluctuations, which make it difficult to see the distinctive features of climatic change.

For studying global climatic change spatial averaging of temperature anomalies is of great importance. This approach is based on the following physical considerations. As seen in Chapter 3, the outgoing longwave radiation is a linear function of surface air temperature. Therefore the averaging of temperature anomalies is similar to the averaging of outgoing emission anomalies. Anomalies of outgoing emission in restricted regions compensate for each other when averaged over the area and their effects on the global climate turn out to be unimportant. Climatic conditions are mainly affected by emission anomalies over vast areas, particularly, over the whole globe. The algebraic sum of this anomaly and the anomaly of shortwave radiation absorbed in the Earth–atmosphere system determines the sign of change in mean air temperature for the entire planet.

Calculations of mean anomalies of air temperature for large areas have been made by Willett (1950, 1974, etc.) and Mitchell (1961, 1963, 1971) using observational data on air temperature from many meteorological stations situated in various geographical regions. In order to determine mean anomalies more accurately, maps of air temperature anomalies, rather than observational data at particular stations, have been utilized. These maps give the distribution of monthly mean temperature anomalies in the Northern Hemisphere for every month since 1881, except for the equatorial zone where observational data are insufficient for compiling anomaly maps (Budyko, 1971).

Figure 10.1 shows the secular trend of air temperature anomalies for the extratropical zone of the Northern Hemisphere and for the 70—85°N latitudinal zone. All data are presented as five-year running means. From this figure it can be seen that, in extratropical latitudes of the Northern Hemisphere, a warming started near the end of the 19th century and reached a weakly expressed maximum during the last years of this century. A temperature decline followed which soon became a rapid increase. This increase in temperature was particularly great at the end of the 1910s and at the beginning of the 1920s. The maximum positive temperature anomaly occurred at the end of the 1930s. In the 1940s the warming changed to cooling. By the mid-1960s the mean temperature of the Northern Hemisphere reached the temperature level of the late-1910s.

The secular temperature trend for extratropical zone of the Northern Hemisphere seems to correspond qualitatively to the trend for the entire globe. The available data (more restricted compared with these for the northern extratropical latitudes) show that, in the equatorial zone and in southern extratropical latitudes, mean air temperature changed, in many regions, in the same way as in the northern extratropics.

Figure 10.1 also shows that the secular temperature trend was larger in

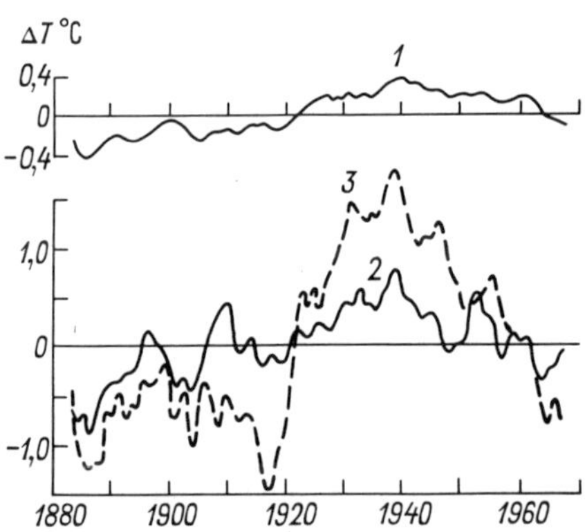


Fig. 10.1. 1, Northern Hemisphere, year; 2, polar zone, warm season; 3, polar zone, cold season.

higher latitudes, and that (in higher latitudes particularly) changes in the cold half-year were stronger than in the warm half-year.

The secular trend for five-year running means of the meridional temperature gradient is presented in Fig. 10.2. Values of the gradient are expressed in temperature degrees per ten degrees of latitude (Budyko and Vinnikov, 1973). As seen in Fig. 10.2, the meridional temperature gradient decreased from the 1880s up to the 1930s. Beginning in the second half of the 1930s it began to increase, with anomalies returning to the values of the 1900s by the mid-1960s.

By the beginning of the 1970s it was known that, over the previous few decades, the climate had been cooling. Since the sign of trend in global mean temperature changes comparatively infrequently, almost all researchers believed that the temperature would continue to drop in the near future. Anxiety was expressed that further cooling would result in a glacial advance and in a deterioration of climatic conditions for agriculture. This point of view related mainly to the mid-latitude countries with moist climates where cooling decreases the growing season and the temperature sums for this period. The available data showed, however, that in the same latitudinal zone in continental regions the cooling was accompanied with an increase in precipitation and a decreased frequency of droughts. This aspect of the change in climatic conditions was favourable for agriculture.

Lamb (1973) has mentioned that there were more than twenty forecasts of climatic change at the beginning of the 1970s, each predicting the continuation of the cooling during the next few decades. Two years later Lamb himself obtained the first data showing a possible climatic change towards warming (Lamb *et al.*, 1975). These data describe the thermal regime in the northern

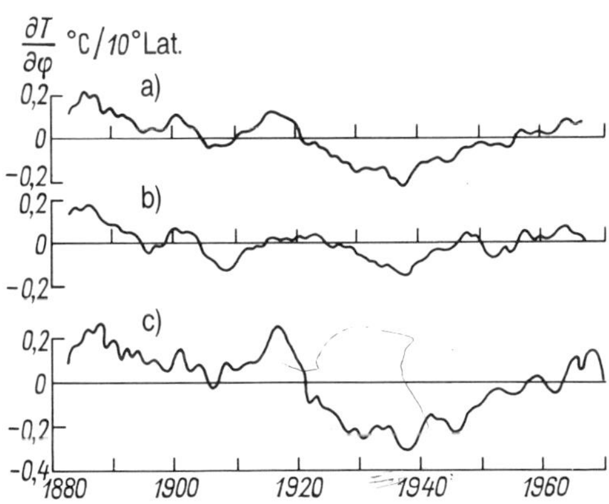


Fig. 10.2. a) year; b) warm season; c) cold season.

Atlantic where, in the region of the Greenland and Norwegian seas, the early tendency to cooling had given place to warming since the 1970—1971 winter. The observation that, at the beginning of the 1970s, the cooling process has changed to a warming was also made by Willett (1974) on the basis of air temperature data. These observations of Willett and Lamb pointed to the necessity of studying the recent climatic change in more detail.

In investigating this question (Borzenkova *et al.*, 1976; Budyko and Vinnikov, 1976) data on changes in mean surface air temperature have been updated up to 1975. Figure 10.3 shows the secular trend of temperature anomalies over the greater part of the Northern Hemisphere (north of 17.5°N; curve 1). The data are expressed as five-year running means. Analysis of the mean annual temperatures reveals that the trend in the Northern Hemisphere has recently changed the sign. The gradual temperature decrease came to a halt in the mid-1960s and was followed by a temperature rise. This warming was later revealed by many researchers. As mentioned above, air temperature variations have especially large amplitudes in high latitudes and during the cold seasons. Figure 10.3 also shows secular variations in the annual mean temperature of the latitudinal zone north of 72.5°N (curve 2). As these curves show, the 1970s warming has been accompanied with a decrease in meridional temperature gradient.

We can also draw important conclusions concerning fluctuations in the thermal regime using observations of the state of polar sea ice. It should be mentioned that sea ice boundaries depend strongly on air temperature (Budyko, 1971), and, at the same time, ice position greatly affects the thermal regime of the atmosphere. At high latitudes the air temperature over the ice-free ocean surface during the cold seasons drops to only a few degrees below zero, since, under such conditions, the ocean releases enough heat into the atmosphere to maintain such a temperature. Under the same conditions, in the presence of ice cover that restricts heat flow from the ocean into the atmosphere to a great extent, the temperature of the lower air

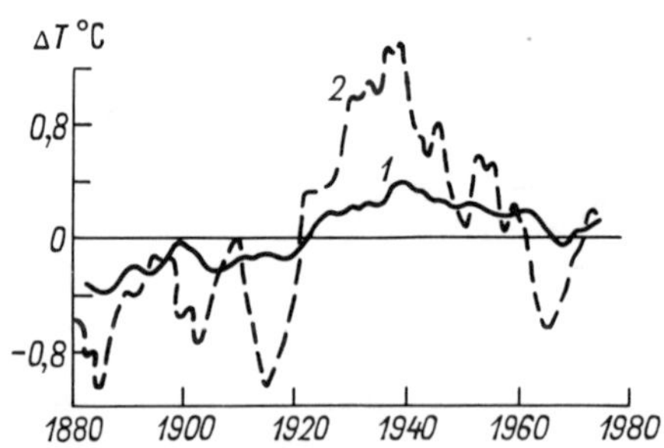


Fig. 10.3.

layer can fall by tens of degrees below zero. Consequently, the migration of the average boundary of polar sea ice has a great influence on the air temperature in relevant areas, especially in the cold seasons. Unfortunately, observational data on the polar sea ice boundary for the period of modern climatic change are rather limited. One hundred years of data on the ice position are available only for the Atlantic sector of the Arctic and adjacent seas. For other parts of the Arctic Ocean, as well as for the Antarctic, data on the polar sea ice boundary cover only the last 20—30 years.

Attempts have been made to elucidate the relationship between climatic change and migration of ice cover in the North Atlantic, in the Barents and Kara Seas. Figure 10.4 shows the secular variations of ice in the Greenland and Barents Seas. Comparison of Figs 10.3 and 10.4 shows that variations in the amount of ice agree well with fluctuations in the mean air temperature in the Northern Hemisphere. In particular, the ice area has decreased rapidly with the development of the warming trend that began in the mid-1960s. Among other investigations of Arctic Sea ice, we mention here the work of Sanderson (1975), who noted a decrease in the amount of ice in most areas of the Arctic over the 1969—1974 period. A description of modern changes in the ice cover of the Arctic Sea is given by Zakharov and Strokina (1978). We present here a diagram from this paper (see Fig. 10.5) showing variations in the ice area in the Arctic Sea over the last four decades for three individual months (July, August and September). From the beginning of the 1940s to the mid-1960s the ice area increased by more than 10%. Since the mid- 1960s it has decreased by ~10%. Such a change in the ice cover is in complete accord with the fluctuations in the mean air temperature at high latitudes. Southern Hemisphere data also show a decrease in sea ice area in the first half of the 1970s (Kukla *et al.*, 1977).

Drozdov, Lamb, and others have established that, during the epochs of warming and cooling, precipitation in various regions of the Earth varied to a great extent. Figure 10.6 presents secular variations in the total precipitation for the cold season of the year (from November to March) in the steppe and forest-steppe zones of the U.S.S.R. The precipitation totals shown are five-year running means based on observations at 21 stations. Comparing Figs.

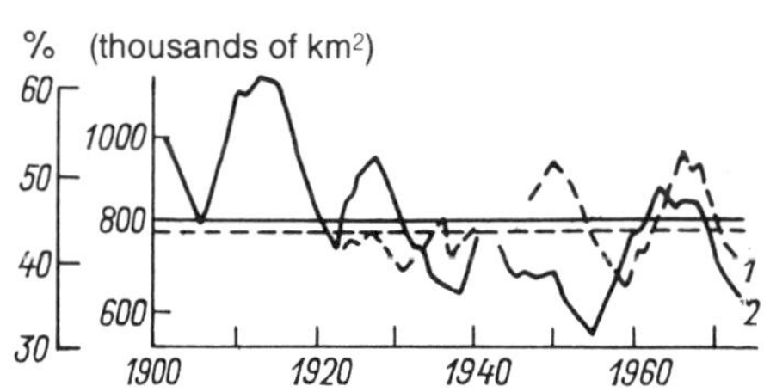


Fig. 10.4. 1, the Greenland Sea; 2, the Barents' Sea.

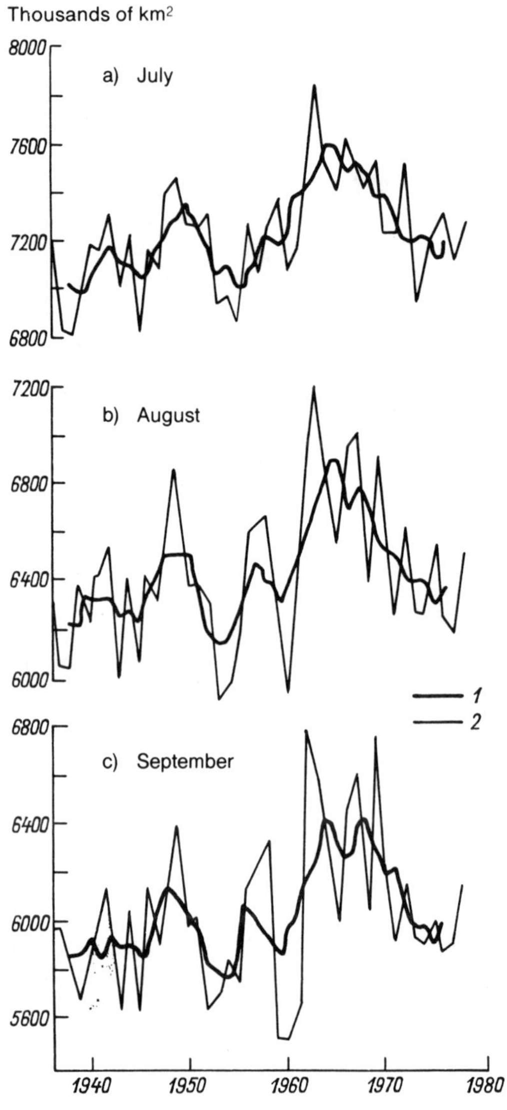


Fig. 10.5. 1, the running means; 2, values for every year.

10.1, 10.2 and 10.6 we can see that a rise in the mean air temperature and a decrease in the meridional tempertaure gradient have been accompanied with a decrease in precipitation in regions of unstable moisture conditions.

Drozdov and Grigorieva (1963, 1971) have noted that, during the period of most rapid warming (for example, the 1930s), droughts covering vast

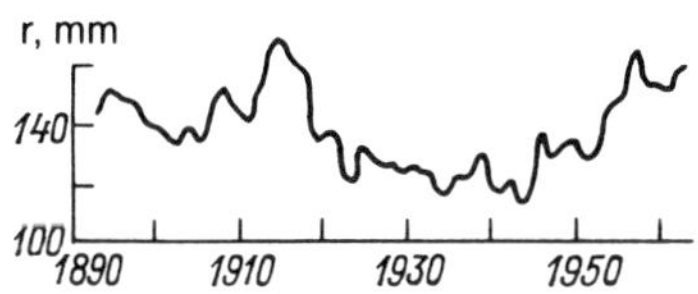


Fig. 10.6.

expanses of the low moisture zones of the U.S.S.R. and North America were more frequent than in previous and subsequent decades. As is well known, in the same period a drastic drop (about 170 cm) occurred in the level of the Caspian Sea, caused by decreased precipitation in the Volga River basin.

Secular variations in precipitation may, to a great extent, depend on variations in the meridional temperature gradient, which in turn affects the nature of the atmospheric circulation. The intensity of water vapour transfer from the ocean to inland regions varies with changes in the meridional temperature gradient. A decrease in the meridional gradient is accompanied with both a weakening of the water vapour flow from the ocean to the interior of the mid-latitudinal belt of the continents and a drop in the amount of precipitation over most of the inland areas. The opposite occurs when the meridional temperature gradient increases.

A detailed study of the effect of variations in the meridional temperature gradient on the precipitation regime was carried out by Drozdov and Grigorieva (1963, 1971). They have found the general characteristics of precipitation variations in high latitudes during warming or cooling periods to be rather complicated, but in mid-latitude areas of insufficient moisture there was a tendency for precipitation to increase as the temperature in the Arctic decreased. This effect was attributed to more intense water vapour transfer into the interior of the continents when the temperature contrast between high and low latitudes is increased.

This concept agrees well with the results obtained by Lamb (1974), who constructed world maps of precipitation anomalies for the periods of increased and decreased mean air temperatures near the Earth's surface. These maps show that, during global cooling, total precipitation increased over most of the continental surface in middle latitudes, decreased in the subtropical and tropical zones of high-pressure belts and increased in the equatorial latitudes. Lamb's conclusions have been corroborated by Borzenkova (1980), who studied the relationships between precipitation anomalies and mean air temperature anomalies in some important agricultural regions of the world. Figure 10.7 shows relative anomalies of latitudinal mean precipitation over the continents during a global cooling period. Curve 1 is constructed from Lamb's data for the entire land surface and curve 2 is based on Borzenkova's data for nine regions on different continents. The agreement of the curves shows that there is a link between

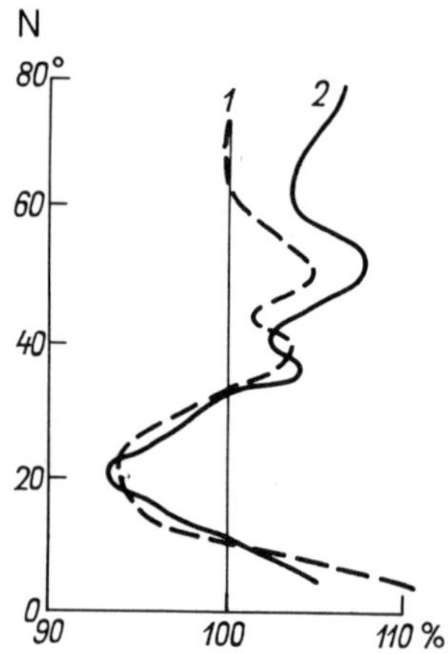


Fig. 10.7.

the distribution of atmospheric precipitation and variations in global mean air temperature.

The relationship between the occurrence of drastic droughts and global climatic fluctuations is of great interest. Rauner (1979) has noticed that droughts in the U.S.S.R. and in the U.S.A. often occur at the same time. This tendency shows the effect of global factors on drought occurrence. Rauner also presents data on the frequency of prolonged and severe droughts embracing most of the crop regions of the U.S.S.R. from 1815 to 1976. For our purpose we divide this interval into two periods: 1815—1919 and 1920—1976. The 5-year running air temperature means of the Northern Hemisphere were, as a rule, lower during the first 5-year period than during the second. According to Rauner, the frequency of severe droughts was 1.1 per 10 years during the first period and 1.9 per 10 years during the second. This conclusion agrees well with the results of other investigations cited above.

The reasons for the present change in climate have only been found comparatively recently as a result of research accomplished during the past few decades. The idea of the relationship between the present change in climate and volcanic activity was proposed by Humphreys (1913, 1929, etc.). His works established that the average amount of direct solar radiation incident on the Earth's surface under cloudless conditions varied noticeably for different years. These variations can be seen quite clearly from plots of the secular variations of the direct radiation based on observational data from actinometric stations. The same curves show that the direct radiation also varies throughout the year on longer time scales such as decades.

It is interesting to compare secular variations in the temperature of the Northern Hemisphere with those of radiation incident on the Earth's surface.

With this aim in mind, actinometric observations over the 1880—1965 period for European and American stations in the 40—60°N latitudinal belt (with the most long-term data series) have been analysed, and a curve of the secular variations in the direct radiation under clear sky conditions averaged over all these stations has been constructed (Budyko and Pivovarova, 1967). Figure 10.8 presents these data smoothed with a 10-year running mean (curve b). As can be seen, solar radiation has varied noticeably.

There are two possible causes for these variations in direct radiation under clear sky conditions, variations in the astronomical solar constant (the luminosity of the Sun) or in the so-called meteorological solar constant (i.e. the amount of radiation incident on the upper boundary of the troposphere). The latter can vary even with stable solar luminosity because of the unstable transparency of the stratosphere.

The first possible cause was proposed in several works; for example, the study carried out by Bossolasco *et al.* (1964). In this work data from three actinometric stations seemed to indicate that the solar constant increases as solar activity (described by Wolf numbers) increases, but only up to a certain limit. With a further increase in solar activity, the solar constant decreases. However, this concept of fluctuations in the astronomical solar constant was not supported by most specialists in the field of actinometry, who believed that the constant did not vary within the limits of accuracy of the measurements (Allen, 1958; Ångström, 1969).

To better identify the causes of the variations in solar radiation revealed in Fig. 10.8, radiation anomalies have been calculated for various altitudes of the Sun (Budyko, 1974). Obviously, if variations in direct solar radiation are connected with fluctuations in the astronomical solar constant, they cannot depend on the altitude of the Sun; whereas radiation changes due to the unstable transparency of the stratosphere must differ considerably with different solar altitudes. An analysis of the available data has shown that the latter dependence is dominant and agrees quite well quantitatively with the results of theoretical culculations of the effect of atmospheric transparency on direct radiation. As a result, we must conclude that the main cause of

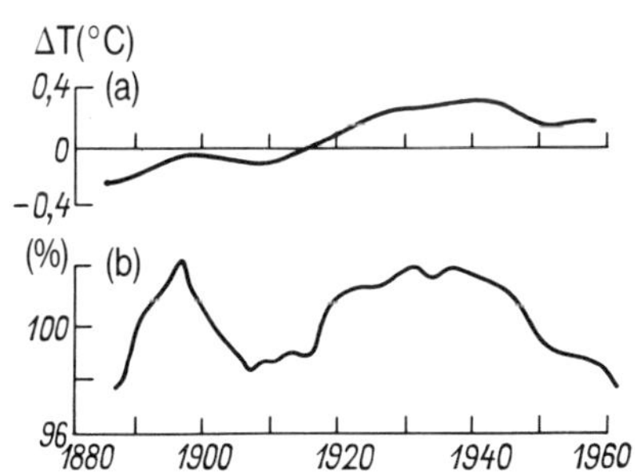


Fig. 10.8.

variations in direct radiation is fluctuations in the meteorological solar constant, while the astronomical solar constant is practically invariable. As indicated in Chapter 2, this conclusion has subsequently been corroborated using satellite data on solar radiation. Note that this conclusion is important in understanding why relationships between solar activity and climatic change are difficult to identify (Khromov, 1973).

To elucidate the mechanism of modern climatic change we compare curve b in Fig. 10.8 with the Northern Hemisphere temperature curve smoothed using a 10-year running mean (curve a). It is evident that curves a and b are very much like. For instance, there are two maxima in each curve, one of them at the end of the 19th century and the other (the major one) in the 1930s. At the same time, there are certain differences between the curves. For example, the first maximum is more pronounced for the secular variations of radiation than for the secular temperature change.

Since curves a and b are similar, they suggest that radiation fluctuations due to changes in atmospheric transparency are an important causal factor for climatic change. To clarify this possibility we need to calculate temperature variations as a function of atmospheric transparency fluctuations for shortwave radiation. To make this calculation, Budyko (1969) found the difference in the mean value of total radiation for the 1888—1917 and 1918—1947 periods (from direct solar radiation data in the Northern Hemisphere and formulas from atmospheric optics). In the second period total solar radiation was 0.3% higher than in the first one. Theoretically, this should correspond to a 0.4 °C increase in mean temperature. The actual temperature difference for these periods is 0.33 °C, which agrees satisfactorily with the calculation results.

The dependence of fluctuations in the radiation regime on volcanic activity has been investigated in many studies. A number of works have studied the radiation effect of the explosive eruption of Agung (on Bali in Indonesia), which occurred in March 1963. To estimate the effect of the Agung eruption on the radiation regime of the U.S.S.R., Budyko and Pivovarova (1967) used observational data on the intensity of direct radiation at noon from a number of actinometric stations in the Soviet Union over the 1957—1966 period. They established that, from 1957 to November 1963, the monthly values of direct radiation averaged over 22 stations situated between 40°N and 60°N varied comparatively little. After December 1963 the values decreased abruptly (see Table 10.1). This was a systematic change, quite distinct from the earlier month to month fluctuations in the intensity of direct radiation that are apparently due to the instability of the atmospheric circulation. Since the decrease in the intensity of direct radiation was much larger in winter than in summer, it is quite evident that these reductions were caused by fluctuations in atmospheric transparency and not by solar constant fluctuations.

TABLE 10.1

Deviation of the direct radiation intensity from the normal over the U.S.S.R. (in %)

| Year | Month | | | | | | | | | | | | Annual |
|---|---|---|---|---|---|---|---|---|---|---|---|---|---|
| | I | II | III | IV | V | VI | VII | VIII | IX | X | XI | XII | |
| 1963 | − 1 | 3 | − 2 | 2 | 2 | 2 | 2 | 1 | 0 | −1 | −2 | −10 | 0 |
| 1964 | −10 | −3 | − 6 | −7 | −2 | −1 | −3 | −2 | −3 | −6 | −8 | −16 | −5 |
| 1965 | −15 | −7 | −11 | −5 | −4 | −3 | −3 | −2 | −5 | −2 | −9 | −12 | −6 |
| 1966 | −10 | −9 | − 7 | −4 | −2 | −1 | 1 | 0 | −2 | −4 | −7 | − 7 | −4 |

This strong dependence of the radiation regime on volcanic eruptions implies that an increase in the radiation which occurred at the end of the 19th century resulted from the removal from the atmosphere of the aerosol particles produced by the Krakatoa eruption. A subsequent reduction in the radiation took place after the eruptions of Mont Pelée and other volcanoes in the early 20th century. The 1915—1920 increase in radiation is apparently explained by an increase in transparency that resulted from the settling of the aerosol after the Katmai eruption. After this, there were no large eruptions for a long time.

The temperature effects of radiation changes after intense eruptions have been dealt with in a number of empirical studies. These have shown that the mean temperature near the Earth's surface decreases by a few tenths of a degree over a period of several months to several years after large eruptions (Humphreys, 1929; Mitchell, 1961, 1963; Angell and Korshover, 1977, Taylor *et al.*, 1980).

Let us consider the effect on temperature of radiation changes due to volcanic eruptions in more detail. It is possible to calculate the effect theoretically, taking into account the heat conductivity and the thermal capacity of ocean water. Since we are interested only in an approximate estimate of this effect, we shall use a simple relationship between radiation and temperature fluctuations that can be established empirically from annual data. We assume that the rate of change in the mean temperature at the Earth's surface is proportional to the difference between the temperature at a given moment ($T$) and the temperature corresponding to stationary conditions ($T_r$), i.e.

$$\mathrm{d}T/\mathrm{d}t = -\lambda(T - T_r), \tag{10.1}$$

where $\lambda$ is a coefficient of proportionality. If the initial temperature is $T_1$, we obtain from (10.1)

$$T - T_r = (T_1 - T_r)\,\mathrm{e}^{-\lambda t}. \tag{10.2}$$

For a rough estimation of $\lambda$, we use data on the annual variations in solar radiation and temperature for the Northern Hemisphere, neglecting the interactions between the Northern and Southern Hemispheres. Since the ratio of the solar radiation at the top of the atmosphere in the Northern Hemisphere during the warm seasons (April—September) to the average annual radiation is 1.29, we find from the relationship of temperature to radiation changes (Chapter 7) that this change in radiation could lead to, approximately, a 40 °C temperature rise in the case where oceanic thermal inertia is absent. The observed difference between the mean temperatures in the Northern Hemisphere for the warm season and for the entire year is 3.5 °C. In this case, if $T - T_1 = 3.5$ °C and $T_r - T_1 = 40$ °C for $t = 1/4$ of the year, we find $\lambda \cong 0.4$ yr$^{-1}$ according to Equation (10.2).

From this result we can determine the annual mean temperature change after a volcanic eruption that has caused an annual mean drop of 10% in direct radiation. In this case, the decrease in total radiation is about 1.5% and $T_r$ drops by ~2 °C. We find, therefore, from (10.2) that the temperature change after the eruption appears to be a few tenths of a degree. This result agrees well with the observed mean annual temperature anomaly after an intensive explosive eruption.

The effect of volcanic activity on present climatic change has been treated in a number of other investigations. Research carried out by Karol (1977) and Karol and Pivovarova (1978) is concerned with the dependence of fluctuations in direct solar radiation and the air temperature of the Earth's surface on stratospheric aerosol loading.

In the first of these studies, aircraft data on stratospheric aerosol at a height of 14—20 km in the Northern and Southern Hemispheres over the 1960—1973 period was compared with anomalies in the direct solar radiation obtained from a number of actinometric stations covering the northern middle latitudes. A close relationship was revealed between the two parameters, corroborating the concept that fluctuations in stratospheric aerosol are extremely important in determining changes in solar radiation received at the surface. This relationship was used to evaluate the effect of stratospheric aerosol fluctuations due to volcanic activity on the thermal regime of the lower atmosphere. The results appear to be similar to the temperature change observed after volcanic eruptions.

The second study compared balloon reports of atmospheric aerosol concentration with fluctuations in the atmospheric transparency based on direct solar radiation measurements from a number of actinometric stations. This investigation found that actinometric observations of direct solar radiation, especially on mountains and at remote stations, reveal the influence of major volcanic eruptions on atmospheric transparency.

Oliver (1976) has calculated air temperature variations in the Northern Hemisphere since the end of the 19th century up to 1968 using a model that allowed him to determine the temperature variations caused by a change in atmospheric transparency after volcanic eruptions. His calculations show that changes in volcanic activity can explain about 70% of the variance of global-scale temperature series.

The same problem has been treated by Sagan and co-workers (Pollack *et al.*, 1975; Pollack *et al.*, 1976; Baldwin *et al.*, 1976), who consider in detail the effect of the products of volcanic eruptions on the radiation regime of the atmosphere. According to observational data, the ejected material seems to comprise solid particles and droplets of sulfuric acid produced from sulphur oxides and water.

In studies of the climatic effects of the Agung eruption, observational data show that the lower stratospheric layers were appreciably heated in the first months after the eruption. According to calculations, this could have been a result of the absorption of longwave radiation by igneous particles. A temperature rise may also affect the troposphere immediately after the eruption when the largest particles of igneous aerosol remain in the air. Later on, this effect disappears and the troposphere and stratosphere become cooler. If the thermal inertia of the Earth–atmosphere system is taken into account, it appears that the mean temperature in the lower air layer drops by several tenths of a degree after a major volcanic eruption. This conclusion agrees reasonably well with the observational records.

The studies in question also treat the problem of how much rising volcanicity contributed to the progression of cooling a few centuries ago, during the period often referred to as the Little Ice Age. The calculations show that the contribution could well be significant.

Hansen *et al.* (1978) have also analysed the effects of the Agung eruption that occurred in 1963. They concluded that all quantitative estimates of changes in the thermal regime after the eruption agree well with the observational data.

The study by Miles and Gildersleeves (1977) is worth noting. This contains an empirical analysis of the relationships between mean air temperature variations in the Northern Hemisphere and climate-forming factors. The authors concluded that variations in mean hemispheric temperature over the last hundred years are mainly related to fluctuations in atmospheric opacity due to volcanic activity and to a rise in carbon dioxide concentration. With these two factors taken into account it is possible to explain 65% of the temperature variability. The amount increases to over 80% if a third factor, i.e. the variation in polar ice area, is taken into consideration.

On the basis of a similar analysis Robock (1978) has established that volcanic activity has an essential effect on the mean air temperature in the Northern Hemisphere whereas solar activity as characterized by the number of sunspots, does not exert such an influence.

Summarizing these ideas, we can set forth a conclusion concerning the reason for recent climatic change. The warming that peaked in the 1930s evidently resulted from a stratospheric transparency increase that enhanced the solar radiation flux coming into the troposphere (the meteorological solar constant). Modifications in the air temperature in different seasons and latitudes were dependent on the optical depth of the stratospheric aerosols and on the shift of the polar sea ice boundary. The retreat of Arctic sea ice due to the warming gave rise, in addition, to an appreciable increase in the air temperature in cold seasons at high northern latitudes. These conclusions are corroborated by calculations based on models of the thermal regime of the atmosphere, the theoretical results being in good agreement with the observational data.

Fluctuations in stratospheric transparency during the first half of the 20th century have probably been due to variations in volcanic activity, in particular, to variations in the amount of ejected materials, especially sulphur-containing gases, released into the stratosphere.

A similar physical mechanism operated in the second half of the 20th century, when the climate was particularly affected by the Agung eruption in the mid-1960s.

In addition to the effects of volcanicity, climatic conditions during the 20th century have also been influenced by man's economic activity, which, in particular, has brought about a rise in the carbon dioxide concentration in the atmosphere. These changes are considered in more detail below.

### 10.1.2. CHANGES IN THE CHEMICAL COMPOSITION OF THE ATMOSPHERE

Man's economic activity affects the chemical composition of the atmosphere which in turn affects its physical state, i.e. climate. Arrhenius (1908) was the first to conclude, at the beginning of this century, that global climate would inevitably change because of the burning of ever-increasing quantities of fossil fuel. Arrhenius thought that the ocean absorbs only a fraction of the carbon dioxide produced by man's economic activity, thus causing a rise in the carbon dioxide concentration of the atmosphere. According to Arrhenius, a doubling of carbon dioxide concentration should lead to a 4 °C increase in mean surface air temperature, an estimate which does not differ much from the modern value (see Chapter 7).

Of great interest are Arrhenius' ideas that the warm climates of the geological past could be explained by an elevated carbon dioxide level in the atmosphere, and that the present growth of carbon dioxide concentration would lead to a renewal of warmer climatic conditions. Arrhenius also suggested that an increased content of carbon dioxide would raise photosynthetic productivity thus exerting a favourable effect on agricultural yields.

Later Callendar (1938) expressed a similar opinion concerning an expected warming due to the growth of carbon dioxide concentration caused by the burning of fossil fuel. The studies of Arrhenius and Callendar were well ahead of the science of their times and so did not attract much attention to the problem of anthropogenic global climate change.

The possibility of noticeable increase in carbon dioxide due to the burning of various kinds of fuel was suggested long before the onset of systematic observations of the carbon dioxide concentration. The reality of this effect was not proven until the 1960s when the first accurate observational data were collected as a result of the International Geophysical Year in 1958. These observational data showed a systematic year-to-year increase in carbon dioxide mass, with the rate being practically the same at stations situated at great distances one from another: Hawaii, Alaska, the South Pole, etc. Figure 10.9 shows, as an example, mean annual carbon dioxide concentrations from observations in Mauna Loa (Hawaii) analysed by Keeling (Climatic Effects . . . , 1982). As seen in the figure, between 1958 and 1981 the carbon dioxide concentration increased by about 7%. Note that the rate of increase in the second half of the period was greater than in the first.

Data presented in Chapter 7 allow us to compare the rates of change in carbon dioxide mass in the geological past with these recent changes. The mean rate of decrease in carbon dioxide concentration in the Tertiary was tens of thousands of times less than the present rate of increase.

The rate of anthropogenic increase of carbon dioxide at present can be compared also with natural gas exchanges between the atmosphere and the lithosphere. It transpires that the anthropogenic carbon dioxide increase is approximately one hundred times its income by natural gas exchange. Such

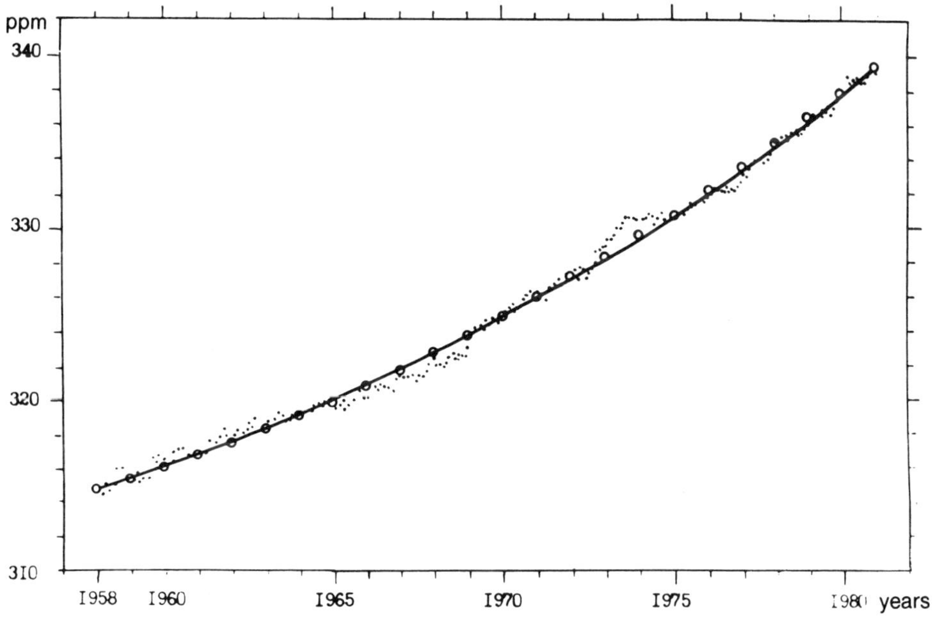


Fig. 10.9.

difference explains a great increase in the rate of change in carbon dioxide concentration at present as compared with the geological past.

Undoubtedly, the present rapid change in atmospheric carbon dioxide depends on anthropogenic factors. Most recently, as a result of burning of coal, petroleum and other fossil fuels, the atmosphere receives annually an amount of carbon dioxide containing about $5 \times 10^{15}$ g of carbon. A small part of this quantity (about 2%) is carbon dioxide formed in producing cement. During last several decades, at least until the early 1970s, the rate of carbon dioxide production (corresponding to the rate of fuel consumption) increased by 4—5% per year.

Atmospheric carbon dioxide concentration may also be influenced by a reduction in carbon amount due to a decrease in the planetary biomass reserve, the effects of deforestation, etc. Unfortunately, available data on this source of carbon dioxide are considerably less accurate than data on the quantity of burned fuel.

According to Bolin (1977), of the total income of carbon dioxide from the burning of fuel, corresponding to a carbon mass gain of $5 \times 10^{15}$ g yr$^{-1}$, about $2.5 \times 10^{15}$ g yr$^{-1}$ is spent on increasing the content of carbon dioxide in the atmosphere and about $1.5 \times 10^{15}$ g yr$^{-1}$ is dissolved in the ocean. What happens to the imbalance of approximately $10^{15}$ g yr$^{-1}$, is not sufficiently clear. Possibly some part of this carbon is absorbed by organic matter on the continents. The fraction that goes into the ocean according to Bolin may be too small. However, the airborne fraction given by Bolin (about 50%) is in accord with most other investigations of the modern carbon balance. The absorption of carbon dioxide by the ocean has been considered in a number of studies. These have established that the rate of absorption is noticeably restricted because of comparatively slow diffusion of carbon dioxide from the atmosphere to the ocean.

The amount of carbon dioxide in the mixed layer of the ocean is about 3—4% of the total carbon dioxide mass in the ocean. A large increase in this amount is impossible, because the size of this reservoir is quite small and because the mixed layer concentration must remain approximately in equilibrium with that in the overlying atmosphere. However, the deep ocean can absorb a considerable mass of carbon dioxide, although the possible flux of carbon dioxide from the atmosphere to the deep ocean is rather limited. The rate of diffusive exchange of water between the mixed layer and deeper layers corresponds to a change in the mixed layer of 2—8% of the total quantity of water per year. In addition to this diffusive exchange, however, the flux of carbon dioxide is noticeably affected by the circulation of oceanic waters, particularly the sinking of cold waters in high latitudes. As a result of these processes the ocean can absorb a significant amount of carbon dioxide. However, the rate of absorption is insufficient to prevent a current rapid growth of atmospheric carbon dioxide concentration.

As mentioned above, according to observations on Mauna Loa and elsewhere the carbon dioxide concentration increased by 7% between 1958 and 1981. The amount of carbon dioxide produced for this period as a result of the burning of fuel and cement production comprises about half of the corresponding value for 1860—1981. We may assume, therefore, that the carbon dioxide concentration has increased by approximately 15% of its 1860 value over the period 1860—1981. By using a similar extrapolation method we may estimate the probable concentration of carbon dioxide in the atmosphere in the near future. Such an estimate has been obtained by Baes *et al.* (1976). They assumed that the carbon dioxide quantity produced by man increases by 4.3% per year and that half of this quantity remains airborne. Their results are given in Table 10.2.

TABLE 10.2

Changes in carbon dioxide concentration

| | 1974 | 2000 | 2025 |
|---|---|---|---|
| Formation of $CO_2$ over a year ($10^{15}$ g of C) | 5 | 14 | 42 |
| $CO_2$ concentration (%) | 0.033 | 0.038 | 0.053 |
| An increase in concentration (% relative to level of 1860) | 12 | 29 | 81 |

The authors of this calculation assume that the modern rate of growth of fuel consumption over the next 50 years seems to be the upper limit for real energy progress. Along with this, probably, the fraction of anthropogenic carbon dioxide absorbed by the oceans will decrease in the future and a considerable component caused by deforestation will be added to the mass of this gas. These factors may compensate for the calculation errors arising from overestimating the rate of fuel consumption.

Baes *et al.* also calculated the total quantity of fuel that would be consumed before 2025 at the assumed 4.3% per year rate of growth. For 1860—2025 this total is close to $10^{18}$ g of carbon, about one-seventh of the present reserves of fossil fuel.

In recent years detailed numerical models have been developed for calculating the expected change in atmospheric carbon dioxide taking into account the main sources and sinks as well as buffer mechanisms influencing carbon dioxide cycle. The development of these models is important in order to better understand the distinctive features of the process of changing carbon dioxide mass in the atmosphere. It is less clear to what extent these detailed models are necessary for estimating carbon dioxide concentrations in the near future. Obviously, such estimations require assumptions about the future levels of fuel consumption and expected changes in biomass. Such assumptions will inevitably be approximate and their probable errors must increase rapidly with prolongation of the forecasting time. It is possible that, as the methods for forecasting future energy use improve, greater accuracy in calculations of the carbon dioxide balance will be needed. However, at present, the need for such accuracy is rather limited.

Some important conclusions regarding future carbon dioxide levels were drawn at the U.S./U.S.S.R. conference on the climatic effect of increasing atmospheric carbon dioxide, held in Leningrad in 1981. The report of this conference (*Climatic Effects* ... , 1982) noted that, for the period of systematic observations (1958—1980), carbon dioxide concentration increased by 22 p.p.m. having reached about 340 p.p.m. The industrial discharge of carbon dioxide for this period amounted to an equivalent concentration change of 39 p.p.m. Studies of the carbon cycle showed that the remaining part of this anthropogenic carbon dioxide input (17 p.p.m.) is absorbed by the ocean, the role of terrestrial biota appearing to be unimportant for the balance of organic carbon. From direct observations carried out in the 19th century and from calculation of the carbon dioxide balance, the preindustrial carbon dioxide concentration has been estimated to be, at most, 290 p.p.m. The increase to 1980, therefore, has been at least 50 p.p.m. This is probably a lower limit because, in recent studies, the preindustrial value of carbon dioxide concentration is assumed to be less than 290 p.p.m., usually in the range 260—280 p.p.m. If the mid-20th century growth rate in the use of fossil fuels is maintained carbon dioxide concentration will reach about 360 p.p.m. by 1990 and 390—400 p.p.m. by 2000.

To estimate changes in carbon dioxide concentration in the 21st century we may use scenarios of energy development which have recently been advanced by a number of scientists in various countries. Among these are calculations by Legasov and Kuzmin (1981), Rotty (1979), and Hafele (1979). In the studies by Legasov and Kuzmin the upper and lower limits of probable industrial discharges of carbon dioxide are indicated, the former being close to Rotty's results and the latter to Hafele's data. In all these scenarios it is assumed that the mean rate of growth of carbon fuel consumption will be less over the next 50 to 70 years than the 4.3% per year that prevailed between 1945 and 1972, probably less than 2.5% per year. Figure 10.10 summarizes expected changes in total anthropogenic carbon dioxide release into the atmosphere. Curves 1*a* and 1*b* are based on the data of Legasov and Kuzmin; curve 2, of Rotty; and curve 3, of Hafele. As can be seen, the four scenarios agree satisfactorily with each other.

The predicted carbon dioxide concentration for 2050 averaged over all scenarios of energy development and all variants of the carbon cycle model is close to 700 p.p.m. Therefore, carbon dioxide concentration is expected to double compared with its pre-industrial value in the mid-21st centry. How long a higher carbon dioxide concentration can remain in the atmosphere is not clear, but this time is not likely to be less than several hundred years.

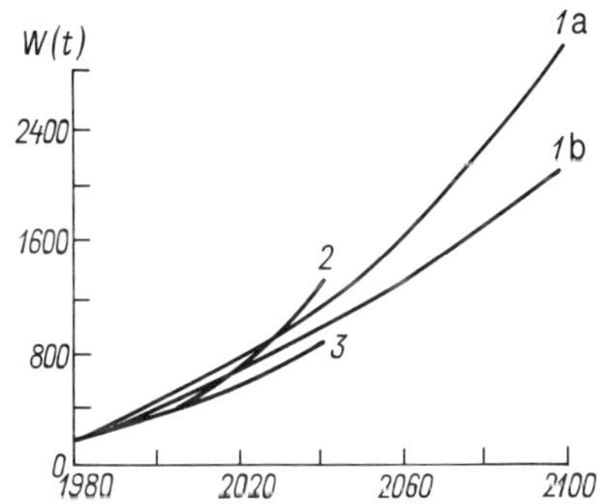


Fig. 10.10. $W(t)$, $CO_2$ release into the atmosphere (in $10^{15}$ g).

If all currently known reserves of carbon fuel are used, carbon dioxide concentration would increase by 6—10 times compared with its pre-industrial value and would reach 0.2—0.3%. As indicated in Chapter 7, such a concentration of carbon dioxide existed throughout a greater part of the Phanerozoic. Thus, the atmosphere in the future could prove to be similar to the atmosphere that prevailed for hundreds of millions of years.

Besides carbon dioxide, the present atmosphere contains a number of gases which, although constituting only a small part of total volume of the atmosphere ('small contaminants') can, nevertheless, affect the climate. A portion of these contaminants is of anthropogenic origin, but some of them existed in small quantities in pre-industrial atmosphere. Among these small contaminants are freons (chlorofluorocarbons), nitrogen oxides,methane and some other gases. The climatic effect of increasing concentration of small contaminants is similar to the effect of carbon dioxide — as a rule it leads to intensification of the greenhouse effect, i.e. a climate change towards warming.

### 10.1.3. ANTHROPOGENIC FACTORS IN PRESENT CLIMATIC CHANGE

As indicated above, along with fluctuations in atmospheric transparency, the growth of carbon dioxide in the atmosphere exerts a noticeable and ever-increasing influence on contemporary climate. Figure 10.11 shows changes in mean surface air temperature in the Northern Hemisphere (five-year running means; curve 2) and thc calculated change in this temperature due to the growth of carbon dioxide amount (curve 3). This calculation is based on the assumption that doubling the carbon dioxide concentration results in an increase of mean surface air temperature by 3 °C. The logarithmic dependence of temperature rise on the growth in carbon dioxide amount is taken

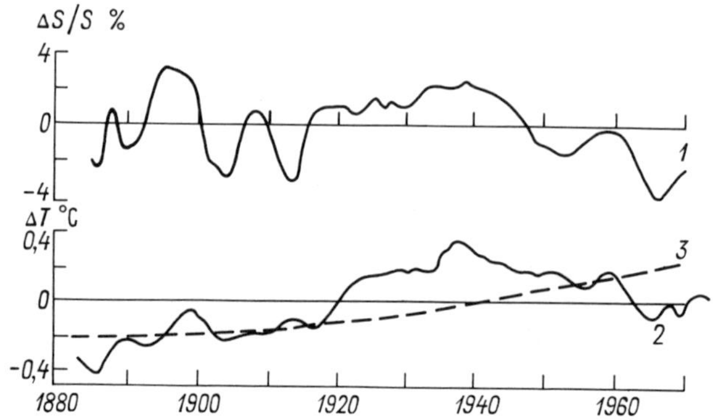


Fig. 10.11.

into account. This figure shows that, although actual changes in temperature differ from the calculated results, observational data confirm the tendency of temperature to rise in the second part of the period in question compared with its first part. For example, before 1920 the temperature was generally below the long-term average, but after this year it was higher, as a rule, than the average value.

The discrepancy between curves 2 and 3 is easy to explain: it is due to fluctuations in atmospheric transparency. As noted above, as atmospheric transparency decreases the temperature lowers, and with increasing transparency the temperature rises. This explanation is confirmed by comparing solar radiation anomalies under cloudless sky obtained at a group of radiation stations, a function of atmospheric transparency, (curve 1) with the secular trend of air temperature (curve 2).

As seen from the figure, since the end of the 19th century increasing carbon dioxide concentration has raised the mean surface air temperature by ~0.5 °C. Although this value is not small, it has been disguised to a great extent by short-term climatic changes associated with variations in atmospheric transparency.

From Fig. 10.11 it can be concluded that, for the period of instrumental observations, the Northern Hemisphere air temperatures averaged over five-year intervals depended largely on the growth of carbon dioxide concentration and fluctuations in atmospheric transparency. To estimate the effect of the latter, Fig. 10.11 shows the direct radiation anomalies under cloudless sky (curve 1) according to Pivovarova (1977). These anomaly values are smoothed using a five-year running mean. The main discrepancy between curves 1 and 2 is the tendency towards rising temperature which is observed over most of the period. The secular course of direct radiation shows no such tendency. Rather, it is characterized by oscillations to both sides of the mean value with a general decrease in radiation throughout the last decades. Apart from the long-term trend discrepancy the curves are similar to each other. The chief maxima and minima of curve 2 correlate with those in curve 1, the temperature variations lagging somewhat behind those in radiation and being smoother.

The relationship between radiation fluctuations and the long-term warming trend which began in the 1920s requires special comment. In 1914—1915, a sharp increase in radiation occurred, following which a long-term positive anomaly in radiation was observed. In 1918—1922, due to increased radiation, the air temperature rose. Then in 1931—1934, radiation increased again, causing a temperature elevation in 1935—1938. The 1953 radiation minimum resulted in the 1956—1957 temperature minimum. The radiation peak in 1959 was followed by the maximum in temperature in 1960. The minimum in radiation in 1966 led to an air temperature minimum in 1967. Thus, in most cases, considerable increases and decreases in radiation were followed by temperature fluctuations that lagged behind the radiation variations by a period of one to five years (more often by three years). Observational data show that variations in radiation income to the troposphere depend chiefly on the mass of stratospheric aerosol, which shows only slight spatial differences between the hemispheres because of intensive horizontal mixing. As a result, for estimating the average conditions of stratospheric transparency, it is sufficient to use actinometric observation data (or observations of aerosol concentration) from only a small number of stations.

The influence of atmospheric transparency variations and changes in carbon dioxide concentration on air temperature can be studied using the afore-mentioned dependences. As noted above, the sign of the difference between the temperature anomalies depicted by curves 2 and 3 coincides in most cases with that for the direct radiation anomalies. This demonstrates the effect of radiation fluctuations on medium time scale air temperature variations in the Northern Hemisphere.

It is interesting to compare the relative values of these temperature anomalies with those of the associated radiation anomalies. Figure 10.11 shows that the ratio of the temperature anomaly to the radiation anomaly increases with the duration of the anomaly. From 1920 to 1945 the longest positive radiation anomaly was observed, its mean value being +1.3%. From calculations based on atmospheric optics it follows that this corresponds to a 0.2% increase in total radiation (Budyko, 1971). Over the period of time under consideration, the temperature anomaly (equal to the average difference between the values presented as curves 2 and 3) is +0.24 °C. The ratio of the second value to the first ($\Delta T_1$) amounts to 1.2 °C for each 1% of total radiation change. Note the agreement between this value and the value of

1.1 °C for each 1% obtained earlier without taking into account the effect of the growth of carbon dioxide concentration on temperature (Budyko, 1969).

By using Equation (10.2), the afore-mentioned relation, and the dependence of global mean temperature on carbon dioxide concentration, a secular trend of mean air temperature for the Northern Hemisphere has been calculated (Budyko, 1977a). For this we assume that $\Delta T_1 = 1.5$ °C, the relation of total radiation anomalies to direct radiation anomalies is 1/6, and $\lambda = 0.1$ yr$^{-1}$ (as established in a number of empirical studies, the value of $\lambda$ declines as the length of the period under consideration increases). Temperature anomalies have been calculated by this method for every year, $T_1$ being taken from observational data for the first year. For successive years the $T_1$ value is the value of $T$ obtained for the previous year. The results are depicted in Fig. 10.12 as curve 2. Comparison of this curve with observational data (curve 1) shows their great similarity. The correlation coefficient between the observed and calculated values is 0.73. This value must be considered encouragingly high, given that the data used are approximate and the numerical model applied is schematic.

We can conclude that air temperature variations averaged over five-year periods in the Northern Hemisphere are largely determinate and depend on causal factors only slightly. The secular temperature trend can be computed with sufficient accuracy using simple climate models, provided the major factors affecting temperature changes are taken into account. It is possible, therefore, not only to compute past climatic changes, but also to forecast future ones.

The second conclusion is that mean air temperature variations in the Northern Hemisphere have been considerably affected by the growth of carbon dioxide concentration over the last hundred years. This effect is reflected by the fact that, as noted above, air temperatures averaged over five-year periods were below the norm in all cases before 1920 and were, as a rule, above the norm after 1920. The question of detecting the effect of the growth of carbon dioxide concentration on modern climate is discussed below in detail.

The influence of other anthropogenic factors on modern climatic conditions is rather limited. Of these factors, the production of energy in various

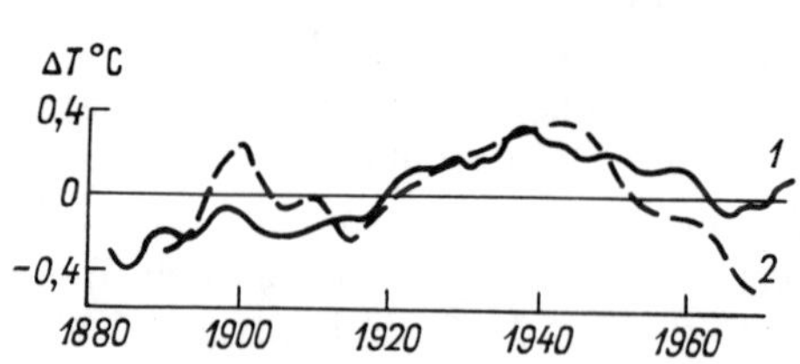


Fig. 10.12.

kinds of economic activities deserves attention. This factor results in additional direct heating of the atmosphere and the Earth's surface. We have calculated (Budyko, 1962a) the quantity of heat produced by man's economic activities. It amounts to about 0.01 W $m^{-2}$ for the entire Earth's surface, and to 2—3 W $m^{-2}$ for highly developed industrial regions with areas of tens to hundreds of thousands of square kilometres. Over large cities (tens of square kilometres) this value increases up to tens and hundreds of watts per square metre. It can be calculated how this additional heat release influences the mean temperature of the Earth.

It was mentioned above that a 1% change in the energy coming to the Earth from the Sun changes the mean temperature near its surface by 1.5 °C. Assuming that the heat produced by man today comprises about 0.006% of the total radiation absorbed by the Earth—atmosphere system, we find that an increase in mean temperature corresponding to this quantity of heat is approximately 0.01 °C. This value is very small. However, in some regions, it must be significantly greater because of the uneven spatial distribution of man-made heat sources.

It was shown in the above paper (Budyko, 1962a) that, in the absence of the atmospheric circulation, the temperature could increase by 1 °C in highly developed industrial regions, and in big cities by tens of degrees. This would probably make life impossible there. The atmospheric circulation considerably weakens the corresponding temperature rise, the attenuation being the greater the smaller the area over which the production of additional heat energy is concentrated.

Sagan *et al.* (1979) have considered the question of variations in mean air temperature due to anthropogenic effects on vegetation cover. Different forms of impact on vegetation are considered in this paper. The authors estimated variations in global albedo and the consequent changes in mean surface air temperature. Of greatest importance, the authors have concluded, is the anthropogenic destruction of savanna vegetation and the conversion of savanna to desert (this occurred over several thousands of years over an area of $9 \times 10^6$ $km^2$ and resulted in an increase in the surface albedo of 0.19). Equally important is the felling of tropical forests (over an area of $7 \times 10^6$ $km^2$ with an increase in albedo of 0.09). Because of these processes, the global albedo of the Earth—atmosphere system has increased by 0.006, resulting in a lowering of the mean air temperature by approximately 1 °C. Sagan *et al.* assumed that, over the last 25 years, as a result of the causes indicated, the global albedo has increased by 0.001, producing a temperature reduction of 0.2 °C. The authors of the paper believe that such albedo changes can explain the cooling that occurred in the Little Ice Age as well as the temperature lowering that took place from 1930 to 1970.

It does seem that, although the estimates made by Sagan *et al.* are somewhat overstated (the actual values of albedo variations seem to be lower than the values in their calculations), albedo variations have actually altered the thermal regime. However, it should be kept in mind that the deforestation must have been accompanied with a considerable release of carbon dioxide, which led to increased mean air temperature. Hence, the albedo effect arising from deforestation was, to a considerable extent, compensated for by increases in carbon dioxide concentration.

In a study of the impact of irrigated lands on global climate (Budyko, 1971), the mean surface air temperature effect of the existing irrigation system has been estimated. The Earth's surface albedo can be lowered by approximately 0.10 over irrigated zones. Considering the relationship between the albedo of the Earth's surface and that of the Earth—atmosphere system (see Chapter 3), it was found that, for little cloudiness, this decrease in the Earth's surface albedo corresponds to a decrease in the Earth—atmosphere albedo by 0.07. The area of irrigated lands is approximately $2 \times 10^6$ km$^2$, that is, about 0.4% of the Earth's total surface area. Irrigation thus decreases the Earth's albedo by approximately 0.0003.

In Chapter 3 it was mentioned that a 0.01 change in the Earth's albedo changes the mean surface air temperature by 2.3 °C. Keeping this in mind, we find that irrigation increases the mean surface temperature by approximately 0.07 °C. This change in temperature is comparatively small, although, as irrigated areas increase, it may become important.

The mean surface air temperature may have been somewhat affected by artificial reservoirs. The mean value of the Earth's surface albedo is decreased in approximately the same way by the construction of reservoirs in regions covered with vegetation as it is by irrigating deserts. Since large artificial reservoirs are built in regions with comparatively moist climate, where cloudiness is considerable, the albedo of the Earth—atmosphere system in this case changes less than in irrigated areas with little cloudiness. In addition, since the total area of artificial reservoirs is significantly less than the area of irrigated lands the effect of the former on the mean surface air temperature proves to be comparatively small.

The urbanization process exerts some influence on climatic conditions as a result of the replacement of forests and fields by constructions, asphalt roads, and so on. Sagan *et al.* (1978) have assessed this effect. They believed that this process had spread over an area of $10^6$ km$^2$, decreasing its albedo from 0.17 to 0.15. The corresponding decrease in global albedo amounts to $2.5 \times 10^{-5}$. This increases the mean air temperature near the Earth's surface by approximately 0.005 °C, i.e. a very small value.

In summary, the material in this section shows that changes in global climate during recent decades seem to depend to a certain extent on man's economic activities.

### 10.1.4. DETECTING ANTHROPOGENIC CLIMATIC CHANGE

The question of detecting anthropogenic climatic change is of considerable value, since obtaining data that confirm man's past impact on climate will also confirm the reality of a rapidly increasing influence of anthropogenic factors on climate today and in the future.

Callendar (1938, etc.) was the first who tried to discern anthropogenic effects on climate. His studies dealt with the question of variations in mean air temperature under the influence of rising carbon dioxide concentration. Callendar was unable to positively identify the climatic effects of carbon dioxide because of several reasons, the most important being: (a) the absence at that time of sufficiently precise systematic measurements of carbon dioxide concentration in the atmosphere; (b) the much smaller increase in carbon dioxide concentration before 1938 compared with more recent time; (c) Callendar had no information on the main cause of natural changes of global climate, variations in atmospheric transparency, and so was unable to take this into account. In fact, the last reason alone is enough to make the problem of detecting anthropogenic climatic change extremely difficult.

In most modern studies of global climatic change it is Northern Hemisphere temperature data that are usually used as an indicator of global conditions. This is partly because the majority of meteorological stations that exist at present are situated in the Northern Hemisphere. Recently, secular time series of mean annual surface air temperature for the Northern Hemisphere have been obtained independently in the U.S.S.R. (Vinnikov *et al.*, 1980), the U.S.A. (Hansen *et al.*, 1981) and England (Jones *et al.*, 1982).

Analysis of these data shows that, over the past hundred years, the surface air temperature of the Northern Hemisphere varied within a range of ±0.5 °C, increasing at a mean rate of about 0.5 °C per 100 yr with 30% of the average annual variance being connected with this linear trend. Up to 1982, the highest value of mean air temperature for the hundred year period was observed in 1981 (*Climate Monitor*, 1981).

In 1977—1983 a series of papers by Budyko (1977b, 1977c, 1980), Vinnikov and Groisman (1981, 1982) and Budyko and Vinnikov (1983) was published analysing these temperature data. These studies are based on the following initial concepts. Anomalies of the mean air temperature can be considered to be the sum of the three components. The first of these is the result of the influence of natural factors on climatic change. The second is called climatic noise (the word 'noise' is used because this component is the sum of temperature changes caused by random variations in the atmospheric circulation, errors in the estimates of all the materials used in the analysis, and variations due to those natural factors that are not taken into account in the calculations). The third component constitutes those changes caused by anthropogenic factors. The aim of detection studies is to identify this component from an analysis of observed changes in mean temperature.

The only natural factor in global climatic change which has been established without doubt by modern theoretical and empirical studies is fluctuations in atmospheric transparency resulting from variations in aerosol concentration in the stratosphere. The calculation of the effects of this factor on the mean air temperature presents no special problems, nevertheless it must be carried out with the highest possible accuracy. The most reliable results of this calculation are obtained using data on atmospheric transparency fluctuations based on observations at available actinometric stations. The application of these data assumes that the amount of tropospheric aerosol for the Earth as a whole can be represented by data from the Northern Hemisphere.

To evaluate the effect of carbon dioxide on mean air temperature, two indicators of its strength may be used: (1) the mean air temperature increase over the last hundred years (or a shorter period of time) due to the growth in carbon dioxide concentration; (2) the air temperature increase due to a doubling of carbon dioxide concentration. The latter can be found from the former if one introduces an additional hypothesis about the character of the dependence of temperature change on changes in carbon dioxide concentation. Both items (1) and (2) are well known independent of observational data on air temperature. As mentioned in Chapter 7, the second value is approximately equal to 3.0 °C based on climate model calculations. As for the first, if we take the carbon dioxide concentration increase over the last hundred years to be 17%, then, using logarithmic dependence $\Delta T$(°C), it is easy to show that its value is 0.6 °C.

To determine these values from observational data on air temperature two approaches may be used. The first one is to calculate the difference between temperature anomalies for various levels of carbon dioxide concentration, either at times when the same conditions of atmospheric transparency existed, or by eliminating the effects of transparency fluctuations on air temperature anomalies by introducing appropriate corrections. In this manner the temperature anomaly attributable to carbon dioxide can be identified provided the value of this anomaly is sufficiently great compared with the 'noise' level discussed above.

In comparing these empirical estimates of the effects of carbon dioxide on temperature with similar values calculated by climate models it should be kept in mind that the latter generally refer to equilibrium conditions, whereas the former are relevant to the transient or time-dependent response of temperature to continually varying carbon dioxide levels. Transient response effects can be estimated using the results of empirical studies related to the thermal inertia of the Earth—atmosphere system. These studies have shown that, for processes occurring over time intervals of several decades, the value of the Northern Hemisphere temperature anomaly due to external climate-forming factors is reduced by approximately 20% due to oceanic thermal inertia (Budyko, 1977b). This empirical value coincides with theoretical estimates of oceanic thermal inertia effects on the climate system (Schneider and Thompson, 1981; Byutner, 1983). Taking this into account, and considering the above value of temperature change, we find that the temperature rise due to the growth of carbon dioxide concentration amounts to 0.5 °C for the last hundred years.

The second approach is based on the energy-balance equation of the Earth—atmosphere system for time-varying conditions. This equation describes the relationship between outgoing longwave emission and carbon dioxide concentration. The parameters of this relationship are determined from observational data on variations in mean surface air temperature. This approach allows us to check hypotheses regarding the contribution of individual factors to climatic change.

Let us now return to studies using the first approach. In one such study the rise in Northern Hemisphere mean temperature due to carbon dioxide concentration growth was estimated by comparing mean temperature anomalies for two periods of time: 1881—1890 and 1961—1970 (Budyko, 1977c). Radiation data showed that the values of mean atmospheric transparency for these time intervals were practically the same. Therefore, the 0.5 °C temperature increase for these periods can probably be attributed to anthropogenic factors. This supposition is strongly supported by the close correspondence between this temperature rise and the value calculated by climate theory as a result of increasing carbon dioxide concentration.

In another study (Budyko, 1977b) a similar type of calculation is carried out using secular temperature variations for the previous eighty years and comparing these variations with those on atmospheric transparency fluctuations and carbon dioxide concentration growth, taking due regard to the effects of oceanic thermal inertia. These data indicate that the value of the parameter $\Delta T_c$ (i.e. the temperature increase due to a doubling of carbon dioxide concentration) is 3.3 °C (Budyko, 1980). This quantity is in good agreement with theoretical estimates of this parameter.

Vinnikov and Groisman (1981, 1982) have detected variations in mean air temperature of the Northern Hemisphere induced by the growth of atmospheric carbon dioxide concentration. They applied a transient model of the Earth's energy balance that considers the thermal inertia of the climate system, albedo variations due to atmospheric transparency fluctuations, carbon dioxide concentration effects on the outgoing longwave emission, and albedo/temperature feedback. Using this model the parameter $\Delta T_c$ turned out to be equal to 2.1—4.2 °C (ref. Table 10.3), which is close to theoretical estimates. The analysis showed that the anthropogenic growth of carbon dioxide concentration has caused a 0.4—0.6 °C increase in annual mean surface air temperature in the Northern Hemisphere at the end of the period in question above the level of the mid-1880s. During individual parts of the period under consideration the rate of warming was intensified or weakened by atmospheric transparency fluctuations. The null hypothesis that there has been no carbon dioxide effect on mean air temperature was rejected at the 1% significance level (in other words, we can be 99% confident that there has been a carbon dioxide effect).

Analysing all the above-mentioned studies on the whole we may conclude that they have shown, with a considerable degree of validity, an anthropogenic change of modern global climate. A similar conclusion can be drawn from Miles and Gildersleeves (1977), Hansen *et al.* (1981) and other similar studies.

TABLE 10.3

Empirical estimates of the parameter $\Delta T_c$ (°C) based on the three series of mean annual air temperature of the Northern Hemisphere: 1 (Vinnikov *et al.*, 1980), 2 (Hansen *et al.*, 1981), 3 (Jones *et al.*, 1982) and the two series of data characterizing atmospheric transparency: A (Pivovarova, 1977), B (Bryson and Goodman, 1980)

| | 1 | 2 | 3 |
|---|---|---|---|
| A | 3.1 | 3.7 | 4.2 |
| B | 2.1 | 2.7 | 2.8 |

## 10.2. Expected Changes of the Biosphere

### 10.2.1. CLIMATIC CONDITIONS OF THE NEAR FUTURE

Man's impact on the biosphere increased during the 19th century and the first part of the 20th century and has grown much more during the scientific—technical revolution that began in the mid-20th century.

Modern changes in the state of the biosphere can be almost entirely attributed to anthropogenic factors. As noted above, these changes hitherto were the sum of various local changes caused by man's impact on different components of the biosphere. At present the global changes associated with man's impact on the Earth's climate must be added to the local ones. These global changes, noticeable even today, should be significantly intensified in the near future.

It appears likely that climatic conditions over the next few decades or longer will depend mainly on the anthropogenic rise in atmospheric carbon dioxide concentration. According to the data presented in Chapter 7, a doubling of the carbon dioxide concentration above the pre-industrial level could occur within a matter of decades and this could lead to an increase in the global average surface air temperature by about 3 °C. There are no other factors, either natural or anthropogenic, that could induce such great fluctuations of the thermal regime within the next decades.

Figure 10.13 shows the expected man-made variations in the mean air

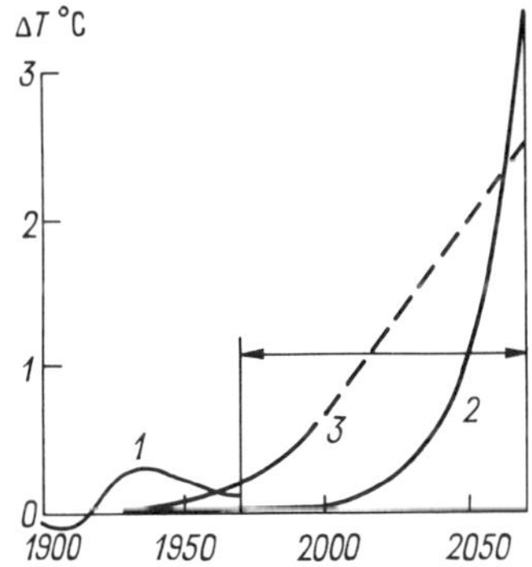


Fig. 10.13. 1, secular change in the mean temperature; 2, temperature change due to energy production growth; 3, temperature change due to $CO_2$ increase.

temperature near the Earth's surface based on a prediction made at the beginning of the 1970s (Budyko, 1972). As seen from this diagram, increasing carbon dioxide concentration in the atmosphere will be the dominant factor in climatic change at the end of the 20th century and in the first half of the 21st century. By the middle of the 21st century increasing carbon dioxide will lead to a mean temperature rise by around 2 °C. Continued growth in energy production could intensify the warming in the second half of the 21st century. This study also made the prediction that a decrease in the meridional temperature gradient will accompany the warming in the second half of the 21st century, leading to a worsening of moisture conditions on the continents in middle latitudes.

The second prediction of anthropogenic climatic change, which appeared a few years after the first one (Budyko and Vinnikov, 1976), is presented in Fig. 10.14. This prediction refers to the time interval up to the year 2000.

At the end of the 1970s and the beginning of the 1980s several estimates of future climatic changes were made by individual scientists (Kellogg, 1977—1978; Flohn, 1977, etc.), by groups of specialists (*Energy and Climate*, 1977; Budyko *et al.*, 1978, 1981, etc.), as well as by international organizations, for example, the U.S./U.S.S.R. Meeting of Experts (Leningrad, June, 1981).

Some of these prognoses presented air temperature, precipitation and other climatic charts for certain time intervals in the next century (Budyko *et al.*, 1978). In the U.S./U.S.S.R. Expert Meeting Report (*Climatic Effects . . .*, 1982), Tables 10.4—10.6, describing future climatic conditions, are given.

These estimates are based on the most probable prediction of forthcoming changes in carbon dioxide concentration in the atmosphere and use a value of the parameter $\Delta T_c$ (the temperature increase corresponding to a doubling of carbon dioxide concentration) equal to 2.5 °C. The calculation takes into account the influence of the thermal inertia of the Earth—atmosphere system

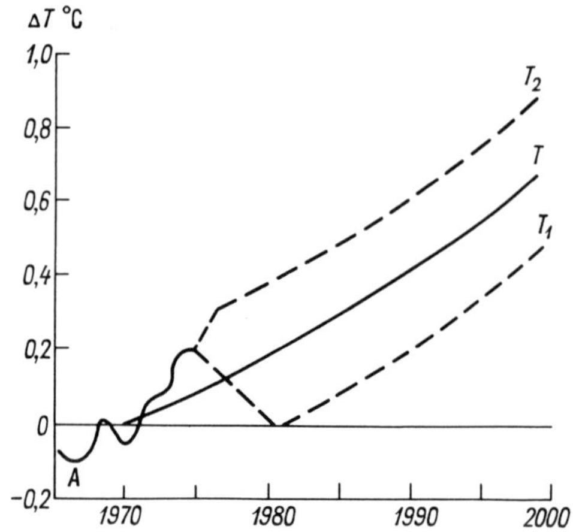


Fig. 10.14. $T$, probable air temperature change in the Northern Hemisphere up to the year 2000 relative to the temperature averaged over the 1881—1975 period; $T_1$ and $T_2$, the upper and lower limits of the expected temperature change; $A$, observed temperature change.

TABLE 10.4

Variations in global mean surface air temperature ($\Delta T$)

| Year | 2000 | 2025 | 2050 |
|---|---|---|---|
| $\Delta T$(°C) | 1—2 | 2—3 | 3—5 |

and the effects of increasing concentrations of small contaminants that also depend on man's economic activities.

Undoubtedly, anthropogenic warming will be considerably stronger in middle and, especially, high latitudes compared with low latitudes. The report mentioned presents the values of expected temperature increase at different latitudes corresponding to an increase in global mean temperature by 3 °C. These values have been obtained by the two independent methods, using atmospheric general circulation models and palaeoclimatic data (Table 10.5). Estimates of accompanying precipitation changes in various latitudinal zones on the continents are shown in Table 10.6.

The state of polar sea ice in the near future is of considerable importance. An air temperature rise in high latitudes should lead to a decrease in the area and thickness of polar sea ice. Polar ice affects the atmospheric thermal regime since the albedo of the oceanic surface covered with ice is much greater than that of an ice-free water surface. In this connection a decrease in

TABLE 10.5

Variations in latitudinal mean surface air temperature (°C) with a 3 °C increase in global mean temperature

| Sources | North latitude | | | | | | | | |
|---|---|---|---|---|---|---|---|---|---|
| | 0—10 | 10—20 | 20—30 | 30—40 | 40—50 | 50—60 | 60—70 | 70—80 | 80—90 |
| Model of general atmospheric circulation (Manabe and Wetherald, 1980) | 1.7 | 2.0 | 2.5 | 3.1 | 3.8 | 4.3 | 5.2 | 6.8 | 7.6 |
| Palaeoclimatic data | 1.2 | 1.4 | 1.6 | 2.4 | 4.0 | 4.6 | 6.8 | 7.9 | 9.5 |

TABLE 10.6

Variations in latitudinal mean total precipitation on the continents (cm $yr^{-1}$) with a 3 °C increase in global mean temperature.

| Sources | North latitude | | | | | | |
|---|---|---|---|---|---|---|---|
| | 10—20 | 20—30 | 30—40 | 40—50 | 50—60 | 60—70 | 70—80 |
| Model of general atmospheric circulation (Manabe and Wetherald, 1980) | 10 | 14 | 3 | −1 | 10 | 9 | 13 |
| Palaeoclimatic data | 12 | 12 | 2 | 2 | 8 | 9 | 13 |

ice area results in an increase of solar radiation absorbed by the Earth—atmosphere system in high latitudes, which induces further warming.

The effects of this positive feedback on the thermal regime, first considered in semi-empirical models of climate theory (Budyko, 1968, etc.), are taken into account in most modern general circulation models. When the air temperature dependence on carbon dioxide concentration is assessed using empirical data on modern climatic change and past climates, the effect of this feedback is included automatically, although its character for past climatic changes somewhat differs from that for recent climatic change.

The effects of carbon dioxide concentration change on polar sea ice can be estimated by several independent methods; either using atmospheric general circulation models, estimating the area and thickness of polar sea ice by semi-empirical means, or by the use of palaeoclimatic data.

Manabe and Stouffer (1980) found, using an atmospheric general circulation model, that ice in the Northern Hemisphere almost completely melts with a 4.4 °C increase in global mean air temperature. However, the parametrization of the ice regime in this model is rather inadequate, as shown by a comparison between the model results for modern ice thickness and observational data. It is possible that ice actually melts completely with a somewhat smaller increase in temperature.

The semi-empirical method for calculating ice cover probably yields more reliable estimates of the effects of global warming. With this method the relative air temperature rise in high latitudes is determined from empirical data on modern climatic change, and the area and thickness of ice is found using the heat balance equation (Budyko, 1971). This method has been used in several studies. The calculations show that ice in the Northern Hemisphere would melt completely (or almost completely) with 2.5 °C increase in global mean air temperature. We can show similarly that, for a 1.2 °C temperature increase in summer, ice in the Northern Hemisphere would retreat northward by approximately 2° of latitude. This would correspond to an area decrease by about 20% and a reduction in ice thickness by nearly one half.

Palaeoclimatic data show that the Arctic polar ice was first formed at the end of the Pliocene when the mean temperature was 2—3 °C higher than the modern one. These data, and the above semi-empirical model results, suggest that the sea ice will probably melt largely or entirely in the second quarter of the 21st century.

This conclusion depends, of course, on the reliability of estimates of future temperature change. The available estimates are presented in Table 10.7, which shows values of the mean air temperature increase for future decades obtained in various studies.

TABLE 10.7

Mean air temperature increase (°C) due to atmospheric chemical composition change (compared with the end of the 19th century).

| Author | 1975 | 2000 | 2025 | 2050 |
|---|---|---|---|---|
| Budyko, 1972* | — | 0.7 | 1.4 | 2.0 |
| Kellogg, 1977—78 | 0.5 | 1.2 | — | 4.0 |
| Budyko, 1979 | 0.5 | 0.9—1.3 | 1.8—2.5 | — |
| Flohn, 1981 | — | 1.0 | 1.5 | 3.3 |
| Budyko *et al.*, 1981 | — | 0.9 | 1.8 | 2.8 |
| *Climatic Effects* . . ., 1982 | — | 1—2 | 2—3 | 3—5 |

* Compared with 1970.

Table 10.7 shows that there are only slight differences between the various estimates of the expected rise in mean surface air temperature due to the influence of variations in atmospheric chemical composition. This indicates that the estimates are rather realistic.

The reliability of available calculations of the spatial distribution of variations in temperature, precipitation and other climate elements for the next decades is more difficult to evaluate. The application of several independent techniques for calculating these variations will, probably, allow us to obtain, in the near future, sufficiently accurate values.

### 10.2.2. THE BIOSPHERE'S EFFECTS OF CLIMATIC CHANGES

The man-induced warming expected in coming decades will bring back climatic conditions that probably took place in the remote past, i.e. in the second part of the Tertiary — the Neogene. To illustrate this, Fig. 10.15(a) shows carbon dioxide concentration variations during the Cenozoic era, up to the modern epoch. Part (b) of the figure shows carbon dioxide concentration changes since 1900 on a larger time scale. The upper and lower curves in part (b) correspond to the results of Keeling and Bacastow (1977) based on extreme alternative for the future development of energy production.

Comparing both sides of the figure, we can easily see that, after a relatively slow rise in the 20th century, the carbon dioxide concentration goes up sharply in the 21st century, reaching a level typical of the end of the Tertiary. This rapid restoration of the ancient atmosphere is accounted for by the fact that the resources of coal, oil, and natural gas that accumulated over millions of years are now being burned within decades in the course of man's economic activities.

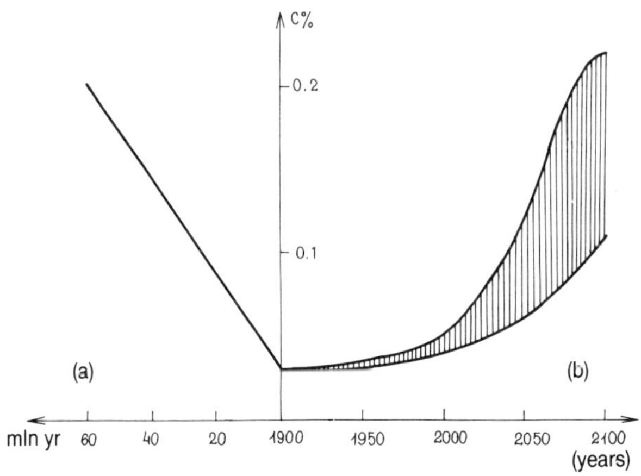


Fig. 10.15.

Figure 10.16 presents calculated variations in the mean surface air temperature (a) for the Neogene and (b) for the 20th and 21st centuries. The values of $\Delta T$ are relative to conditions prevailing at the beginning of the 20th century. The Neogene data are based on both palaeoclimatic information and climate model calculations, the results of which are in good agreement. Two methods were also used to estimate $\Delta T$ for the late 20th and 21st centuries, model calculations and calculations based on the empirical relationship between $\Delta T$ and carbon dioxide concentration (both of which yielded similar results).

It can be seen from the figure that at the mid-21st century the mean surface air temperature will reach the level of the Pliocene and will, probably, continue to rise.

Both model calculations and palaeoclimatic data show that there are great differences in carbon dioxide-induced temperature increases at various latitudes. At high latitudes the temperature rise is considerably higher than in low latitudinal belts. Therefore, anthropogenic climate change will cause natural conditions to be modified most drastically in high latitudes and to a lesser extent in middle latitudes.

One of the most important consequences of anthropogenic climate change will be a shift in the geographical zones since these depend on climatic factors. With warming, the geographical zones of moist climate move northward. Changes in the position of the arid climate zones are more difficult to predict because these changes depend considerably on variations in the potential evaporation/precipitation ratio.

The data given above on possible changes in climate at the beginning of the next century show that both precipitation and potential evaporation rates will increase over the greater part of the continents. Their ratio, therefore,

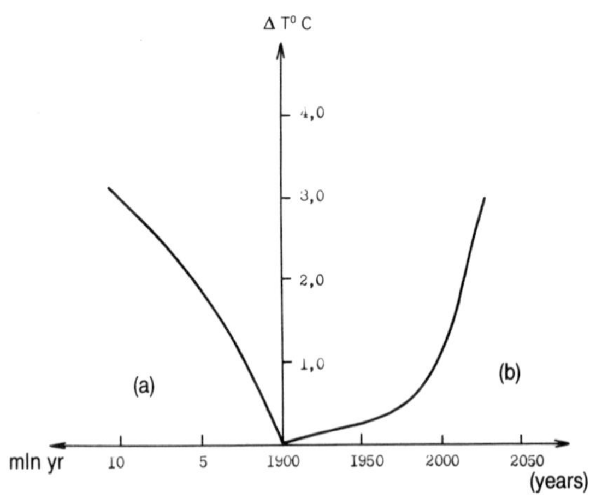


Fig. 10.16.

will not change much. One might think that, for this reason, the positions of arid geographical zones will not undergo large changes. The boundaries of humid climatic zones will, however, show much larger shifts.

The results of our calculation for the latitudinal belt 40—70°N are presented in Fig. 10.17. Part (a) shows the change in the average latitudinal boundary of the geographical zones in the Northern Hemisphere during the Neogene: part (b) depicts the same factor for the 20th and 21st centuries. It can be seen that, at the end of the Tertiary, the humid climate zones moved southward, first at a slow rate and then more and more quickly. This conclusion is confirmed by palaeogeographical data which show that natural conditions in the Neogene differed markedly from present conditions.

During the Miocene the central parts of western Europe were occupied by forests of evergreen plants including palms. In the north of Europe, Spitsbergen included, there were rich coniferous and deciduous forests comprising birch trees, beeches, oaks, pines, fir trees and other plants. In the present steppe zone of southeastern Europe, beech and oak groves existed together with some evergreen trees. Later, savanna vegetation developed here. Coniferous and deciduous forest covered the northern part of Asia. In the Pliocene the major geobotanical zones shifted southward, though at that time they occupied higher latitudes than during the present epoch (Sinitsyn, 1965).

Part (b) of Fig. 10.17 shows that geographical zones would shift towards higher latitudes by 15° by the middle of the 21st century. As mentioned above, the more arid geographical zones change their location to a far lesser extent, since the growth of potential evaporation due to the warming trend will be compensated for by an increase in precipitation. But, even so, the

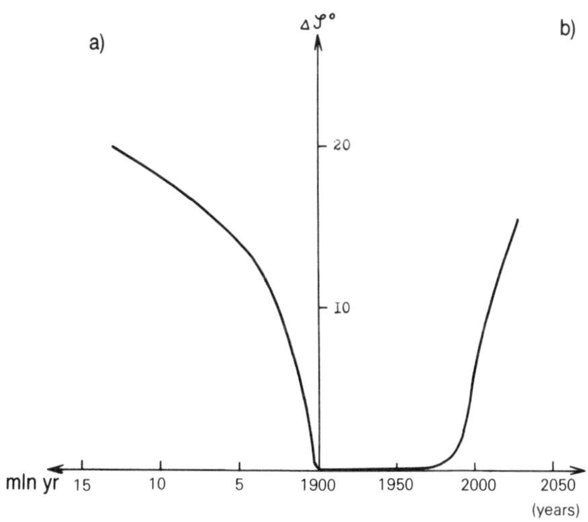


Fig. 10.17.

geographical zonality of the 21st century will differ greatly from that of the last centuries. Note, however, that a change in the components of the natural environment cannot immediately follow a rapid climatic change. The change in the surface water regime will follow the climatic change closely, but changes in vegetation cover, and especially in the soil, will be considerably slower.

Obviously, in a period as short as a few decades, the components of the environment can adapt only partially to new climatic conditions. At the same time we should not underestimate the flexibility of many of these components to respond quickly to global climatic change. For instance, Grigoriev (1956) pointed out that, during the relatively short-term warming of the 1920—1930s, the northern margins of forests in some forest-tundra regions moved noticeably further into higher latitudes.

The direct physiological influence of the expected increase in atmospheric carbon dioxide on autotrophic plant productivity is of essential importance. This problem has been investigated using numerical models of photosynthesis as mentioned in Chapter 4 and described by Menzhulin (1976), Menzhulin and Savvateev (1980) and other authors. Menzhulin found that, for a carbon dioxide concentration increase of several tens of a per cent above its modern value, the relative increase in wheat productivity amounts to about half of the corresponding relative carbon dioxide increase. It should be borne in mind that this conclusion applies to vegetation under conditions of sufficient light, heat, moisture, and mineral nutrition. Under less favourable conditions the dependence of photosynthesis on carbon dioxide increase is weaker.

It is clearly of great interest to estimate the magnitude of this effect for the entire biosphere at present-day rates of carbon dioxide growth. This is, however, a difficult task. Keeling and Bacastow (1977) think that, for small variations in carbon dioxide concentration, the relative increase in productivity is 0.27 of the relative change in carbon dioxide concentration. Since most of the natural vegetation cover exists under conditions that are far from entirely favourable, their estimate does not contradict the result obtained by Menzhulin.

Taking the above estimate of Menzhulin as a first approximation, and assuming that the carbon dioxide concentration has increased by 15% since the end of the 19th century, it is possible that this change is responsible for an increase in wheat yield (and probably that of other crops) of about 7—8%. The actual increase in the average wheat yield over the indicated period of time has been far greater than this value. It is impossible to verify our estimate using data on changes in yields *because of* the far stronger impact on yields which has resulted from progress in agricultural technology and various other factors. Nevertheless, the carbon dioxide effects mentioned here may be of vital importance in the economy.

It must be emphasized, however, that an increase in carbon dioxide concentration favourable for plant productivity could be more or less compensated for by a worsening of climatic conditions in some regions. As mentioned in the previous section, during a warming trend, average precipitation decreases and droughts become more frequent over a considerable part of the continental areas in the middle latitudes of the Northern Hemisphere. It is rather difficult to estimate quantitatively the effects of such climatic change on average yields. Undoubtedly, over areas subjected to a severe drought, the adverse weather would affect crop yield more than the carbon dioxide increase. However, if we assume that, as a result of climatic change, droughts that decrease the crop yield by 15% will occur 10—30% more often, and that, for the remaining years the yield will be about 10% higher due to the increase in carbon dioxide, the rise in total crop productivity over approximately a decade could appreciably outweigh the damage caused by droughts.

We might suppose that, overall, global vegetation productivity due to an increase in atmospheric carbon dioxide concentration must become higher, since photosynthesis is directy affected by carbon dioxide variations, due to an increase in the total precipitation over some part of the continents, and due to increasing the area of the more fertile zones because of their spread to higher latitudes. It is very difficult to take all these factors into consideration. Some rough estimates concerning these relationships, obtained by Yefimova, are presented in Fig. 10.18. Part (a) shows the variation in total productivity of continental vegetation in the late Neogene. Part (b) shows the possible change in future productivity with a continuation of global warming.

It is likely that the 50% increase in total productivity shown in the figure will not actually occur in a few decades, since the productivity of the natural vegetation could not reach its maximum potential in such a short period of time in a rapidly changing environment. However, we expect that the average

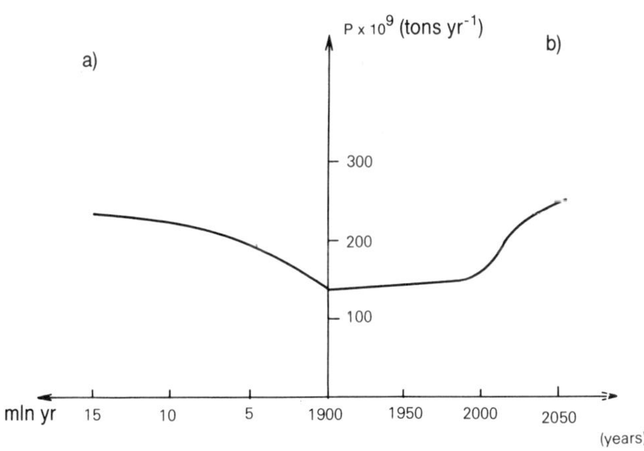


Fig. 10.18.

productivity of both natural vegetation and arable lands will increase considerably as the carbon dioxide concentration goes up.

By burning the coal and oil reserves accumulated for hundreds of millions of years, man is rapidly restoring the chemical composition of the ancient atmosphere. From the above estimates, it will take only a few decades to reach the level of carbon dioxide that existed in the late Tertiary period, i.e. several million years ago. The process going on now can be considered as 'rejuvenation' of the biosphere, i.e. the restoration of natural conditions with a warm climate at all latitudes, without polar ice cover, and with high productivity of autotrophic plants, which can sustain a far greater mass of heterotrophic living organisms compared to the modern epoch.

This change in natural conditions could be especially favourable for countries with a cold continental climate where the development of agriculture and many other branches of economic activity is difficult.

If a return to the climatic conditions of the Tertiary occurred rather slowly, this process could be favourable for mankind. In fact, the changes will occur rapidly, and this rapidity creates a number of problems whose solutions may be difficult to find. An opinion that has been expressed repeatedly is that any sudden change in climatic conditions embracing vast areas must lead to severe economic consequences involving large investments in order to reorganize all the branches of economic activity that are dependent on climate. At the same time, even a slow inadvertent climatic change could be harmful in some regions.

Returning to the question of the practical importance of the warming trend, we should reiterate that some consequences may be favourable to man's economic activities and some, unfavourable. The impact of the warming trend on agricultural productivity will evidently be of practical importance. As mentioned above, this impact will differ in different climatic zones. For example, crop productivity will rise in regions with a comparatively cold and humid climate where a longer growing season and increased temperature sums over this season could result in a considerably higher crop yield. At the same time, during certain periods of the warming trend, precipitation could decrease in a number of regions with unstable moisture conditions. This would have an adverse effect on productivity. However, since the carbon dioxide growth will contribute directly to yield increases, it might be thought that the overall, global-scale impact of the warming trend on productivity will be positive.

One of the most important results of the expected climatic change is associated with the river run-off rate, since fresh water resources even now, in a number of regions, do not satisfy existing demands. Our calculations have shown that, by the end of the 20th century, moisture conditions could worsen in some territories, but somewhat later, during the Arctic warming, river run-off will increase considerably. This increase in water resources will only occur, however, in zones of sufficiently humid climate.

Among other important practical problems associated with the warming effects, we should mention the melting of the surface layers of permafrost soils which cover almost half of the U.S.S.R. Some difficulties could therefore arise concerning the preservation of buildings in these regions.

The possibility of a contraction of interior continental glaciers, which influence the water regime of arid regions, could also be of importance. Grosswald and Kotlyakov (1978) have treated the question of the impact of global warming on glaciers and have concluded that all Arctic glaciers, except those of Greenland, could disappear rather quickly. They also think that the mountain glaciers in middle latitudes will respond more slowly to the warming.

Changes in the World Ocean level associated with changes in global ice mass are also noteworthy. As we know, over the past few million years the World Ocean level repeatedly rose and fell following the decay and development of land glaciers. During the greatest Quaternary glaciations the ocean level decreased by over 100 m compared to its present-day position as a result of the considerable amount of water that was transferred from the ocean to the continents in the formation of ice cover. During the warmest interglacial epochs the ocean level was up to several metres higher than at present.

The complete melting of the present glaciers would cause a 90 m rise in the ocean level. Antarctic glaciers and the Greenland ice sheet comprise almost the whole water mass of present-day glaciers. (The melting of the former would increase the ocean level by about 80 m; that of the latter by 7 m.) A 90 m rise in ocean level would result in a reduction in total land area by several per cent. The magnitude of this area change may be comparatively small, but the land involved includes a number of areas where the largest cities of the world are situated.

Even the consequences of a partial melting of present-day glaciers, which would lead to an ocean level rise by several metres, would be disastrous for countries situated in lowlands, such as the Netherlands, and for some large coastal cities. Even coastal towns that are situated at a higher level would suffer greatly, for their port facilities would be flooded. Therefore, even a relatively insignificant rise in the ocean level will present global problems.

The question of the fate of the Antarctic and Greenland glaciers during the warming trend has been discussed only recently. The first work in this field was presented by Mercer (1978) who concluded that the western part of the Antarctic ice mass could melt comparatively soon (perhaps even in 50 years) as a result of the forthcoming warming trend. Such an event would raise the ocean level by $\sim 5$ m. This possibility arises from the specific nature of the West Antarctic ice sheet, which rests on the sea bottom unlike the East Antarctic ice sheet, which rests on land. Such glaciers are very sensitive to fluctuations in sea water temperature and can easily succumb to a temperature rise.

Grosswald and Kotlyakov (1978) have reached a conclusion similar to Mercer's on the basis of a more detailed treatment of the problem. According to Thomas, Sanderson and Rous, the destruction of the West Antarctic ice sheet as a result of the warming trend could take a few hundred years. At the same time they do not exclude the possibility of a fast development of this process (Thomas *et al.*, 1979). The necessity for a detailed treatment of this vital issue is quite obvious.

One of the most acute problems that will arise with a rapid climatic change is associated with the task of preserving many species of wild animals and plants. The destruction of specific ecological environments will threaten their existence. The question of the preservation of food-fish reserves, which are sometimes very sensitive to climatic changes, is a part of the same problem. There are a great number of similar problems in this field because a global climatic change will affect so many components of the environment.

In the early 1970s the possibility of great global climate change due to man's economic activities was supported by only a few scientists. However, in the second half of the 1970s it has been corroborated by the conclusions of various scientific organizations, subsequently attracting the attention of governmental institutions in many countries to the problem of anthropogenic climatic change.

Beginning in 1976, several reports on the implications of the growth of carbon dioxide concentration were published. In 'WMO Statement on Climatic Change' (1976) and 'Technical Report By the WMO Executive Committee Panel of Experts on Climatic Change' (1977), an appreciable change in climatic conditions as a result of the burning of ever-growing amounts of coal, oil, and other kinds of fuel was considered possible.

In 1977 a commission of the U.S. National Academy of Sciences issued the report *Energy and Climate.* This report discussed the influence of anthropogenic factors on climate (including aerosols, gas injections, and heat production), the state and goals of climate modelling and carbon cycle modelling, and substantiation of the methods for observation of climate change. The introduction to the report begins: "Worldwide industrial civilization may face a major decision over the next few decades — whether to continue reliance on fossil fuels as principal sources of energy or to invest the research and engineering effort, and the capital that will make it possible to substitute other energy sources for fossil fuels within the next 50 years. The second alternative presents many difficulties, but the possible climatic consequences of reliance on fossil fuels for another one or two centuries may be so severe as to leave no other choice."

This report indicates that increasing carbon dioxide should be the major determinant of climatic change in coming centuries. Increasing carbon dioxide could lead to an increase of more than 6 °C in global mean surface air temperature with an increase several times greater in high latitudes. Variations in climate and atmospheric composition are of particular importance for agriculture because they can result in an increase in the rate of photosynthesis, as well as movements of agroclimatic regions to higher latitudes, shifts of the boundaries of arid zones, and other factors.

The First World Conference on Climate was held by the World Meteorological Organization and other international institutions in Geneva in 1979. Among the participants at the conference were leading Soviet specialists in the field of geography and geophysics: I. P. Gerasimov, Ye. K. Fyodorov, Yu. A. Izrael, and G. I. Marchuk. In his report, Fyodorov (1979) said: "Future climatic changes are inevitable. They will become apparent and probably irreversible during the nearest decades . . . it is necessary to develop a certain set of strategies i.e. to have a set of measures planned beforehand which would help mankind in eliminating undesirable consequences of expected climatic changes . . . What are the main elements of this strategy? The first and most important is to forecast climatic changes . . . The second is to assess the effects of natural or man-made climate changes . . . The third is to recommend measures that would allow us to avoid the adverse outcomes of climate changes or to prevent the occurrence of changes themselves".

Gerasimov (1979), Fyodorov (1979), Izrael (1979a) and Marchuk (1979) have substantiated major sections of the U.S.S.R. climate research programme, which is aimed at assessing, in particular, present-day climatic changes and towards forecasting their future tendencies.

The World Climate Conference accepted a declaration that appealed to all the countries of the world. This stated, in particular, that the burning of fossil fuel, deforestation and soil cultivation have increased the carbon dioxide concentration in the atmosphere by 15% over the past hundred years, an increase that is continuing at present. Therefore, the development of a warming trend, which would be most significant in middle and high latitudes, is expected. Since man-made climate change would be favourable in some regions and unfavourable in others, complex social and technological problems of international significance emerge. To solve these problems, unprecedented efforts are necessary to organize studies of climatic change and to create new forms of international cooperation.

At present the World Meteorological Organization and meteorological services of many countries are carrying out widespread investigations on the problems outlined in the World Climate Programme. Among these problems is the calculation of future climatic change and the estimation of the effect of this change on the biosphere and on man's economic activities. Similar programmes are operating in many countries, with special scientific groups to study the current climate change.

Global climate change is not the only way that man's activities can affect the biosphere over the coming decades. Since the mid-20th century many ecological investigations have been devoted to studying the effects on the biosphere of technological progress accompanied with the growth of population on the Earth. As noted in Chapter 9, among the major social and ecological problems facing mankind today are: (a) the difficulties in food production for the rapidly growing populations of many developing countries; (b) exhaustion of oil and other minerals; (c) an increase in environmental pollution noticeable particularly in most developed countries. The question of preserving modern floras and faunas is also of great importance and causes anxiety in many countries where nature is threatened.

It should be recognized that, although solving these ecological problems presents great difficulties, progress in science and technology over recent decades shows that ways of overcoming these difficulties can be found. For instance, Fyodorov (1977) said: ". . . The possibilities of creation of new chemicals and production of energy increase faster than the sources of raw materials and energy exhaust . . . the use of renewable natural resources enables us to raise many times the natural productivity as a whole . . . environmental pollution could be reduced to a negligibly small value . . ."

However, Fyodorov emphasizes: "The matter is whether mankind be able soon to control its development." In another study he says: "Socialism is the only social system that will be able to solve, in the future, the ecological problem completely. However, through peaceful co-existence and close co-operation between the countries with different state systems, we must take immediate measures to stave off the ecological crises. To do this it is necessary to decrease the war budgets and to spend the additional money for the reconstruction of industry, and the protection of the natural environment and resources" (Fyodorov, 1974).

### 10.2.3. THE DISTANT FUTURE OF THE BIOSPHERE

In studying this problem we need to consider natural climatic changes that have occurred over long periods of time. These changes can depend on the following factors:

1. *Astronomical factors.* A change in the position of the Earth's surface relative to the Sun produces climatic changes on time scales of tens of thousands of years.
2. *Composition of the atmosphere.* In Tertiary and Quaternary times the climate was definitely affected by variations in carbon dioxide concentration in the atmosphere. Taking into account the rate at which these vatiations have occurred we may conclude that there could be a gradual decrease in carbon dioxide concentration over time periods exceeding 100 000 years which may have important climatic effects. At the same time the climate might be influenced by shorter term variations in the amount of carbon dioxide on time scales of tens of thousands of years.
3. *The structure of the Earth's surface.* A change in the surface relief and related alterations in the positions of ocean coastlines can noticeably

affect climatic conditions over vast areas on time scales of hundreds of thousands or millions of years.

4. *The solar constant.* One should take into account the inevitability of variations in solar radiation caused by the Sun's evolution. These variations can considerably affect climatic conditions for the periods exceeding 100 million years.

Let us now look at the effects that these factors may produce on future climatic conditions. Future changes in the position of the Earth's surface relative to the Sun (orbital variations) can be estimated quite precisely. Calculations in this field allow us to evaluate the fluctuations in solar radiation income during the warm half-year, the time which is most influential in causing changes in the polar ice boundary. According to Milankovich (1930), radiation fluctuations at latitudes close to the polar circles produce a particularly strong effect on the ice boundary. The available calculations show that, in approximately 5000 years and subsequently over intervals of tens of thousands of years the amount of radiation will drop noticeably in the region of the 'critical latitudes'. These changes could lead to the onset of new ice ages, i.e. to the development of great continental glaciations. Some time after each of these epochs the amount of radiation would increase resulting in the destruction of glaciers.

Decreases in radiation in the 'critical latitudes' of the Northern Hemisphere would repeat with increasing amplitude eventually reaching the value of radiation decrease that caused the last Würm glaciation (Berger, 1977). A comparatively great radiation decrease in the 'critical latitudes' of the Northern Hemisphere will occur within the next hundred thousand years, resulting, at this time, in the development of a great glaciation.

The effects of astronomical factors on future climate may well be accentuated in the future by a further reduction in carbon dioxide concentration in the atmosphere. Over many millions of years the tendency has been for carbon dioxide concentration to decline so it is probable that it would continue to drop in the future in the absence of man's influence. The rate of this drop can be estimated approximately from data presented in Chapter 7 on decreasing carbon dioxide concentration at the end of the Cenozoic era. By extrapolating these data, we can estimate when the atmospheric carbon dioxide concentration might fall to a level low enough to lead to complete glaciation of the Earth. This time is of the order of several million years (Budyko, 1980). With this conclusion in mind, we may paint a picture of the possible future evolution of the biosphere.

A continuing reduction in carbon dioxide concentration will be followed by a gradual decrease in autotrophic plant productivity and in the total mass of living organisms on the Earth. At the same time, the polar ice zone will gradually expand advancing to lower latitudes with the onset of an ice age. In several million years, ice cover will reach a critical latitude and, after this, it will spread to the equator as a result of positive feedback. This will produce a complete and highly stable glaciation of the planet, with the stability ensured by low negative temperatures at all latitudes of the globe.

It is possible that complete glaciation would lead to the cessation of all biological processes on our planet. This assumption is based on the fact that, over the long period of the existence of the Antarctic glaciation, no living organisms have appeared that can maintain their life in its central regions. With a spreading of central Antarctic climatic conditions over the entire globe even for a compartively short period of time, it is difficult to expect living organisms to be capable of adapting to such an unfavourable climate.

Calculations by semi-empirical climate models show that the indicated glaciation could only be destroyed by increasing the global mean air temperature by several tens of degrees. This temperature increase could occur due to the accumulation of atmospheric carbon dioxide ejected by volcanic eruptions. Such an accumulation is likely because there would be no carbon dioxide loss in the formation of carbonate rocks because of the freezing of all water bodies. However, if the level of volcanic activity at the time of the formation of a 'white Earth' were sufficiently low, the planetary glaciation would be maintained for a very long period. Another extreme possibility that would lead to preservation of the planetary glaciation is if the temperature lowering were sufficient to cause freezing of any carbon dioxide coming from the depth of the Earth's crust.

The conclusion that the entirely ice covered planet could exist was based on simplified theoretical calculations, which were made before the exploration of the Solar System by means of spacecraft. These researches have shown that several satellites of the large planets are entirely covered with ice: they constitute analogues of 'white Earth'. Europa — the moon of Jupiter — is the best known of these.

Looking further forward into the far distant future, due to the general long time scale increase in solar radiation, the ice covers of these satellites as well as the Earth's ice cover (if it were to become completely glaciated) should disappear. In this way the hypothetical ice cover of the Earth would be destroyed in several billion years, but after that the surface air temperature will rise to about 80 °C (Budyko, 1974). As the solar constant continues to grow further increases in temperature will lead to the disappearance of the World Ocean, which in turn will produce a rapid accumulation of carbon dioxide in the atmosphere and a corresponding considerable temperature increase. As a result of this, the chemical composition of the atmosphere and the Earth's climate will eventually approach the conditions existing now on Venus.

In the far distant future the emission of solar radiation will cease after which the temperature on our planet will drop to near zero degrees Kelvin.

Returning to the less distant future, let us consider an alternative scenario in which the expected carbon dioxide concentration reduction is insufficient for the entire Earth to become ice covered. One or several cycles in the development of the biosphere may occur connected with increasing carbon dioxide and oxygen concentrations in the atmosphere. In the more immediate future, epochs in which the atmosphere is enriched with these gases could be followed by new qualitative changes in plants and animals similar to those changes in animate nature which repeatedly took place during the Phanerozoic.

Thus, the natural development of the biosphere would end in its destruction, in several billion years time, due to the air temperature increase caused by the growth of the solar constant. This possibility is in stark contrast with 'white Earth' scenario discussed above in which the biosphere ceases to exist far earlier — in a few million years. There is, of course, a third possibility that cannot be ruled out because it has existed throughout the entire biosphere's history; namely the biosphere's partial or complete destruction under the action of the factors that caused the critical epochs of geological history.

The duration of the biosphere's existence may well be critically dependent on man's impact. If this impact leads to the creation of the noosphere, the duration of the biosphere's existence can be expected to increase considerably. The creation of the noosphere will be completed after the formation of systems that control the state of the biosphere, and after methods for controlling large-scale biospheric processes have been fully developed. One of the first indications of the creation of the noosphere will be the elimination of the danger of global ecological crises, such as those discussed in Chapter 9. Among the problems that creating the noosphere should solve is climate control: the development of methods for controlling climate and optimizing climatic change in the interests of mankind as a whole. As has been noted, this problem in its simplest form is accessible even to modern technology.

In previous studies (Budyko, 1974, etc.) a method for climate control was described which could prevent, for a number of years, the development of a global warming trend. From estimates of the effects of stratospheric aerosols on the shortwave radiation flux reaching the Earth's surface, it follows that comparatively small variations in aerosol mass in the lower stratosphere can change the 'meteorological solar constant' (and hence the mean surface air temperature) noticeably. The burning of sulphur on stratospheric aircraft could lower the atmospheric transparency to solar radiation appreciably and lead to a reduction in surface air temperature.

There are also technologically realizable methods which could increase mean surface air temperature (spreading water vapour, freons and other gases which absorb longwave radiation in the stratosphere).

Undoubtedly, in the future, the number of possibilities for affecting climate will increase. This will make it possible to control climate in the interests of the entire human population.

Although some methods of climate control may already be technically feasible, their use is still not practical. This is because any application of these methods will result in complicated and contradictory consequences that could be unfavourable for individual countries. Thus, implementation of such methods in the near future is hardly probable. Similar methods can, however, be widely used in the future when the creation of the noosphere is complete. The creation of the noosphere should increase the stability of the biosphere markedly and so increase the time period of its existence considerably.

Among the main problems to be solved in creating the noosphere is the need for unlimited use of the Earth's resources and space, not only to meet the needs of mankind, but also to maintain constant scientific and technological progress. Another important problem is the protection of the biosphere from the possibly unfavourable or even catastrophic effects of external factors that could do great damage to the biosphere or even cause its complete destruction, examples of which have been described above.

With regard to this latter point, even now, when the noosphere is far from complete, the biosphere can still be protected from some unfavourable influences of global character. For instance, present-day changes of the atmospheric chemical composition are an unpremeditated consequence of the burning of ever-increasing amounts of fossil fuel. The current rapid growth of atmospheric carbon dioxide will soon make it impossible for ice ages, which repeatedly caused great damage to the biosphere in the past, to develop. More distant, but not fantastic, is the promise of creating a system preventing the bombardment of the Earth by large asteroids, a feature of the past record responsible for mass extinctions of organisms. The production of energy required for the comparatively small change in the orbit of an asteroid in order to divert it from collision with the Earth, will, apparently, be technologically possible in the not too distant future.

It is not possible to enumerate here all of the problems that must be solved in order to prolong the existence of the biosphere. However, a question of special significance is that of the spreading of the noosphere outside the Earth. This question is associated with the restricted size of our planet, and with the difficulty of creating completely closed technological processes. Such closed technological systems may well be essential because of the Earth's finite resources. Clearly, the utilization of the resources of other planets will help to maintain the prolonged scientific and technological progress of mankind.

At the same time space studies could answer the question of the presence of other biospheres in the universe. Contacts with other noospheres could possibly be of importance for the future of mankind. In this connection it is desirable to continue to investigate the place of the Earth's biosphere in the universe as a whole.

## 10.3. Man in the Universe

### 10.3.1. LIFE IN THE UNIVERSE

As we know, many of the philosophers of antiquity believed in the existence of life on various celestial bodies. This concept was widely accepted among prominent scientists in the field of astronomy in the 18th and, particularly, in the 19th centuries. The supporters of the concept thought that the inhabitants of the planets and stars were reasoning creatures and that there were numerous civilizations in the universe.

The launching of artificial satellites held out new prospects for studying the question of extraterrestrial life. During flights to the Moon and Mars attempts were made to find living organisms there. But even before these flights, many scientific and popular books that were devoted to this question appeared, among them books by Shklovsky (1963) and Sagan (1974).

In addition to the investigations carried out to find life on other planets, the last few years have witnessed attempts to receive information that might be transmitted by representatives of extraterrestrial civilizations and to transmit information into space by means of radio. Several international conferences concerned with this matter were also held, indicating a rising interest in the question of extraterrestrial civilizations.

Some scientists are sceptical when treating this problem. For example, the famous biologist G. G. Simpson expressed the following view in 1963 before the implementation of the U.S. programme to investigate the existence of life on the Moon and Mars: "In the case of extraterrestrial life, what real evidence there is pertains only to our solar system and it is strongly, although not quite conclusively, opposed to the presence of life on any other planet of this system. When scientists argue for the existence of such life, they are thus discussing a partly scientific question, but are on the wrong side. Outside of our solar system there is absolutely no objective evidence at all, either pro or con. When scientists discuss that issue they are at best extrapolating probabilities beyond any presently testable point. At worst, and usually, they are simply fantasizing" (Simpson, 1969, pp. 73—74).

According to Simpson, it is practically impossible to obtain any information on the existence of life outside the Solar System and there is no hope at all of discovering an extraterrestrial civilization.

Not only biologists, but also some specialists in the exact sciences concerned with space investigations, are rather sceptical. They regard the search for extraterrestrial civilizations as useless.

The absence of a common opinion on this subject does not preclude the organization of works on the problem of searching for extraterrestrial civilizations. Such research has been carried out in the U.S.S.R. and in other countries. The outcomes of these works are cited in a number of publications among which are the collections *Communication with Extraterrestrial*

*Intelligence* (1973) and *The Problem of the Search for Extraterrestrial Civilizations* (1981). These publications describe the results of the first attempts to detect radio signals from extraterrestrial civilizations. Although these attempts gave negative results, scientists who support the continuation of such studies, express their confidence in the great probability of detecting, in the future, signals from extraterrestrial civilizations. This confidence is based on the fact that such a small number of studies has so far been carried in the search for artificial signals, on the insufficiently high level of present technology for receiving signals, and on the assurance that the Earth is not a unique object in the universe. The latter consideration seems to us to be incorrect, an opinion which we discuss below in detail.

Some recent studies concerning the existence of extraterrestrial civilizations deserve attention. Hart (1975) has stated that, if reasoning creatures were in space, they would already be present on the Earth. Hart considers all ideas concerning the existence of extraterrestrial civilizations which explain why their representatives have not visited the Earth to be unconvincing. Hart therefore concludes that the Earth's civilization is the only one in our galaxy.

Similar conclusions were drawn several years ago by Shklovsky (1976) who was, earlier, one of the active supporters of the search for extraterrestrial civilizations. Shklovsky noted that the widespread opinion, that a considerable fraction of stars have planetary systems, is not confirmed by recent astronomical investigations. At the same time he emphasized that, although the mechanism for the origin of life is not yet known, the probability of this event occurring on any other planet is very small. Since any sufficiently developed civilization should be capable of exploring not only its own planetary system but also the galaxy, Shklovsky indicated that the results of this activity would be clearly seen on Earth. In summary, he concluded that the Earth's civilization is most likely alone, if not in the entire universe then, in any case, in our galaxy or the local galactic system.

These considerations of Shklovsky were discussed in some of the studies presented in the afore-mentioned collection *The Problem of the Search for Extraterrestrial Civilizations.* The authors of these studies disagreed with Shklovsky's opinions. Such disagreements are natural because so much of the subject is based on assumptions whose validity is totally unknown. Thus, for example, we simply do not know the probability of the appearance of life on a planet similar to the Earth at the early stages of its existence. Although Shklovsky's views on this are more probable compared with the contrary point of view, it is difficult to draw any definite conclusions about the existence of extraterrestrial civilizations on the basis of this particular assumption.

As mentioned above, the search for extraterrestrial civilizations is being continued in the U.S.S.R., U.S.A. and in other countries. For instance, observations of artificial radio signals from space have recently begun on a specially equipped radio telescope in Harvard (Eberhart and Thomsen,

1983). The supporters of the search for extraterrestrial intelligence strongly believe that, in the future, these investigations should be considerably enlarged.

Since direct methods for studying the problem of extraterrestrial life have not yet been effective and are often connected with large capital investments (flights of spacecrafts), the possibility of using other approaches is of interest. One such approach is to study those features in the evolution of the Earth's biosphere which could be important in understanding the more general conditions for the existence of life in the universe. At the same time it is necessary to understand what particular circumstances have allowed life on Earth to survive for such a long period of time.

Palaeontological evidence shows that living organisms have continually existed on our planet for almost four billion years. This is a very long period, being little less than the time elapsed since the Earth's origin and amounting to a considerable part of the time since the 'big bang', when the universe itself began to form. Since there is no information on the distribution of life in outer space, we have no way of assessing the probability that life might exist on a planet for such a long period of time. This probability could be either relatively high or quite low.

This problem has several other aspects. What is the probability that the biosphere could exist for a period long enough for reasoning creatures to originate, develop, and begin to investigate their environment by scientific methods? The fact that the time elapsed from the first emergence of life on the Earth to the development of modern civilization is great, even on a cosmic time scale, can be easily explained by the slowness of the development of the higher nervous activity in the course of the natural evolution of living organisms. That this slow development follows from the essence of the evolution process is clear from the following remark: "In fact what natural selection favors is always and only the genetic characteristics of those organisms that leave the most descendants under existing conditions in any one period. If those leaving the most descendants are the fittest in some sense, for example the most intelligent, then natural selection does maintain or increase fitness in that special sense. But if . . . those less fit in one special sense, less intelligent, say, have more descendants, then natural selection will decrease fitness in that sense" (Simpson, 1969, pp. 56—57).

These words do not contradict the well-known fact that, in the course of evolution, the tendency towards complication of living organisms prevails, the development of the nervous system in animals being a particular example. This question is discussed in detail in Chapter 9.

The ancient age of the Earth's biosphere should be compared with the very short-term 'space era' of the Earth's civilization. The duration of this era corresponds to the time during which both the technique of space flight and the use of radioastronomy as a means for detecting signals from extraterrestrial civilizations were mastered. In both cases, this time period is only a few decades. It is only over the same short period that radio and television transmission has been on a large scale, creating the possibility that others might detect the Earth's civilization from space by means of radioastronomy. These time intervals are short, not only compared with the total duration of the biosphere's existence, but also with the time required for radio waves to travel from the Earth to the majority of stars in our galaxy. The short duration of the 'space era' strongly limits the possibility of discovering extraterrestrial civilizations by direct methods.

### 10.3.2. THE BIOSPHERE'S STABILITY

At present there is no widely accepted explanation of the great age of the Earth's biosphere. In fact, this question is not discussed at all in scientific literature and, apparently, is considered more or less clear. In reality, it is not at all easy to explain the long existence of the Earth's biosphere; particularly when one recalls that, throughout its history, the biosphere has been subjected to significant external effects that have considerably changed its state. Since the 'zone' of life (i.e. the range of physical and chemical conditions in which organisms can live) is very limited compared to the overall variability of these conditions in, for example, the Solar System, the maintenance of life on our planet over billions of years requires special study. This study is intimately connected with the high sensitivity of most organisms to even comparatively small changes in the environment.

Regardless of widespread opinion about the extensive and long-lasting distribution of life on Earth, organisms have actually spread over only a very restricted part of our planet. The total mass of living substance is of the order of $n \times 10^{18}$ g, which corresponds to a surface density of $\sim 1$ g cm$^{-2}$. Since wood, which has a very low metabolic rate, is the main component of the total biomass, the amount of biologically active living substance probably does not exceed 0.1 g cm$^{-2}$. This value is negligibly small compared to the mass of the atmosphere, hydrosphere, and the upper layers of the lithosphere contained in a vertical column of 1 cm$^2$ section, i.e. compared with the main components of the environment surrounding living organisms.

Assuming that the mean thickness of the layer where organisms exist is of the order of 10 km, it is evident that living substance (with an average specific weight close to 1) occupies not more than one-millionth of the volume of this layer.

As mentioned earlier living substance could be considered as a very thin film situated on the surface of the continents, in the soil and in the upper layers of water basins. This film is partly broken: it vanishes completely or almost completely in the zones with permanent ice and snow cover, in the driest continental regions, and in those ocean regions with only small amounts of the mineral matter necessary for plants. Thus, even over billions of years of evolution, living organisms have still not adapted to conditions which exist over a considerable part of our planet's surface. Life has adapted still less to the conditions in other regions of the biosphere. The quantity of organisms in the atmosphere, in the deep layers of the oceans, and in the lithosphere beyond the soil layer is, as a rule, negligibly small, even compared to their relatively small mass in the surface film.

It is worth noting that the areas occupied by individual species of animals and plants are usually very limited. These areas, for the great majority of species, cover only a small part of the Earth's surface, and temporal fluctuations in the natural conditions within these zones frequently lead to a sharp change in the number of organisms, and sometimes to their extinction.

For most of the history of the biosphere, the boundaries of the life zone were largely determined by the need to maintain the photosynthesis of autotrophic plants. This is because almost all living substance (particularly multicellular organisms) depends on the redistribution of the solar radiation energy consumed by autotrophic plants. The study of photosynthesis shows that this process takes place within a rather narrow temperature range, requires readily accessible water resources, definite carbon dioxide concentrations, and necessary mineral substances. A vital condition of photosynthesis is the absorption by the plants of sufficient amounts of radiation in a definite range of wavelengths. The combination of these essential conditions is far from ubiquitous. Indeed, over a considerable part of the Earth's surface, it only occurs for part of the year. The combination of natural conditions that provides high productivity of autotrophic plants, and therefore results in the existence of a large mass of heterotrophic organisms per unit of area, is even more rare.

Although, today the productivity of natural vegetation reaches rather high values in a number of regions with a humid tropical climate, it is much less than the productivity expected with a higher $CO_2$ concentration and a lower $O_2$ concentration in the atmosphere.

The productivity of photosynthesis is a crucial factor in explaining the relatively small amount of living substance in the modern biosphere. In Chapter 4 results were presented on the relationship among plant productivity, the biomass, and the factors influencing photosynthesis. Considering this relationship, one can establish that the total biomass depends appreciably on carbon dioxide concentrations in the atmosphere and in water basins. An increase in this concentration should lead to a growth in the total biomass.

The efficiency with which vegetation uses available energy and water resources was also estimated in Chapter 4. Plants consume only about 0.1% of the total solar radiation reaching the Earth's surface and approximately the same fraction of total precipitation is used by land plants in the synthesis of organic matter. Under favourable conditions, even with the present-day composition of the atmosphere, these figures could increase considerably.

In summary, then, we have shown that the life zone covers only a small part of the range of natural conditions that exist on our planet. In addition, the amount of living substance existing within this zone is negligibly small compared to the volume of its surroundings. The range of conditions spanned by the life zone is extremely narrow compared with the range of physical and chemical conditions that exist on the planets of the Solar System (to say nothing of the conditions on other planets of the universe). Thus, when viewed in this context, the flexibility of living organisms (i.e. their adaptability to various surroundings) appears to be very limited.

Given the narrowness of the life zone and the limited amount of living substance in the external envelope of the Earth, the antiquity of the biosphere seems to be rather remarkable, a fact that requires some explanation. To do so we need to study the mechanisms that maintain the biosphere's stability. This stability must depend both on the factors that govern its development under more or less constant external conditions, and on the variability of the external factors that influence it.

Let us consider first the biosphere's stability in the absence of changing external factors. The typically balanced character of the ecology of Earth's biosphere is to some extent a paradoxical feature. As is known, heterotrophic organisms are related to one another and to autotrophic organisms by trophic relations that frequently result in the annihilation of one organism by another. If predators, which obtain food most effectively by killing other animals, leave a greater number of descendants, it would seem that natural selection would tend to favour predatory skills. The same inference can be drawn concerning the ability of herbivorous animals to consume increasing amounts of living plant matter. Carrying these arguments one step further, it appears that natural selection would result in the quick destruction of the most prosperous species because they would soon annihilate their sources of food.

Of course, ecology has long been concerned with the complex mechanisms that act to prevent the mutual destruction of species in the course of trophic interrelations; but this study is still far from completion. The use of numerical models of ecological systems is essential for progress in such investigations. These studies have established that a system of negative feedbacks exists that decrease the variability of the size of populations incorporated into ecological systems. Such feedbacks are of great importance in increasing the stability of ecological systems. The efficiency of these feedbacks is clearly demonstrated by the antiquity of many present species and genera. Some of these have existed for tens to hundreds of millions of years.

Another stabilizing factor is the rather slow process of evolution under constant external conditions. This is revealed by the existence of various 'living fossils'. For example, a great number of ancient forms have remained in the open oceans, where environmental conditions have been almost constant over long time periods. This observation leads us to the assumption that, if the external factors (such as solar radiation and atmospheric carbon dioxide level) that influence the biosphere in general remained almost constant, i.e. if the abiotic components of the biosphere changed comparatively slowly, then the rate of evolution would be slow and a host of ancient forms of organisms would persist for long periods of time. However, even under such conditions, changes in the abiotic environment would still occur. In particular, under the action of photoautotrophic plants, the amount of oxygen in the atmosphere would gradually increase. Such a change in the environment would have to stimulate the progressive development of organisms, although the rate of the progress under these conditions would probably have been comparatively low.

Although, in the absence of changes in external factors, the biosphere would be more stable, this stability would not necessarily guarantee its unlimited existence.

Let us now consider the question of the biosphere's stability under changing external conditions. We begin with a comparison of global and local ecological systems. The latter usually exist under varying environmental conditions. It has often been proposed that ecological systems possess certain features of integrity: as a result they are sometimes considered to be 'superorganisms'. Is it possible to use such a term for the global ecological system — the biosphere? The use of this term seems to be warranted because all components of the biosphere, including living organisms, are closely connected to each other. Living organisms in turn exert a considerable effect on other abiotic components of the biosphere. Their influence on the atmosphere is particularly great. Autotrophic plants play a central role in the balance of carbon dioxide and in the production of oxygen. The deposits of the upper layers of the lithosphere are also largely created by living organisms.

However, the use of the term 'superorganism' for the global ecological system would be justified only if there were a comparatively high level of stability, sufficient to provide a prolonged existence of the system even under the influence of variable external factors. As is known, a more or less high level of the stability is achieved in the course of natural selection. This was mentioned in Chapter 5 with reference to local ecological systems, which are characterized by a high level of stability achieved in the course of natural selection. The question of the stability of global ecological system — the Earth's biosphere — has not been studied in depth. However, it is probable that this stability is comparatively small. In this case, the use of the term 'superorganism' for the global biosphere is not justified. As a further consequence of this argument, the persistence of the Earth's biosphere over billions of years must be largely due to the coincidence of unconnected favourable circumstances. The probability that a similar chain of favourable coincidences would be repeated for other external bodies is extremely low.

Let us consider the reasons for this conclusion. The way some organisms affect their environment seems to improve the conditions of their existence, but mainly in local ecological systems over comparatively short periods of time. In these short time scales, local effects probably do not influence the global ecological system from the point of view of adaptation, because the time period during which the biosphere undergoes noticeable changes is long compared to the period of the existence of most species.

Of course, not all organisms affect their environments to their own benefit. For instance, phototrophic plants produce oxygen, but the high concentration of oxygen in the present atmosphere reduces plant productivity. The utilization of carbon dioxide by plants has resulted in the lowering of carbon dioxide concentration in air and in water. This has not only decreased plant photosynthetic productivity, but it even threatens the existence of the biosphere because of the increased probability of the Earth's glaciation.

Thus the effects of organisms on the Earth's biosphere are not to increase its stability. Although the present biosphere has evolved under varying external conditions, it is likely that a similar situation would emerge under constant external conditions.

It has repeatedly been suggested that great changes in the biosphere can be induced by comparatively insignificant external factors. Examples cited include the critical epochs of geological history considered in Chapter 8, which led to the extinctions of many animal groups at the end of several of the geological periods. The extinction of numerous animal groups in South America at the end of the Tertiary is an obvious case of a large biospheric change resulting from a relatively small change in external factors. This extinction episode was caused by the connection of South and North America as a result of the uplift of land in the region of the Panama isthmus. The penetration of highly developed mammals into South America led to a rapid disappearance of many representatives of the original fauna that had existed there throughout the whole Tertiary period when the continent was isolated from other continents.

The balanced ecology of the biosphere was considerably disturbed as a result of man's activities, which can be considered as an external factor relative to the biosphere's various components. Large, man-induced changes in the biosphere's ecology appeared in the early stages of the existence of modern man, who succeeded in annihilating many species of large animals (see Chapter 9). Man's impact on animate nature has sharply increased today. As a result of the inadvertent consequences of economic activities, many species of organisms have been annihilated and the populations of a great number of species have declined drastically. The ease with which man has annihilated living organisms over vast areas is one of the proofs of the great sensitivity of the biosphere to changes in external factors.

The duration of the biosphere's existence under the influence of variations in external factors depends on its stability, which in turn is determined by a set of direct relations and feedbacks between the system's components and external factors. Ignoring for the moment the question of man's activities, among the external factors whose changes can influence the global ecological system are solar radiation and exchanges of matter between the deep layers of the lithosphere and the biosphere.

The importance of the first of these factors is obvious. To take two extreme cases, when solar irradiance becomes too great or when it ceases (this could happen in several billion years) the biosphere will cease to exist. As a less extreme case, the results described in Chapter 7 show that climatic conditions can vary considerably even with comparatively small fluctuations of solar radiation. This high sensitivity of the thermal regime to solar radiation fluctuations leads us to the assumption that the biosphere's existence may depend on even relatively small changes in the radiation of the Sun.

The question of the effects of solar constant variations on climate in the geological past, treated in Chapter 7, is of considerable importance. The Sun's evolution is assumed to result in a gradual reduction of its diameter and an increase in its luminosity. Other things being equal, this change in solar radiation should correspond to a very cold climate in the Precamrian and in the Palaeozoic, which conflicts with the geological record. This conflict can be resolved by the fact that, in the past, the chemical composition of the air differed fundamentally from the present one. For instance, the atmosphere probably contained a great quantity of carbon dioxide. The question therefore arises: are these changes in external factors (solar radiation and chemical composition of the atmosphere), whose effects on the thermal regime proved to be opposite and mutually compensatory, mere coincidence? The mutual compensation cannot be attributed to deterministic relationships between the two processes. As mentioned in Chapter 7, the carbon dioxide concentration level in the atmosphere depends mainly on the rate of upper mantle degassing. Throughout the history of the biosphere, this rate has varied widely and, apparently, independently of the atmospheric thermal regime. Although the long-term decrease in carbon dioxide concentration with declining degassing rate seems to have been retarded by a fall in the rate of carbonate formation, this factor did not change the tendency towards a lowering of atmospheric carbon dioxide amount which has prevailed throughout the history of the biosphere.

Data presented in Chapter 7 show a ten-fold variation in atmospheric carbon dioxide concentration during the Phanerozoic, from several tenths to several hundredths per cent of the atmospheric volume: and a five-fold variation in oxygen amount, between $\frac{1}{3}$ and 1.5—2 relative to its modern mass.

Undoubtedly, variations in atmospheric chemical composition had considerable effects on animate nature. But is it possible for variations in atmospheric chemical composition to go beyond the limits under which organisms can survive? Autotrophic plants, the main source of energy for the modern biosphere, can live in the range of atmospheric carbon dioxide concentration from about 0.01 to several per cent. For most animals there is an upper limit of carbon dioxide concentration above which they cannot survive. In many cases this limit amounts to a concentration of several per cent by volume.

Although many plants and animals have only limited ranges of tolerance for oxygen partial pressure (these ranges differ markedly for aerobic and anaerobic organisms), variations in atmospheric oxygen mass are less important for the stability of the biosphere than changes in carbon dioxide concentration. This is largely because the oxygen concentration in the atmosphere is much greater than that for carbon dioxide. The source of free oxygen —the photosynthesis of autotrophic plants — is less variable than the source of atmospheric carbon dioxide. As a result, carbon dioxide concentration in Phanerozoic has varied much more widely than oxygen concentration over the lifetime of the biosphere (see in Chapter 7). The actual difference between relative changes in $CO_2$ and $O_2$ concentrations should be much greater than the values obtained in Chapter 7, since these values are averages for geological periods or epochs. They do not characterize fluctuations in atmospheric gas composition for shorter time periods. It is unlikely that atmospheric oxygen concentration could vary markedly over relatively short time intervals, but large variations are a possibility for carbon dioxide. This greater variability for carbon dioxide is demonstrated by the present rapid growth of carbon dioxide concentration due to man's economic activities.

Atmospheric carbon dioxide has a considerable influence on temperature through the greenhouse effect. A concentration of carbon dioxide considerably less than the modern one would result in a major temperature drop, which, as some calculations show, could induce complete glaciation of the Earth. Thus, fluctuations in carbon dioxide mass have both direct and indirect (due to climatic change) effects on living organisms.

In the above discussion we mentioned the possibility of ecological catastrophes associated with the Sun's evolution. Catastrophes resulting from changes in atmospheric chemical composition might take place much earlier, possibly leading to the destruction of all living organisms. Variations in carbon dioxide concentration are the most dangerous in this sense. As noted above, the range of this concentration at which most autotrophic plants can live is relatively narrow, corresponding to an increase or decrease in atmospheric carbon dioxide mass by approximately one order of magnitude relative to its average value for the Phanerozoic. At present time this mass is much less than its mean value in the past and currently is quite close to the lower limit of the range.

As the data of Chapter 7 show, the process of decreasing atmospheric carbon dioxide began in the Cretaceous and accelerated in the late Tertiary period. The probable cause of decreasing carbon dioxide mass was the attenuation of volcanic activity, which in turn may have resulted from the exhaustion of radioactive element resources that provide the heating in the deep layers of the Earth. We can make some very approximate estimates of the consequences of a continuation of this process. Over a period of several million years, i.e. a short period of time from the point of view of the Earth's history, one of the two major ecological catastrophes discussed above could have taken place. Complete glaciation of the Earth could occur, or autotrophic plants could become extinct. The latter would also mean the extinction of heterotrophic organisms because of the impossibility of maintaining the proper level of photosynthesis required to compensate for the expenditure of organic matter on respiration and other physiological functions of plants. In this connection it is useful to recall a statement by Ronov (1976) that life on the Earth can exist only with a sufficiently high level of radioactive decay in the lithosphere.

The precariousness of life on Earth is further demonstrated by a consideration of the glacial—interglacial cycles of the past few million years. It is likely that the Earth's biosphere was close to catastrophe during the most well-developed Quaternary glacial periods when ice advanced almost to the critical latitude beyond which the climate becomes unstable and ice spreads to the equator as a result of positive feedback. In addition, we should remember that, throughout the entire history of the biosphere, the possibility (considered in previous chapters) existed of its complete or partial destruction through the collision of the Earth with a sufficiently large external body. Large impact craters can be seen distinctly on the planets without an atmosphere. Some of these visual signs show also that large parts of planetary surfaces were destroyed by secondary events after their collisions with large asteroids. For instance, a great part of the surface of Mercury is occupied by the Caloris basin formed, apparently, 3—4 billion years ago when the solid crust of Mercury was destroyed by a large asteroid that produced magma withdrawal. It is quite possible that the collision of such an asteroid with the Earth would have annihilated life on our planet.

In order to make a quantitative estimate of the probability of retaining life on Earth over the period of the biosphere's existence, it is necessary to study the stability of the biosphere relative to external effects. This question has been discussed in a number of studies including the monograph *Global Ecology* (Budyko, 1977a). Here it is pointed out that the comparative stability of climatic conditions required for a prolonged existence of life is achieved as a result of the action of several feedbacks between different elements of the meteorological regime. The most important feedback is the relationship between outgoing longwave radiation and surface air temperature. With growing heat inflow the temperature rises, thus increasing outgoing emission.

It is interesting to note that this relationship, which is important for retaining life, is much weaker for our planet than for the Moon (which has no living organisms): in other words, a change in solar constant will affect the temperature of the Earth more than that of the Moon. It can easily be shown using the Stefan—Boltzmann equation (see Chapter 3) that, on the airless Moon, a 1% increase in the solar constant would cause the mean surface temperature to rise by 0.6—0.7 °C. On the Earth, an increase in heat income results in intensified evaporation and growing water vapour content of the atmosphere. This strengthening of the greenhouse effect means that the corresponding temperature rise for our planet is up to 1.5 °C. Actually, this value is close to 2 °C due to the effects of warming causing melting of the polar ice and additional absorption of solar radiation.

As mentioned above, retaining life on the Earth is possible only for limited variations in atmospheric carbon dioxide and oxygen concentration. In the study of changes in atmospheric chemical composition treated in Chapter 7, it was assumed that the loss of carbon dioxide due to carbonate formation is proportional to its concentration in the atmosphere. A similar assumption has been made for the expenditure of atmospheric oxygen on oxidizing minerals. It is obvious that these relationships decrease the possible range of variations in atmospheric $CO_2$ and $O_2$ concentration. These relations can be compared with similar dependences found in other investigations of the stability of atmospheric chemical composition. For instance, Van Valen (1971) distinguished three feedbacks that can stabilize the amount of oxygen in the atmosphere. These are based on the fact that, with the growth of oxygen concentration, (a) the productivity of phytotrophic plants producing oxygen decreases, (b) the expenditure of oxygen on oxidizing various minerals increases, and (c) the fraction of deposited non-oxidized organic carbon decreases. Van Valen pointed out that these relationships are important for restricting variations in oxygen concentration, but he did not present them in quantitative form.

The factors that stabilize the level of carbon dioxide concentration in the atmosphere and restrict variations in mean surface air temperature have recently been studied by Walker *et al.* (1981). They noted that the expenditure of carbon dioxide on carbonate formation increases as the weathering of silicates intensifies due to rising temperature. Using the implied temperature/carbon dioxide concentration relation, the authors calculated the way changes in solar radiation influence global mean temperature when due account is taken of the damping effect resulting from this carbon dioxide balance feedback. As mentioned in the above paper, this approach helps to explain the warm climates in the early history of the Earth when solar luminosity was low, but it still requires us to assume that a volcanic activity was more intense during the early history of the Earth.

The dependence of the rates of change in $CO_2$ and $O_2$ concentration on their respective concentrations, as assumed in our calculations of the evolution of atmospheric chemical composition, coincides qualitatively with the dependence accepted in the above study. The dependence in the study by Walker *et al.* is stronger than in our calculations, so it has a more pronounced restriction on variations in carbon dioxide concentration in the geological past. Nevertheless, as seen from Chapter 7, the use of this dependence still leads to the conclusion that the amount of carbon dioxide in the past atmosphere varied within wide limits.

For estimating the biosphere's stability to variations in external factors the following fact is of considerable importance. The range of variation of mean air temperature and $CO_2$ and $O_2$ concentrations during the Phanerozoic coincides roughly with the range of these parameters within which the biosphere can survive. Of particular importance is the fact that the modern, comparatively low concentration of carbon dioxide and low mean air temperature are much closer to the values at which phototrophic plants will disappear and the Earth will become ice-covered than they are to the maximum values these elements reached during the Phanerozoic. Thus, data on the past variations in air temperature and atmospheric chemical composition lead us to conclude that the stability of the biosphere to changes in external factors is very restricted.

The biosphere's stability is probably much lower for comparatively short term variations in air temperature and atmospheric chemical composition. Data on the modern growth in carbon dioxide concentration show that there are no strong natural negative feedbacks that could considerably attenuate an increase in atmospheric carbon dioxide content with strengthening its sources. The current variations in atmospheric chemical composition result from a man-made carbon dioxide source that is about two orders of magnitude stronger than modern volcanism and other natural sources. However, the present-day level of volcanic activity is far below than that which is possible.

During the Phanerozoic it is likely that there was permanent danger of global poisoning of aerobic organisms by carbon dioxide and other products of volcanic eruptions. This undoubtedly took place repeatedly on local scales in regions where explosive eruptions occurred. Apart from known cases of mass death of people poisoned by eruption products, we note observations of naturalists who have found the remnants of dead animals poisoned by volcanic gases from eruptions in tropical countries. Similar finds have been made by Simpson in his research on ancient fauna in South America (Simpson, 1978). Since only a small part of the remnants of large animals is found by palaeontologists, the finding of this site of mass mortality indicates that such events were frequent.

Another danger for living organisms must have been the decreased atmospheric transparency associated with periods of intense volcanic activity. As observational data and calculations show, variations in volcanic activity over the last hundred years have led to changes in surface air temperature by several tenths of a degree. Over hundreds of millions of years far greater temperature variations must have occurred due to the simultaneous eruptions of many volcanoes in some periods. This has been discussed in Chapter 8 where approximate calculations showed that such events could have caused coolings of 5—10 °C over individual decades. Such a temperature decrease could have been a sufficient cause for the extinction of many species of stenothermal organisms (i.e. organisms that could not adapt to large temperature variations in the environment). In this connection it has been hypothesized that an increase in volcanic activity may have caused the dramatic faunal changes that occurred in critical epochs of geological history. The impact of a comparatively large asteroid would have played the same role and had similar consequences.

It is worth considering whether an increase in volcanic activity or an asteroid impact could result in a large enough cooling for all living organisms to be destroyed. Although we know that this hasn't happened, this does not answer the question of its probability of occurrence; just as the longevity of the biosphere does not explain why the biosphere has existed for so long. We need to estimate the probability of changes in external factors that could have large impacts on the biosphere. Obtaining such estimates is a matter for the future. However, the restricted data available now suggest that, over the very long period of the existence of the biosphere, this probability was very high.

Note that the considerations described in this section differ noticeably from the views of the authors of the 'Gaia hypothesis' who supposed that the activities of living organisms are directed towards increasing the stability of the biosphere and guaranteeing its prolonged existence (Margulis and Lovelock, 1974; Lovelock, 1979, etc.). It seems to us that the authors of the 'Gaia hypothesis' have not based their hypothesis on firm grounds.

### 10.3.3. THE UNIQUENESS OF THE EARTH'S BIOSPHERE

It follows from the above considerations that life in the universe is a rare phenomenon. This refers particularly to the highest life forms, whose development could have resulted in the creation of extraterrestrial civilizations.

The only known case when civilization has emerged is the result of evolutionary changes in organisms on our planet over four billion years. This period of time far exceeds the time of the formation and dissemination of the first organisms on the Earth. Therefore, the probability of civilizations occurring on other planets is many times less than the probability that living organisms might appear on them.

Variations in natural conditions that accelerated the evolutionary process seem to be of great importance in the emergence of the highest forms of organisms, those responsible for the creation of civilizations. It is possible that, on a hypothetical planet on which conditions were favourable for life but were constant for billions of years, the slowness of evolution would exclude the possibility of the emergence of sufficiently complicated organisms and, consequently, the creation of a civilization.

Thus, the emergence of highly developed organisms apparently requires a definite changeability of external conditions. However, these changes would always have to be within the limits of the narrow life zone of organisms. Such a condition lowers the probability of the emergence of extraterrestrial civilizations.

The method widely used in assessing the number of planets in the universe where civilizations might exist is to multiply the number of stars by several coefficients, each less than unity. These coefficients characterize the probabilities of (a) the existence of planets near a star, (b) the presence on these planets of conditions suitable for life, (c) the emergence of life on these planets, (d) the progress of organisms up to the level of the emergence of highly developed civilizations, and (e) the long-term survival of these civilizations. The authors of these assessments acknowledge the great difficulty of obtaining any valid estimates of these coefficients (with the possible exception of the first of them). Nevertheless, given the enormous number of stars in the universe, they can, by using very small (in their opinion) values for every coefficient, still obtain a rather large number for the civilizations existing now. For example, Sagan (1974) estimates that in our galaxy approximately one million planets have highly developed civilizations.

We are sceptical of these calculations and agree with Simpson that such a method of computation is, to a large measure, arbitrary. In addition, there may well be other factors that considerably lower the probability of the existence of life (and, particularly, intelligent life) in the universe.

From our studies of the stability of the Earth's biosphere we conclude that this stability is not very high, and that the probability of the existence of life on a planet similar to the Earth for billions of years is extremely low. Comparatively small variations in solar radiation and in volcanic activity seem to be sufficient to destroy most or all the life on the Earth. Furthermore, other external factors that could exert a fatal influence on life have been cited above. For example, the fall of large asteroids could produce variations in the state of the biosphere, approaching the limits of the life zone.

Without going into this and other similar possibilities in more detail, we note that, from our point of view, the independent evolution of the Sun, and the weak dependence of the evolution of the deep layers of the lithosphere on the state of the biosphere are factors of utmost importance for the problem under consideration. Because the biosphere is so sensitive to variations in solar radiation and to atmospheric composition changes resulting from degassing of the lithosphere, the absence of feedback stabilizing the 'Sun-biosphere' system and the weakness of such feedbacks in the 'deep layers of the lithosphere—biosphere' system, considerably decreases the probability of the prolonged existence of the biosphere that is required for a highly developed civilization to emerge. These considerations suggest that the probability of life in the universe, particularly in its highest forms associated with the appearance of civilizations, is insignificant. This conclusion is confirmed by the negative results of attempts to detect life on the Moon and Mars, by the failure of attempts to pick up signals from extraterrestrial civilizations, and by the absence of reliable information on past contacts between mankind and the inhabitants of other planets.

This conclusion is based on studies of the Earth's biosphere, the results of which have been published previously in works by the author. A similar conclusion has recently been reached by Hart (1978, 1979), who evaluated the possibility of maintaining conditions favourable for life on planets revolving around different stars.

In the first of these articles Hart constructed a numerical model for the evolution of the chemical composition of the Earth's atmosphere. Using this model he showed that it is only possible to maintain conditions favourable for the existence of organisms for long periods for a specific range of values of distance between the Earth and the Sun. An increase in this distance by a small value could have led to complete glaciation of the planet in the remote past. A slight decrease in this distance could lead to an increase of the planet's surface temperature to the boiling point of water, after which the disappearance of the hydrosphere, the main sink for carbon dioxide emitted from the Earth's crust, would result in a rapid further temperature rise, as has occurred on Venus.

Hart's second article treats the question of the existence of 'zones of life' (i.e. places where the existence of living organisms is possible on planets similar to the Earth) for stars of different types. Using the above model, he found that, near many stars, zones of life could not exist for long. For the stars that could have long-term zones of life, the volume of these zones, which is determined by the closest and farthest approach of the planet to the star, is very small in every case.

Although Hart's model of the evolution of the atmosphere is rather schematic, it correctly simulates many features of this process. In addition, Hart has shown that his conclusions regarding the properties of the life zone are comparatively insensitive to the possible errors in the parameters introduced into his model. For this reason, Hart's results can be considered as evidence that life in the universe is a very rare phenomenon.

In view of these considerations, the Earth's biosphere is unique in two respects. First, the very fact of its long existence under conditions that repeatedly could have led to the destruction of living things on our planet is extraordinary. Second, there is the unusual possibility that has appeared in our own time, that of directed influence on the biosphere by one of its components, man, towards active maintenance of the biosphere's existence and prevention of the processes threatening its further existence. The presence of this possibility offers promise for a long future existence of the biosphere, and the distribution of the noosphere in space to other external bodies.

## 10.4. Preservation of the Biosphere

The scientific and technological revolution occurring now offers the possibility of creating a controlled global ecological system, i.e. completing the conversion of the biosphere into the noosphere. One of the main characteristics of the noosphere will be its high stability to the effects of external factors that could cause considerable damage to, or even total destruction of the biosphere.

Factors that might lead to partial or complete destruction of the biosphere have been described above. Almost all of these refer to the future, to millions or even billions of years distant from the present time. An exception to this is the possibility of a new glacial age, a possibility which could occur within the next few thousands of years. Of course, modern economic activities are influencing climate in precisely the opposite direction, increasing the global mean surface air temperature. In the coming decades and centuries the expected temperature rise could result in melting a considerable part of the available permanent ice. This has one reassuring implication; the problem of preventing a new ice age should present no difficulties to future technology.

A far closer and more real danger for the biosphere is the possibility of anthropogenic ecological crises affecting mankind himself, so destroying any chance that we might create a fully developed noosphere. There are two forms of possible anthropogenic ecological crisis. Economic development carried out without a proper synthesis of ecological and economic planning, may have unfavourable consequences that could lead to the crisis phenomena considered in Chapter 9. As noted in this chapter, this crisis is not inevitable. It can be prevented by increasing the level of cooperation between various countries. The second possibility is that of unlimited nuclear conflict. This presents today's greatest danger for mankind and the biosphere. Several calculations have shown the large number of people that can die as a result of nuclear war. In addition to deaths caused directly by nuclear explosions we must add massive mortality from epidemics because of the deterioration of medical service. In the course of a nuclear war and soon after a considerable, or perhaps the greater part of mankind will die (Blokhin, 1983).

This conclusion, however, does not take into account more distant after-effects of nuclear warfare, effects that could cause the death of a far greater number of people. These consequences of nuclear war, which are inseparable from the question of the fate of the biosphere after a nuclear conflict, will now be considered.

The estimation of the effects of unlimited nuclear warfare on the biosphere has been discussed repeatedly. In a number of cases it has been suggested that these effects would be restricted and that, after a nuclear conflict, the biosphere would be little changed. This is the conclusion reached, in particular, in the report of the U.S. National Academy of Sciences (*Long-Term Worldwide Effects* . . . , 1975). However, the grounds given for this conclusion are unconvincing, and the question of the after-effects of nuclear warfare requires further study.

According to the above report, in the mid-1970s the total number of warheads was 50000 with a total explosive power of roughly $2 \times 10^{10}$ tonnes of TNT. This energy exceeds the energy of the atomic bomb that destroyed Hiroshima by more than a million times. In the earlier sections of this chapter we have demonstrated the limited stability of the biosphere to external effects. It is natural, therefore, to pose the question: could the explosion of the available warheads produce a global ecological crisis similar to those crises that resulted in repeated mass extinctions of organisms throughout the history of the biosphere?

Undoubtedly, the biospheric effects of unlimited nuclear conflict will be varied, with many different effects on the conditions vital to the existence of living organisms. However, it is a change in global climate produced by nuclear warfare that seems to present the greatest danger for animate nature. This change would be similar in some respects to the change following the fall of large asteroid, discussed in Chapter 8.

As in the case of an asteroid impact, numerous nuclear explosions would produce a great number of aerosol particles in the atmosphere. These would considerably reduce the transparency of the atmosphere to solar radiation. The residence time of optically active particles ejected to the stratosphere would be several months, similar to that of fine dust ejected to the stratosphere during explosive volcanic eruptions. This residence time is somewhat less than that of the sulphate aerosol particles that are formed from sulphur-containing gases after a volcanic eruption. Since the formation of these particles is a relatively slow process, their residence time is usually of the order of one or two years.

This difference in residence time for aerosols produced by volcanic and nuclear explosions restricts the possibility of using data on climate effects of volcanic eruptions for modelling climatic changes after nuclear conflict. A better analogue would be to use data related to the impact of large asteroids; although, unfortunately, these data are very limited.

Chapter 8 treats a previous estimate of the global cooling after a large asteroid impact in the late Cretaceous. It has been found that the temperature lowering amounted to 5—10 °C (Budyko, 1980), apparently enough to cause the extinction of most animal species at the Mesozoic—Cenozoic boundary. This value should be comparable with the temperature change after an unlimited atomic war.

The difficulty in estimating the nuclear war effect directly is basically associated with uncertainties in the amount of submicron aerosol particles that would be ejected in this case into the stratosphere. In the report of the U.S. National Academy of Sciences it is anticipated that the mass of these particles will be up to $\sim 10^8$ tonnes (*Long-Term Worldwide Effects . . .* , 1975). This amount will be supplemented with a great quantity of highly dispersed aerosol particles formed as a result of fires that would inevitably embrace vast areas. These particles would mainly consist of soot and ash. The total quantity of these particles is difficult to assess, but it would probably amount to around 100 million tonnes (Crutzen and Birks, 1982). As a result, the increase in the total mass of optically active particles in the stratosphere will be of order $10^8$ tonnes, i.e. $\sim 10^4$ g cm$^{-2}$ for the entire surface of the Earth. From calculations using the formulas of atmospheric optics it follows that, with this concentration of stratospheric aerosol, the amount of solar radiation coming to the troposphere would decrease by more than half (Budyko, 1974).

Using current knowledge of the dependence of air temperature on solar radiation income we find that the asymptotic, or steady-state cooling due to such a decrease in solar radiation would be 50—100 °C. The actual temperature decrease would be noticeably less than this due to thermal inertia effects of the climate system. These effects could easily be taken into account by using Equation (10.1). As a result, if we assume that the residence time of aerosol in the atmosphere is of the order of several months, the lowering of global mean surface air temperature would amount to 5—10 °C. This value coincides with the estimate of the mean temperature lowering after a large asteroid impact.

The expected temperature decrease in different regions of the globe would differ noticeably from this average value. Since an increase in aerosol mass in the Northern Hemisphere would be larger and since the ocean area is less than in the Southern Hemisphere, the mean temperature decline in the Northern Hemisphere would probably be somewhat larger, around 10—15 °C. Analysis of data on annual temperature variations shows that the expected mean temperature lowering over the northern continents would amount to about 30 °C, this value being greater in the continental regions remote from the oceans. It is probable that the most drastic temperature decrease would only last for several months, with less severe cooling persisting noticeably longer.

There would, of course, be many other dramatic changes in climate after a nuclear war which cannot be discussed here. We note, however, that the estimated temperature change is extremely large, comparable with the climatic catastrophes of the geological past which have been described in Chapter 8.

Although the above calculation of the expected cooling after a nuclear war is very schematic, this simplification is justified by the essential uncertainty of a number of parameters of the problem under consideration, particularly estimates of the amount of aerosol entering the stratosphere. In spite of these simplifications the estimates presented here seem to admit the possibility of catastrophic after-effects of an unlimited nuclear war for the entire biosphere.

In the course of this catastrophe a host of animal and plant species would die. At the same time new calamities would arise for those people who survived the nuclear conflict. For example, it is probable that a drastic cooling (particularly in the regions with warm climate where houses are not heated) would cause many deaths from cold. This, however, would not be the major danger the global cooling would create for mankind. Far greater consequences would arise from the disturbance in the food production system. The lowering of temperature would mean that photosynthetic productivity would decrease significantly, possibly leading to complete or almost complete destruction of crops. As Crutzen and Birks (1982) noted, photosynthetic productivity might also lower because of the noticeable decrease in solar radiation reaching the Earth's surface. Since the world's existing food reserves are insignificant compared with annual consumption, any drastic lowering of food production would lead to mass mortality.

The eruption of the volcano Tambora in 1815, one of the major eruptions over the last several hundred years, can give us some idea of the after-effects of such a climatic change. Although after this eruption mean global temperature lowered by only 0.5—1 °C, the cold weather still resulted in a drastic decrease in crop yields in a number of tropical regions where a part of the population died from starvation (Ganapathy, 1980). It is obvious that a nuclear climatic catastrophe would have immeasurably heavier consequences.

It is highly probable that the explosions of many nuclear bombs would cause other environmental changes that would be disastrous for life. Injection of great amounts of nitrogen oxides may lower the ozone concentration in the stratosphere considerably (Crutzen and Birks, 1982; Izrael *et al.*, 1983). According to calculations of the latter study, the ozone content after a nuclear conflict would decrease by 30—60%, and would remain at this level for many months. As is well known, any attenuation of the ozone screen would cause a considerable increase in solar UV-radiation reaching the Earth's surface. This could result in death for some organisms and would cause a sharp rise in the frequency of some diseases (for instance, carcinoma of the skin in man).

Thus it can be seen that nuclear war could result in a global ecological catastrophe comparable with ecological crises of the past. However, unlike past crises, the consequences of which were restricted to the extinction of many animals, a nuclear catastrophe could, in addition to mass extinction of organisms, completely crush the hopes for the creation of the noosphere, and with this the chances of prolonged existence for the biosphere. Since a nuclear war would destroy the major centres of modern civilization, it is unlikely that the small part of mankind surviving the catastrophe would be able to recover the present-day level of science and technology, a level which is necessary in order to gain control over (and so preserve) the global ecological system. In addition, there could well be a loss of incentive for renewed scientific and technological development after the catastrophic consequences of nuclear war.

The need to prevent anthropogenic ecological crises is not restricted to the possibility of premeditated onset of nuclear war. The very fact of preparation for such a war brings severe trials to the peoples of the world. There is real danger of occasional inadvertent uses of modern weapons which could escalate to an unlimited nuclear conflict. Furthermore, the high level of defense budgets in developed countries reduces the funds available for work directed towards preventing other global ecological–economic crises, such as those associated with population growth, the exhaustion of mineral resources, evironmental pollution, and global climatic change. A drastic decrease in the amount of weapons is required if we are to create a controlled global ecological system and so ensure the prolonged existence of the biosphere.

The Conference of scientists for deliverance of mankind from a threat of nuclear war, for disarmament and peace was held in Moscow in May 1983. At this Conference a resolution was adopted that says: "We are firmly convinced that to maintain peace at present is the primary and imperative need of mankind. Today there is no problem more important and urgent for the peoples and governments than removing the threat of nuclear death . . .

"We believe that mankind will actually be delivered from the burden of weapons. We are convinced that the cessation of senseless expenditure of energy and resources, and the transference of this huge industrial and scientific and technological potential to peace industries will offer wide possibilities for producing material wealth and raising people's standard of life. We clearly understand what beneficial results would be brought through disarmament and improvement of international relations that will save people from distrust and fears, through development of fruitful scientific cooperation, and joint solutions to global problems of the present . . ." (*The Appeal . . .*, 1983).

It is beyond any doubt that the question of removing the possibility of nuclear conflict is the central problem at the present time.

# CONCLUSION

The information given in this book allows us to draw the following principal conclusions.

The examination of the energy regime of the biosphere, and of the cycles of different kinds of matter within it, allows us to understand the biosphere's history better, to properly evaluate its present state, and to clarify the prospects for the future development of the biosphere.

In studying the evolution of the biosphere we have shown that the life zone (i.e. the range of physical and chemical conditions under which the existence of organisms is possible) is very narrow compared to the variability of these conditions on celestial bodies. This zone is even narrower for individual species of organisms or their higher taxonomic groups. Since the physical and chemical conditions in the external envelopes of the Earth are constantly varying as a result of changes in the solar radiation, fluctuations in tectonic processes (particularly degassing of the Earth's mantle) and bombardments of the Earth's surface by celestial bodies, these changes must be an important factor in the evolution of organisms. Some of these changes have led to the destruction of numerous species of living things and threatened the existence of the biosphere.

Since the external factors which influence the evolution of the biosphere act independently of the existence of life on the Earth, the preservation of the biosphere for billions of years and, moreover, its long progressive development seem to be events whose probability is very low. It follows from this that the possibility of discovering extraterrestrial life (and particularly extraterrestrial civilizations) is practically absent.

The appearance of modern man was the most important mark in the biosphere's evolution. It has already been shown today that man's economic activities are a decisive factor in causing not only diverse local changes in the biosphere, but also global changes manifested by changes in the atmospheric chemical composition and climate.

Man influences the biosphere in many ways. In some cases, man's economic activities have led to local ecological crises which could possibly develop into global crises. The most grave of future possibilities is that of an ecological crisis produced by nuclear war, an event which might threaten the further existence of mankind and the biosphere.

On the other hand, there are numerous examples showing that directed human activity can eliminate ecological crises. This gives hope for the possibility of developing a controlled global ecological system, immune to both anthropogenic ecological crises and crises evoked by natural factors. In this case the prolonged existence of the biosphere will be ensured.

Because the probability that extraterrestrial intelligent life exists elsewhere in the universe is so small, we must recognize our high responsibility to preserve and further develop the Earth's civilization, so that it will lead to the formation of the noosphere. It is probable that the next few decades will be a decisive period, a time during which it will become clear whether or not mankind is capable of maintaining and distributing intelligent life in the universe.

# REFERENCES


Adhémar, J. F.: 1842, *Les révolutions de la mer*, Déluges périodiques, Paris, 184 pp.

Aizenshtat, B. A.: 1965, 'A Method for Calculation and the Results of Determining Certain Bioclimatic Characteristics', *Proc. SANIGMI* **22**(37), 3—41 (R).

Allen, C. W.: 1958, 'Solar Radiation', *Quart. J. Roy. Met. Soc.* **84**(362), 307—318.

Alpatiev, A. M.: 1950, 'Rational Use of Precipitation Makes up a Basis for Overcoming Droughts', in *Agroclimatic Conditions of the Ukrainian Steppe and the Ways Making Them Better*, Izd. AN Ukr. S.S.R., Kiev, pp. 75—93 (R).

Alpatiev, A. M.: 1954, *Moisture Cycle in Cultivated Plants*, Gidrometeoizdat, Leningrad, 248 pp. (R).

Alt, E.: 1929, 'Der Stand des meteoroligischen Strahlungsproblems', *Met. Z.* **46**(10), 504—513.

Alvarez, L. W. *et al.*: 1980, 'Extraterrestrial Cause for the Cretaceous — Tertiary Extinctions: Experiment and Theory', *Science* **208**, 1095—1108.

Alvarez, W. *et al.*: 1982, 'Current Status of the Impact Theory for the Terminal Cretaceous Extinction', in *Geological Implications of Impacts of Large Asteroids and Comets on the Earth*, The Geological Society of America, Special Paper No. 190, pp. 305—315.

Andrews, P.: 1982, 'Hominoid Evolution', *Nature* **295**, 185—186.

Angell, J. K. and Korshover, J.: 1977, 'Estimate of the Global Change in Temperature, Surface to 100 mb, between 1958 and 1975', *Monthly Weather Rev.* **105**(4), 375—385.

Antonov, V. S.: 1968, 'A Possible Cause for the Pulsations of Water Exchange between the Arctice and Atlantic Oceans', *Problemy Arktiki i Antarktiki* **29**, 13—18 (R).

Armand, D. L.: 1966, *To Ourselves and Our Grandchildren* (2nd edn), Mysl', Moscow, 254 pp. (R).

Armand, D. L.: 1967, 'Some Problems and Methods of the Physics of Landscapes', in *Geophysics of the Landscape*, Nauka, Moscow, pp. 7—24 (R).

Arrhenius, S.: 1896, 'On the Influence of the Carbonic Acid in the Air upon the Temperature of Ground', *Philos. Mag.* **41**, 237—275.

Arrhenius, S.: 1903, *Lehrbuch der Kosmischen Physik*, Hirzal, Leipzig, 1026 S.

Arrhenius, S.: 1908, *Das Werden den Welten*, Leipzig, 108 S.

Asaro, F., *et al.*: 1982, Geochemical Anomalies near the Eocene-Oligocene and Permian-Triassic Boundaries, in *Geological Implications of Impacts of Large Asteroids and Comets on the Earth*, The Geological Society of America. Special Paper No. 190, pp. 517—528.

*Atlas of the Heat Balance*. 1955 (Ed. M. I. Budyko), Izd. GGO, Leningrad, 41 pp. (R).

*Atlas of the Heat Balance of the Earth*. 1963 (Ed. M. I. Budyko), Moscow, 63 pp. (R).

*Atlas of the Oceans. 1974—1980.* Moscow. Vol. 1: The Pacific Ocean, 302 pp; vol. 2: The Atlantic and Indian Oceans, 306 pp; vol. 3: The Arctic Ocean, 184 pp. (R).

*Atlas of the World Water Balance*. 1974. Gidro-meteoizdat, Moscow-Leningrad, 69 pp. (R).

Augustsson, T. and Ramenathan, V.: 1977, 'A Radiative-Convective Model Study of the $CO_2$-Climate Problem', *J. Atm. Sci.* **34**, 448—451.

Axelrod, D. I. and Baily, H. P.: 1969, 'Paleotemperature Analysis of Tertiary Floras', *Palaeogeogr., Palaeoclimatol., Palaeoecol.* **6**, 163–195.

Baes, C. F., *et al.*: 1976, *The Global Carbon Dioxide Problem* (C. F. Baes, H. E. Goeller, J. S. Olson *et al.*), Oak Ridge National Lab., 78 pp.

Bakker, R. T.: 1971, 'Dinosaurs Physiology and the Origin of Mammals', *Evolution* **25**(4), 636—658.

Bakker, R. T.: 1974, 'Dinosaurs Bioenergetics', *Evolution* **28**(3), 497—503.

Baldwin, B., *et al.*: 1976, 'Stratospheric Aerosols and Climatic Change' (B. Baldwin, J. B. Pollack, A. Sommers, S. B. Toon), *Nature* **263**, 551—555.

Baur, F. and Philipps, H.: 1934, 'Der Wärmehaushalt der Lufthülle der Nordhalbkugel', *Gerlands Beitr. Z. Geophys.* **42**, 160—207.

Bazilevich, N. I. and Rodin, L. Ye.: 1967, 'Schematic Maps of Productivity and Biological Circulation of Basic Types of Land Vegetation on the Earth', *Bull. Geogr. Soc. USSR* **3**, 190—194 (R).

Bazilevich, N. I. and Rodin, L. Ye.: 1969, 'Geographic Laws of Productivity and Circulation of Chemical Elements in the Earth's Basic Types of Vegetation', in, *General Theoretical Problems of Biological Productivity*, Nauka, Leningrad, pp. 24—32 (R)

Bazilevich, N. I., Rodin, L. Ye, and Rozov, N. N.: 1970, *Geographical Aspects of Studying Biological Productivity*, Izd. Geogr. Obshch. S.S.R., Leningrad, 28 pp. (R).

Bazilevich, N. I., Rodin, L. Ye., and Rozov, N. N.: 1971, 'How Much Does the Living Matter of the Planet Weigh?', *Priroda* **1**, 46—53 (R).

Bedgley, F. I.: 1966, 'Heat Budget at the Surface of the Arctic Ocean', *Proc. Symp. on the Arctic Heat Budget and Atmospheric Circulation*, Rand Corporation, Santa Monica, California, pp. 267—278.

Beeva, I. M. and Vinnikov, K. Ya.: 1981, 'On the Dependence of the Out-Going Longwave Radiation of the Earth—Atmosphere System on Temperature and Cloudiness', *Proc. State Hydrol. Inst.* **271**, 77—82 (R).

Berger, A. L.: 1973, ***Théorie astronomique des paléoclimats***, Lou-vain, vol. 1, 197 pp.; vol. 2, 108 pp.

Berger, A. L.: 1975, 'La paléoclimatologie quantitative du Quaternaire. P. 1—2', *Revue des Questions Scientifiques* **146**(2), 167—179; (3), 278—293.

Berger, A. L: 1976, 'Obliquity and Precession for the Last 5,000,000 Years', *Astron. Astrophys.* **51**, 127—135.

Berger, A. L.: 1978, 'Théorie astronomique des paléoclimats, une nouvelle approche', ***Bull. Soc. Belge Géologie* 87,** 9–25.

**21**, 227—235.

Berger, A. L.: 1978, 'Théorie astronomique des paléoclimat, une nouvelle approche', *Bull. Soc. Belge Géologie* **87**, 9—25.

Berkner, L. V. and Marshall, L. C.: 1965a, 'On Origin and Rise of Oxygen Concentration in the Earth's Atmosphere', *J. Atm. Sci.* **22**(3), 225—261.

Berkner, L. V. and Marshall, L. C.: 1965b, 'Oxygen and Evolution', *New Scient.* **28**(469), 415—419.

Berkner, L. V. and Marshall, L. C.: 1966, 'Limitation on Oxygen Concentration in a Primitive Planetary Atmosphere', *J. Atm. Sci.* **23**(2), 133—143.

Bernal, J. D.: 1951, *The Physical Basis of Life*, Routledge and Kegan Paul, London, 80 pp.

Bibikov, S. N.: 1969, 'Some Aspects of Paleoeconomic Modelling of the Paleolithic', *Sovietskaya Archeologiya* **4**, 5—22 (R).

Blinova, Ye. N.: 1947, 'On the Problem of the Average Annual Distribution of Temperature in the Earth's Atmosphere With Due Account of the Continents and Oceans', *Bull. Acad. Sci. USSR, Ser.: Geogr. Geophys.* **11**(1), 3—13 (R).

Blokhin, N. N.: 1983, 'Reports of the Participants of the Conference', *Inf. Bull. Acad. Sci. USSR* **9**, 32—37 (R).

Bolin, B.: 1977, 'Changes in Land Biota and its Importance for the Carbon Cycle', *Science* **196**, 613.

Bolin, B. and Keeling, C. D.: 1963, 'Large-Scale Atmospheric Mixing as Deduced from the Seasonal and Meridional Variations of Carbon Dioxide', *J. Geophys. Res.* **68**(13), 3899—3920.

Borzenkova, I. I.: 1980, 'Fluctuations in the Moisture Conditions of the Sahara Desert and the Adjacent Regions for the Last Twenty Thousand Years', *Bull. Acad. Sci. USSR, Ser.: Geogr.* 3, 36—45 (R).

Borzenkova, I. I. *et al.*: 1976, 'Changes in the Air Temperature of the Northern Hemisphere over the 1881—1975 Period', *Meteorol. Gidrol.* 7, 27—35 (R).

Bossolasco, M. *et al.*: 1964, 'Global-strahlung und Sonnenaktivität (M. Bossolasco, C. Ciconi, J. Dagnino *et al.*), *Pure and Applied Geophys.* **57** 221—224.

Braidwood, R. I. and Reed, C. A.: 1957, 'The Achievement and Early Consequences of Food Production', *Cold Spring Harbor Symp. Quant. Biol.*, vol. 22, Population Studies, Animal Ecology and Demography, NY, pp. 19—31.

Brinkman, R. A.: 1969, 'Dissociation of Water Vapour and Evolution of Oxygen in the Terrestrial Atmosphere', *J. Geophys. Res.* **74**(23), 5355—5368.

Brooks, C. E. P.: 1950, *Climate Through the Ages* (2nd edn), Benn, London, 395 pp.

Brückner, E.: 1901, 'Die Herkunft der Regens', *Gäa*, Natur und Leben, Leipzig und Cologne, Bd 36.

Bryan, K. *et al.*: 1982, 'Transient Climate Response to Increasing Atmospheric Carbon Dioxide', *Science* **215**, 56—58.

Bryson, R. A. and Goodman, B. M.: 1980, 'Volcanic Activity and Climatic Change', *Science* **207**, 1041—1044.

Buchardt, B.: 1978, 'Oxygen Isotope Paleotemperatures' from the Tertiary Period in the North Sea Areas', *Nature* **275**, 121—123.

Budyko, M. I.: 1948, *Evaporation Under Natural Conditions*, Gidrometeoizdat, Leningrad, 136 pp. (R); OTS 63—11061, Isr. Program Sci. Transl., Jerusalem, 1963; MGA 16.2—17.

Budyko, M. I.: 1949, 'On the Theory of Intensity of Physico-Geographical Process', *Voprosy Geog.* **15**, 25—45 (R).

Budyko, M. I.: 1950, 'On the Theory of Hydro-Meteorological Effectiveness of Shelterbelts', in *Problems of Hydro-Meteorological Effectiveness of Shelter Belts*, Gidrometeoizdat, Leningrad, pp. 69—84 (R).

Budyko, M. I.: 1951, 'On the Influence of Reclamation Practices on Potential Evaporation', *Bull. Acad. Sci. USSR, Ser.: Geogr.* **1**, 16—35; **2**, 57—61 (R).

Budyko, M. I.: 1956, *Heat Balance of the Earth's Surface*, Gidrometeoizdat, Leningrad, 255 pp. (R).

Budyko, M. I.: 1958, Transl. by N. A. Stepanova, MGA BE—286, V. S. Weather Bureau, Washington, D.C., 11B—25.

Budyko M. I.: 1958, 'Temperature of the Active Surface and Its Bioclimatic Significance', in *Current Problems of Meteorology of the Surface Air Layer*, Gidrometeoizdat, Leningrad, pp. 201—211 (R).

Budyko, M. I.: 1959, 'On the Heat Balance of Living Things', *Bull. Acad Sci. USSR, Ser.: Geogr.* **1**, 29—35 (R).

Budyko, M. I.: 1961, 'On the Thermal Zonation of the Earth', *Meteorol. Gidrol.* **11**, 7—14 (R).

Budyko, M. I.: 1962a, 'Some Ways of Influencing the Climate', *Meteorol. Gidrol.* **2**, 3—8 (R).

Budyko, M. I.: 1962b, 'Polar Ice and Climate', *Bull. Acad. Sci. USSR, Ser.: Geogr.* **6**, 3—10 (R).

Budyko, M. I.: 1964, 'On the Theory of the Influence of Climatic Factors on Photosynthesis', *Proc. Acad Sci. USSR* **158**(2), 331—337 (R).

Budyko, M. I.: 1966, 'The Possibilities of Changing Climate by Forcing Polar Ice', in *Current Problems of Climatology*, Gidrometeoizdat, Leningrad, pp. 347—357 (R).

Budyko, M. I.: 1967, 'On the Causes of the Extinction of Animals at the End of the Pleistocene', *Bull. Acad. Sci. USSR, Ser.: Geogr.* **2**, 28—36 (R).

Budyko, M. I.: 1968, 'On the Origin of Ice Ages', *Meteorol. Gidrol.* **11**, 3—12 (R).

Budyko, M. I.: 1969, *Climatic Changes*, Gidrometeoizdat, Leningrad, 35 pp.

Budyko, M. I.: 1971, *Climate and Life*. Gidrometeoizdat, Leningrad, 472 pp. (R); 1974, English edition, ed. D. H. Miller, Academic Press, New York.

Budyko, M. I.: 1972, *Man's Impact on Climate*, Gidrometeoizdat, Leningrad, 47 pp. (R).

Budyko, M. I.: 1974, *Climatic Change*, Gidrometeoizdat, Leningrad, 280 pp. (R); 1977, English Translation: American Geophysical Union, Washington, D.C.

Budyko, M. I.: 1975. 'The Dependence of Mean Air Temperature on Changes in Solar Radiation', *Meteorol. Gidrol.* **10**, 3—10 (R).

Budyko, M. I.: 1977a, *Global Ecology*, Mysl', Moscow, 328 pp. (R); 1980, English Translation: Progress Publishers, Moscow, 323 pp.

Budyko, M. I.: 1977b, 'Studies of Modern Climatic Change', *Meteorol. Gidrol.* **11**, 42—57 (R).

Budyko, M. I.: 1977c, *Present-Day Climate Change*, Gidrometeoizdat, Leningrad, 47 pp. (R).

Budyko, M. I.: 1978, 'The Thermal Regime of Dinosaurs', *Zhurnal obshchey biologii* **2**, 179—188 (R).

Budyko, M. I.: 1979, 'Semiempirical Model of Atmosphere's Thermal Regime and Real Climate', *Meteorol. Gidrol.* **4**, 5—17 (R).

Budyko, M. I.: 1980, *The Earth's Climate: Past and Future*, Gidrometeoizdat, Leningrad, 352 pp. (R); 1982, English edition: Academic Press, New York, 307 pp.

Budyko, M. I.: 1981, 'Changes in the Thermal Regime of the Atmosphere in the Phanerozoic', *Meteorol. Gidrol.* **10**, 5—10 (R).

Budyko, M. I.: 1982, *Changes in the Environment and in the Successive Faunas*, Gidrometeoizdat, Leningrad, 77 pp. (R).

Budyko, M. I. and Drozdov, O. A.: 1950, 'On Water Cycle on a Limited Land Territory', in *Problems of Hydro-Meteorological Effectiveness of Shelter Belts*, Gidrometeoizdat, Leningrad, pp. 14—23.

Budyko, M. I. and Drozdov, O. A.: 1953, 'The Laws of Water Cycle in the Atmosphere', *Bull. Acad. Sci. USSR, Ser.: Geogr.* **4**, 167—170 (R).

Budyko, M. I., Kotlyakov, V. M., and Meshcheryakov, Yu. A.: 1970, *On the Application of Quantitative Methods in Physico-Geographical Studies*, Izd. Geogr. Obsheh. SSSR, Leningrad, 17 pp. (R).

Budyko, M. I. and Pivovarova, Z. I.: 1967, 'The Effects of Volcanic Eruptions on the Solar Radiation Incident on the Earth's Surface', *Meteorol. Gidrol.* **10**, 3—7 (R).

Budyko, M. I. and Ronov, A. B.: 1979, 'The Evolution of the Atmosphere in the Phanerozoic', *Geochimiya* **5**, 643—653 (R).

Budyko, M. I. and Tsitsenko, G. V.: 1960, 'Climatic Factors of Man's Heat Regime', *Bull. Acad. Sci. USSR, Ser.: Geogr.* **3**, 3—11 (R).

Budyko, M. I. and Vasishcheva, M. A.: 1971, 'The Influence of Astronomical Factors on Quaternary Glaciations', *Meteorol. Gidrol.* **6**, 37—47 (R).

Budyko, M. I. and Vinnikov, K. Ya.: 1973, 'Present-Day Changes in Climate', *Meteorol. Gidrol.* **9**, 3—13 (R).

Budyko, M. I. and Vinnikov, K. Ya.: 1976, 'Global Warming', *Meteorol. Gidrol.* **7**, 16—26 (R).

Budyko, M. I. and Vinnikov, K. Ya.: 1983, 'The Problem of Revealing Anthropogenic Changes in Global Climate', *Meteorol. Gidrol.* **9**, 14—26 (R).

Budyko, M. I., Vinnikov, K. Ya., and Yefimova, N. A.: 1983, 'The Dependence of Air Temperature and Precipitation of Carbon Dioxide Content of the Atmosphere', *Meteorol. Gidrol.* **4**, 5—13 (R).

Budyko, M. I. and Yefimova, N. A.: 1968, 'The Use of Solar Energy by Natural Plant Cover on the USSR Territory', *Botanichesky Zhurnal* **23** (10), 1384—1389 (R).

Budyko, M. I. and Yefimova, N. A.: 1981, 'The Effects of Carbon Dioxide on the Climate', *Meteorol. Gidrol.* **2**, 5—17 (R).

Budyko, M. I. and Zubenok, L. I.: 1961, 'The Determination of Evaporation from the Surface of Land', *Bull. Acad. Sci. USSR, Ser.: Geogr.* **6**, 3—17 (R).

Budyko, M. I., *et al.*: 1978, 'Impendent Changes in Climate', *Bull. Acad. Sci. USSR, Ser.: Geogr.* **6**, 5—10 (R).

Budyko, M. I., *et al.*: 1981 'Anthropogenic Climatic Change', *Meteorol. Gidrol.* **8**, 5—14 (R).

Burton, A. C. and Edholm, O. G.: 1955, *Man in a Cold Environment*, Arnold, London, 273 pp.

Butzer, K. W.: 1964, *Environment and Archeology*, Aldine, Chicago, 524 pp.

Büttner, K.: 1938, *Physikalische Bioklimatologie*, Akad. Verl.-Ges., Leipzig, 155 S.

Büttner, K.: 1951, 'Physical Aspects of Human Bioclimatology', *Compend: Meteorol*, (Ed. T. Malone), Amer. Meteorol. Soc., Boston, pp. 1112—1125.

Byutner, E. K.: 1961, 'On the Time of Fixation of a Stable Amount of Oxygen in the Atmospheres of the Planets Containing Water Vapour', *Rep. Acad Sci. USSR* **138**, 1050—1052 (R).

Byutner, E. K.: 1983, 'The Oceanic Response to Changes in Greenhouse Effect of the Atmosphere', *Bull. Acad. Sci. USSR, Phys. Atmos. Ocea.* **8**, 892—895 (R).

Callendar, G. S.: 1938, 'The Artificial Production of Carbon Dioxide and Its Influence on Temperature', *Quart. J. Roy. Met. Soc.* **64**(27), 223—240.

*Carbon Dioxide and Climate: A Scientific Assessment*, 1979, Nat. Acad. Sci., Washington, D.C., 22 pp.

*Carbon Dioxide and Climate: A Second Assessment*, 1982, Nat. Acad. Sci., Washington, D.C., 72 pp.

Cess, R. D.: 1976, 'Climate Change: An Appraisal of Atmospheric Feedback Mechanisms Employing Zonal Climatology', *J. Atm. Sci.* **33**(10), 1831—1843.

Cess, R. D.: 1978, 'Biosphere-Albedo Feedback and Climate Modeling', *J. Atm. Sci.* **35**, 1765—1768.

Cess, R. D. and Wronka, J. C.: 1979, 'Ice Ages and the Milankovich Theory: A Study of Interactive Climate Feedback Mechanisms', *Tellus* **31**(3), 185—192.

Chamberlin, T. C.: 1897, 'A Group of Hypotheses Bearing on Climatic Changes', *J. Geology* **5**, 653—683.

Chamberlin, T. C.: 1898, 'The Influence of Great Epochs of Limestone Formation upon the Consitution of the Atmosphere', *J. Geology* **6**, 609—621.

Chamberlin, T. C.: 1899, 'An Attempt to Frame a Working Hypothesis of the Cause of Glacial Periods on an Atmospheric Basis', *J. Geology* **7**, 545—584.

CIAP Monograph 4, 1975, *The Natural and Radiatively Perturbed Troposphere*. U.S. Dep, Transport, 1352 pp.

CLIMAP Project Members, 1976, 'The Surface of Ice-Age Earth', *Science* **191**, 1131—1137.

*Climate monitor*, 1981, vol. 10, (5), 113—136.

'Climatic Effects of Increased Atmospheric Carbon Dioxide' (Proc. Soviet-American Meeting on Studying Climatic Effects of Increased Atmospheric Carbon Dioxide, Leningrad, 15—20 June 1981). 1982 Gidrometeoizdat, Leningrad, 56 pp. (R).

Cloud, P. E.: 1974, 'Atmosphere, Development of', *Encyclopaedia Britannica*, 15th ed., pp. 313—319.

Colbert, E. H.: 1958, *Evolution of the Vertebrates*, John Willey and Sons, N. Y.-L., 479 pp.

Colbert, E. H.: 1965, *The Age of Reptiles* Weidenfeld and Nicolson, London, 228 pp.

Colbert, E. H., Cowels, R. B., and Bogart, C. M.: 1964 'Temperature Tolerances in the American Alligator and Their Bearing on the Habits, Evolution and Extinction of the Dinosaurs', *Bull. Amer. Mus. Nat. History* **26**, (No. 7).

Commoner, B.: 1971, *The Closing Circle*, Knopf, N.Y., 326 pp.

*Communication with Extraterrestrial Intelligence*, 1973, (Ed. C. Sagan), MIT Press, Cambridge, Mass. and London, 428 pp.

Cronin, J. E. *et al.*: 1981, 'Tempo and Mode in Homind Evolution', *Nature* **292**, 113–122.

Crutzen, P. J. and Birks, J. W.: 1982, 'The Atmosphere After a Nuclear War', *Ambio* **11**, 115—125.

Cuvier, G.: 1812, *Discours sur les révolutions de la surface de globe*.

Darwin, C.: 1859, *The Origin of Species by Means of Natural Selection*, Murray, London, 426 pp.

Darwin, C.: 1871, *The Descent of Man and Selection in Relation to Sex*, Murray, London, 688 pp.
Davidson, J. L. and Philip, J. R.: 1958, 'Light and Pasture Growth. Arid Zone Research. Climatology and Microclimatology', *Proc. of the Canberra Symp., UNESCO*, pp. 181—187.
Davitashvili, L. Sh.: 1969, *The Causes for the Extinction of Organisms*, Nauka, Moscow, 440 pp. (R).
Desmond, A. J.: 1976, *The Hot-Blooded Dinosaurs. A Revolution in Paleontology*, The Dial Press, New York, 238 pp.
De Wit, C. T.: 1965, 'Photosynthesis of Leaf Canopies', *Agric. Res. Rep.*, No. 663, Wageningen, 57 pp.
Dietz, R. S.: 1961, 'Continent and Ocean Basin Evolution by Spreading of the Sea Floor', *Nature* **190**(4779), 854—857.
Dines, W. H.: 1917, 'The Heat Balance of the Atmosphere', *Quart. J. Roy. Met. Soc.* **43**(182), 151—158.
Dobrodeev, O. P. and Suyetova, I. A.: 1976, *The Living Matter of the Earth*, Izd. MGU, Moscow, pp. 26—58 (R).
Dobzhansky, T.: 1951, *Genetics and the Origin of Species* (3rd edn), Columbia Univ. Press, New York, 364 pp.
Dokuchaev, V. V.: 1948, *On Natural Zones*, Moscow, Geografgiz, 63 p. (R).
Donn, W. L. and Shaw, D. M.: 1966, 'The Heat Budget of an Ice-Free and Ice-Covered Arctic Ocean', *J. Geophys. Res.* **71**(4), 1087—1093.
Dorodnitsyn, A. A., Izvekov, B. I., and Shvets, M. Ye: 1939, 'A Mathematical Theory of General Circulation', *Meteorol. Gidrol.* **4**, 32—42 (R).
Doronin, Yu. P.: 1963, 'On the Heat Balance of Central Arctic', *Proc. Arct. Antarct. Inst.* **253**, 178—184 (R).
Dorst, J.: 1965, *Avant que nature meure*, Switzerland.
Drozdov, A. V.: 1969, 'The Productivity of Zonal Land Plant Communities and Indices of Hydrothermal Regime of the Territory', in *General Theoretical Problems of Biological Productivity*, Leningrad, Nauka, pp. 33—38 (R).
Drozdov, O. A. and Grigorieva, A. S.: 1963, *The Water Cycle in the Atmosphere*, Gidrometeoizdat, Leningrad, 314 pp. (R).
Drozdov, O. A. and Grigorieva, A. S.: 1971, *Long-Term Cycle Fluctuations in Atmospheric Precipitation over the USSR Territory*, Gidrometeoizdat, Leningrad, 153 pp. (R).
Duncan, W. G., *et al.*: 1967, 'A Model for Simulating Photosynthesis in Plant Communities', *Hilgardia* **38**(4), Berkely, pp. 181—205.
Duvigneaud, P. and Tanghe, M.: 1967, *Ecosystèmes et Biosphère*, Ministère de l'Eduction Nationale et de la Culture, Bruxelles.
Dymnikov, V. P., *et al.*: 1980, 'The Study of Climate Sensitivity to $CO_2$ Doubling by Means of Zonally Averaged Model of Atmospheric General Circulation', in *Mathematical Modelling of the Dynamics of the Atmosphere and the Ocean*, 2, Novosibirsk, pp. 39—50 (R).
Eberhart, J. and Thomsen, D. E.: 1983, 'The Search for Extraterrestrial Intelligence', *Science News* **123**, 149.
*Ecological Physiology of Animals. P. 3*, 1982, Nauka, Leningrad, 504 pp. (R).
Eldredge, N. and Gould, S. J.: 1972, 'Punctuated Equilibria: An Alternative to Phyletic Gradualism', in Schopf, T. J. M. (ed.), *Models in Paleobiology*, Freeman Cooper, San Francisco, pp. 82—115.
Ellis, J. S. and Vonder Haar, T. H.: 1976, 'Zonal Average Earth Radiation Budget Measurement from Satellite for Climate Studies', Atm. Sci. Paper, 240, Colorado St. Univ., Fort Collins, Colo., 50 pp.
Elton, C.: 1930, *Anomal Ecology and Evolution*, Clarendon Press, Oxford, H. Milford, London, 96 pp.

*Energy and Climate*, 1977, Studies in Geophysics, Nat. Acad. Sci., Washington, 158 pp.
Fairbridge, R. E.: 1967, 'Ice-Age Theory', *The Encyclopaedia of Atmospheric Science and Astrogeology*. Reinhold Publ. Corp., New York, Amsterdam, London, pp. 462—474
Farlow, J. O.: 1976, 'A Consideration of the Trophic Dynamics of a Late Cretaceous Large-Dinosaur Community', *Ecology* **57**, 841—857.
Feduccia, A.: 1973, 'Dinosaurs as Reptiles', *Evolution* **27**(1), 160—169.
Fifield, R.: 1983. 'Mere Hiccups in the History of Time', *New Scient.* **98**, 704—706.
Fletcher, J. O.: 1966, 'The Arctic Heat Budget and Atmospheric Circulation', *Proc. of the Symposium on the Arctic Heat Budget and Atmospheric Circulation* Rand Corporation. Santa Monica, Calif., pp. 23—43.
Flint, R. F.: 1957, *Glacial and Pleistocene Geology*, J. Wiley, New York; Chapman and Hall, London, 571 pp.
Flohn, H.: 1977, 'Climate and Energy. A Scenario to a 21st Century Problem', *Climatic Change* **1**(1), 5—20.
Flohn, H.: 1981, *Life on a Warmer Earth*, IIASA, 75 pp.
Forrester, J. W.: 1971, *World Dynamics*, Wright-Allen Press, Cambridge, Mass., 142 pp.
Frakes, L. A.: 1979, *Climates Throughout Geologic Time*, Elsevier, Amsterdam; Oxford, New York, 310 pp.
Frolov, I. T.: 1979, *The Perspectives of Man*, Politizdat, Moscow, 336 pp. (R).
*Fundamentals of Forest Biogeocenology*, 1964 (Eds. V. N. Sukachev, N. V. Dylis), Nauka, Moscow, 574 pp. (R).
Fyodorov, Ye. K.: 1972, *Interactions of the Society and Nature*, Gidrometeoizdat, Leningrad, 88 pp. (R).
Fyodorov, Ye. K.: 1974, 'Concluding Remarks', in B. Commoner: *Closing Circle*, Gidrometeoizdat, Leningrad, pp. 248—278 (R).
Fyodorov, Ye. K.: 1977, *Ecological Crisis and Social Progress*, Gidrometeoizdat, Leningrad, 176 pp. (R).
Fyodorov, Ye. K.: 1979, 'Climatic Change and Strategy of Mankind', *Meteorol. Gidrol.* **7**, 12—24 (R).
Ganapathy, R.: 1980, 'A Major Meteorite Impact on the Earth 65 Million Years Ago: Evidence from Cretaceous-Tertiary Boundary Clay', *Science* **209**, 921—923.
Ganapathy, R.: 1982, 'Evidence for a Major Meteorite Impact on the Earth 34 Million Years Ago: Implication on the Origin of North American Tektites and Eocene Extinction', in *Geological Implications of Impacts of Large Asteroids and Comets on the Earth*, The Geological Society of America, Special Paper No. 190, pp. 513—516.
Garrels, R. M.: 1975, *The Circulation of Carbon, Oxygen and Sulphur Throughout Geologic Time*, Nauka, Moscow, 47 pp. (R).
Garrels, R. M. and Mackenzie, F. T.: 1971, *Evolution of Sedimentary Rocks*, Norton, New York, 397 pp.
Gates, W. L.: 1976, 'Modeling the Ice-Age Climate', *Science* **191**, 1138—1144.
Gause, G.: 1934, *The Struggle for Existence*, Williams and Wilkins, Baltimore, 163 pp.
Geoffroy, St.-Hilaire: 1833, 'Le degré d'influence du monde ambiant pour modifier les formes animales', *Mem. Acad. Sci., Paris* **12**, 64—93.
*Geology of Astroblems*, 1980, Nedra, Leningrad, 231 pp. (R).
George, W.: 1964, *Biologist Philosopher. A Study of the Life and Writings of Alfred Russel Wallace*, Abelard-Schuman, London, 320 pp.
Gerasimov, I. P.: 1945, 'The World Map of Soils and General Laws of the Geography of Soils', *Pochvovedeniye* **3—4**, 152—161 (R).
Gerasimov, I. P.: 1960, 'Hydrothermal Factors of Soil Formation', in *The Heat and Water Regimes of the Earth's Surface*, Gidrometeoizdat, Leningrad, pp. 144–161 (R).
Gerasimov, I. P.: 1970, 'Nature and the Development of Primitive Society', *Bull. Acad. Sci. USSR, Ser.: Geogr.* **1**, 5—8 (R).

Gerasimov, I. P. (ed.): 1971, *The Resources of the Biosphere on the USSR Territory*, Nauka, Moscow, 295 pp (R).
Gerasimov, I. P.: 1979, 'Climates of the Past Geological Epochs', *Meteorol. Gidrol.* 7, 37—53 (R).
Gerasimov, I. P. and Markov, K. K.: 1939, *Ice Ages on the Territory of the USSR*, Izd. Ak. Nauk SSSR, Moscow, Leningrad, 462 pp. (R).
Gershanovich, D. Ye. and Muromtsev, A. M.: 1982, *Oceanological Basis for Biological Productivity of the World Ocean*, Gidrometeoizdat, Leningrad, 320 pp. (R).
Gilyarov, M. S.: 1966, 'The Evolution of Insects in the Transition to Passive Distribution and the Principle of Feedback in Phylogenetic Development', *Zool. Zh.* **43**(1). 3—23 (R).
Gilyarov, M. S.: 1975, 'The General Trends in the Evolution of Insects and Higher Vertebrates', *Zool. Zh.* **54**(6), 822—831 (R).
Golyitsyn, G. S.: 1974, 'Modern Concepts on the Nature of Dust Storms on Mars', *Inf. Bull. Acad. Sci. USSR* **1**, 24–31 (R).
Gould, S. I.: 1981, 'The Chance that Shapes our End', *New Scient.* **89**(1239), 347—350.
Gould, S. J. and Eldredge, N.: 1977, 'Punctuated Equilibria: the Tempo and Mode of Evolution Reconsidered', *Paleobiology* **3**, 115—151.
Greenberg, J.: 1982, 'Fossils Trigger Questions of Human Origins', *Science News* **121**, 84.
Gribbin, J.: 1982, 'Carbon Dioxide, Ammonia and Life', *New Scient.* **94**(1305), 413—416.
Grigoriev, A. A.: 1936, 'On Certain Interrelations Among Basic Elements of Physico-Geographical Environment and their Evolution', *Prob. fiz. geog.* **3**, 3—30 (R).
Grigoriev, A. A.: 1946, 'Some Results of the Development of New Ideas in Physical Geography', *Bull. Acad. Sci. USSR, Ser.: Geogr. Geophys.* **10**(2), 139—168 (R).
Grigoriev, A. A.: 1954, 'Geographical Zonality and Some of Its Regularities', *Bull. Acad. Sci. USSR, Ser.: Geogr.* **5**, 17—39; **6**, 41—50 (R).
Grigoriev, A. A.: 1956, *The Sub-Arctic*, Geografgiz, Moscow, 223 pp. (R).
Grigoriev, A. A.: 1966, *The Laws of the Structural Properties and the Development of Geographical Environment*, Mysl, Moscow, 382 pp. (R).
Grigoriev, A. A.: 1970, *The Types of Geographic Environment*, Moscow, Mysl', 472 pp. (R).
Grigoriev, A. A. and Budyko, M. I.: 1956, 'On the Periodic Law of Geographic Zonality', *Rep. Acad. Sci. USSR* **110**(1), 129—132 (R).
Grigoriev, A. A. and Budyko, M. I.: 1962, 'On Seasonal Variations in the Climatic Factors of Geographic Zonality', *Rep. Acad. Sci. USSR* **143**(2), 391—393 (R).
Grossvald, M. G. and Kotlyakov, V. M.: 1978, 'The Forthcoming Change in Climate and the Fate of Glaciers', *Bull. Acad. Sci. USSR, Ser.: Geogr.* **6**, 21—32 (R).
Gvishiani, D. M.: 1978, 'Methodological Problems of Modelling Global Development', *Voprosy filos.* **2**, 14—28 (R).
Hafele, W.: 1979, 'A Global and Long-Range Picture of Energy Developments', *Science* **209**, 174—182.
Hallam, A. (ed.): 1977, *Patterns of Evolution. As Illustrated by the Fossil Record*, Elsevier, Amsterdam, Oxford, New York, 591 pp.
Hameed, S., *et al.*: 1980, 'Response of the Global Climate to Changes in Atmospheric Chemical Composition Due to Fossil Fuel Burning', *J. Geophys. Res.* **85**, (C12), 7537—7545.
Hammer, C. U., Clausen, H. B., and Dansgaard, W.: 1980, Greenland Ice Sheet Evidence of Post-Glacial Volcanism and Its Climatic Impact', *Nature* **288**(5788), 230—235.
Hansen, J. E. *et al.*: 1978 'Mount Agung Eruption Provides Test of Global Climatic Perturbations' (J. E. Hansen, W. Ch. Wang, A. A. Lacis), *Science* **199**, 1065–1068.
Hansen, J. E., *et al.*: 1979, *Proposal for Research in Global Carbon Dioxide Source/Sink Budget and Climate Effects*, Goddard Inst. for Space Studies, New York, 60 pp.
Hansen, J. E., *et al.*: 1981, Climate Impact of Increasing Atmospheric Carbon Dioxide', *Science* **213**(4511), 957—966.

Hart, M. H.: 1975, 'An Explanation for the Absence of Extraterrestrials on Earth', *Quart. J. Roy. Astro. Soc.* **16**, 128—135.

Hart, M. H.: 1978, 'The Evolution of the Atmosphere of the Earth', *Icarus* **33**, 23—29.

Hart, M. H.: 1979, 'Habitable Zones about Main Sequence Stars', *Icarus,* **37**, 351—357.

Hays, J. D. *et al.*: 1976, 'Variations in the Earth's Orbit: Pacemaker of the Ice Ages', *Science* **194**, 1121—1132.

Herbert, W.: 1982a, 'Fossil Raises Question About Earliest Primates', *Science News* **121**, 372.

Herbert, W.: 1982b, 'Into the Pliocene Gap', *Science News* **121**, 389.

Herman, Y.: 1981, 'Causes of Massive Biotic Extinctions and Explosive Evolutionary Diversification throughout Phanerozoic Time', *Geology* **9**, 104—108.

Hess, H. H.: 1962, 'History of Ocean Basins', in *Petrologic Studies*, Geol. Soc. Amer., pp. 599—620.

Holdane, J.: 1929, *The Origin of Life*, Rationalist Annual.

Holland, H. D.: 1980, 'Atmosphere, Evolution of', *Encyclopaedia of Ocean and Atmospheric Sciences*, McGraw Hill, New York, pp. 33–45.

Holloway, J. L. and Manabe, S.: 1971, 'A Global General Circulation Model with Hydrology and Mountains', *Monthly Weather Rev.* **99**(5), 335—370.

Houghton, H. G.: 1954, 'On the Annual Heat Balance of the Northern Hemisphere', *J. Met.* **11**,(1), 1—9.

Hsü, K. J.: 1980, 'Terrestrial Catastrophe caused by Cometary Impact at the End of the Cretaceous', *Nature* **285**, 201—203.

Hsü, K. J. *et al.*: 1982, 'Mass Mortality and its Environmental and Evolutionary Consequences', *Science* **216**, 249—256.

Humphreys, W. J.: 1913 'Volcanic Dust and Other Factors in the Production of Climatic Changes and Their Possible Relation to Ice Age', *J. Franklin Inst.* **176**, 131 pp.

Humphreys, W. J.: 1929, *Physics of the Air* (2nd edn), McGraw-Hill, New York, 654 pp.

Huxley, J.: 1942, *Evolution. The Modern Synthesis* (3rd edn), Allen and Unwin, London, 1974, 750 pp.

Huxley, J.: 1962, 'Eastern Africa: the Ecological Base', *Endeavour* **21**(82), London, pp. 98—107.

Izrael, Yu. A.: 1974, *Peaceful Nuclear Explosions and the Environment*, Gidrometeoizdat, Leningrad, 135 pp. (R).

Izrael, Yu. A.: 1979a, 'Climate Monitoring and the Service of Collecting Climatic Data for the Determination of Climatic Changes and Fluctuations', *Meteorol. Gidrol.* **7**, 54—67 (R).

Izrael, Yu. A.: 1979b, *Ecology and Monitoring of the State of Natural Environment*, Gidrometeoizdat, Leningrad, 375 pp. (R).

Izrael, Yu. A., Petrov, V. N., and Severov, D. A.: 1983, 'On the Impact of Nuclear Detonations on the Ozone Content of the Atmosphere', *Meteorol. Gidrol.* **9**, 5—13 (R).

Jerison, H. J.: 1973, *Evolution of the Brain and Intelligence*, Academic Press, New York and London, 482 pp.

Jones, P. D., *et al.*: 1982, 'Variations in Surface Air Temperatures', *Monthly Weather Rev.* **110**, 59—70.

Kalesnik, S. V.: 1970, *General Geographical Regularities of the Earth*, Mysl', Moscow, 283 pp. (R).

Kamshilov, M. M.: 1979, *The Evolution of the Biosphere* (2nd edn), Nauka, Moscow, 256 pp. (R).

Kandror, I. S., *et al.*: 1966. 'An Experimental Test of Applying the Budyko-Tsitsenko Heat Balance Equation for Determining the Heat State of Man Under Real Conditions', *Gigiena i sanitariya* **3**, 51—57 (R).

Karol, I. L.: 1977, 'Variations in the Global Content of Stratospheric Aerosols and Their Relation to Fluctuations in the Mean Direct Solar Radiation and Temperature Near the Earth's Surface', *Meteorol. Gidrol.* **3**, 32—40 (R).

Karol, I. L. and Pivovarova, Z. I.: 1978, 'The Relationship Between Changes in the Stratospheric Aerosol Content and Fluctuations in the Solar Radiation', *Meteorol. Gidrol.* **69**, 35—42 (R).

Kashin, K. I. and Pogosyan, Kh. P.: 1960, 'On the Water Cycle in the Atmosphere', in *Problems of Hydro-Meteorological Effectiveness of Shelter Belts*, Gidrometeoizdat, Leningrad, 5—13 (R).

Keeling, C. D. and Bacastow, R. B.: 1977, 'Impact of Industrial Gases on Climate', in *Energy and Climate. Studies in Geophysics*, Nat. Acad. Sci., Washington, D.C., pp. 72—95.

Keller, H.: 1906, 'Niederschlag, Abfluss and Verdunstung in Mitteleuropa', *Besondere Mitteil. z. Jahrb. f. Gewässerkunder Norddeutschlands*. Bd 1, H. 4.

Kellogg, W. W.: 1977—1978, 'The Impact of Human Activities on Climate', *Bull. WMO*, 1977, **26**(4), 285—299; 1978, **27**(1), 3—12 (R).

Khromov, S. P.: 1973, 'Solar Cycles and Climate', *Meteorol. Gidrol.* **9**, 93—111 (R).

Kimball, H. H.: 1918, 'Volcanic Eruptions and Solar Radiation Intensivities', *Monthly Weather Rev.*, **46**(8), 355—356.

Kleiber, M.: 1932, '. . . Body Size and Metabolism. Berkeley, California, Univ. of California print. off Hilgardia . . .', vol. 6, No. 11, pp. 315—353.

Kleiber, M.: 1961, 'The Fire of Life: An Introduction to Animal Energetics', Wiley, New York and London, 454 pp.

Klige, R. K.: 1980, *The Level of the Ocean in Geological Past*, Nauka, Moscow, 111 pp. (R).

Koblents-Mishke, O. L. and Sorokin, Yu. I.: 1962, 'Primary Production of the Ocean', *Oceanology* **2**(3), 506—510 (R).

Kochin, N. Ye.: 1936, 'The Construction of the Model of Zonal Circulation of the Atmosphere', *Proc. Main Geoph. Observ.* **10**, 3—27 (R).

Kondratiev, K. Ya. and Moskalenko, N. I.: 1980, 'The Evolution of the Atmosphere and Greenhouse Effect', *Bull. Acad. Sci. USSR, Phys. Atmos. Ocean.* **16**(11), 1151—1162. (R)

Köppen, W. und Wegener, A.: 1924, *Die Klimate der geologischen Vorseit*, Berlin, 256 S.

Kovda, V. A.: 1969, 'Problems of Biological and Economic Productivity of Land', in *General Theoretical Problems of Biological Productivity*, Nauka, Leningrad, 8—23. (R)

Kovda, V. A.: 1973, *Fundamentals of Soil Science*, Nauka, Moscow, 1—447 pp., 2—468 pp. (R)

Krapivin, V. F., Svirezhev, Yu. M. and Tarko, A. M.: 1982, *Mathematical Modelling of Global Processes*, Nauka, Moscow, 272 pp. (R)

Krasilov, V. A.: 1977, *Evolution and Biostratigraphy*, Nauka, Moscow, 256 pp. (R)

Kukla, G. J., Angell, J. K., and Korshover J.: 'New Data on Climatic Trends', *Nature* **270**(5638), 573—580.

Kurtén, B.: 1965, 'Evolution in Geological Time', in *Ideas in Modern Biology* (Ed. by J. A. Moore), Natural History Press, Garden City, NY, 563 pp.

Kurtén, B.: 1971a, *The Age of Mammals*, Columbia Univ. Press, NY, 250 pp.

Kurtén, B.: 1971b, 'Time and Hominid Brain Size', *Commentationes Biologicae,* **36**, 3—7.

Lack, D.: 1954, *The Natural Regulation of Animal Numbers*, Clarendon Press, Oxford, 344 pp.

Lamb, H. H.: 1964, 'The Tole of Atmosphere and Oceans in Relation to Climatic Changes and the Growth of Ice Sheets on Land', in *Problems in Paleoclimatology*, Intersci, London, New York, Sydney, pp. 332—348.

Lamb, H. H.: 1966, *The Changing Climate*, Methuen, London, 236 pp.

Lamb, H. H.: 1970, 'Volcanic Dust in the Atmosphere: With a Chronology and Assessment of its Meteorological Significance', *Philos. Trans. Roy. Soc.* London Ser. A and *Phys. Sci.* **266**(1179), 425—433.

Lamb, H. H.: 1973, 'Climate Change and Foresight in Agriculture: the Possibilities of Long-Term Weather Advice', *Outlook on Agriculture* **7**(5), 203—210.

Lamb, H. H.: 1974, *The Current Trend of World Climate*, University of East Anglia, Norwich, 28 pp.

Lamb, H. H.: 1977, *Climate: Present, Past and Future V.2*, Methuen, London 835 pp.

Lamb, H. H., Malmberg, S. A. and Colebrook, J. M.: 1975, 'Climatic Reversal in Northern North Atlantic', *Nature* **256**(5517), 479.

Laplace, P. S.: 1796, *Exposition du systeme du monde* v.2, Paris.

Lavrenko, Ye. M.: 1949, 'On the Phytogeosphere', *Voprosy geog.* **15**, 53—66. (R)

Lavrenko, Ye. M.: 1971, 'The Major Problems of Biogeocenology and the Purposes of Biogeocenological Studies in the USSR', *Zhurnal obshchey biologli* **4**, 395—408. (R)

Lavrenko, Ye. M., Andreev, V. N. and Leontiev, V. A.: 1955, 'The Profile of Productivity of the Overground Part of Natural Vegetation Cover in the USSR from Tundras to Deserts', *Botanichesky Zhurnal* **40**(3), 415—419. (R)

Legasov, V. A. and Kuzmin, I. I.: 1981, 'The Problems of Energetics', *Priroda* **2**, 8—23. (R)

Leonov, A. K.: 1947, 'An Attempt of Qualitative Consideration of Water Masses, Heat and Salts that are Introduced by the Atlantic and Pacific Currents into the Arctic Basin', *Meteorol. Gidrol.* **5**, 92—106 (R).

Lettau, H.: 1954, 'Study of the Mass, Momentum and Energy Budget of the Atmosphere', *Arch. Met. Geophys. Bioklimat.*, Ser. A, Bd 7.

Lieth, H.: 1963, 'The Role of Vegetation in the Carbon Dioxide Content of the Atmosphere', *J. Geophys. Res.* **68**(13), 3887—3898.

Lieth, H.: 1964, 'Versuch einer kartographischen Darstellung der Produktivität der Phlanzendecke auf der Erde', in *Geogr. Taschenbuch* (E. Meynen, ed.), F. Steiner Verlag, Wiesbaden, pp. 72—80.

Liopo, T. N.: 1966, 'Statistical Probability Methods of Medico-Geographical Estimation of Climate', in *Summary and Perspectives of Introducing New Methods for Geographical Studies*, Irkutsk, pp. 62—64 (R).

Liopo, T. N.: 1968, 'The Nomogram for the Determination of Characteristics of Heat-Isolating Properties of Clothing for Thermal Comfort', *Proc. Main Geoph. Observ.* **233**, 156—160 (R).

*Long-Term Worldwide Effects of Multiple Nuclear-Weapons Detonations*, National Academy of Sciences, 1975.

Lovelock, J. E.: 1979, *Gaia*, Oxford Univ. Press, Oxford, 157 pp.

Lvovich, M. I.: 1974, *The World Water Resources and Their Future*, Mysl', Moscow, 448 pp. (R).

Lyell, C.: 1830—1833, *Principles of Geology*, vol. 1, 1830, 511 pp.; vol. 2, 1832, 330 pp.; vol. 3, 1833, 398 pp. London.

Lyubimenko, V. N.: 1935, *Photosynthesis and Chemosynthesis in the Plant World*, Selkhozgiz, Moscow, Leningrad, 320 pp. (R).

Maksimov, N. A.: 1926, *Physiological Basis of Drought Resistence of Plants*, GIZ, Leningrad, 436 pp. (R)

Maksimov, N. A.: 1944, *A History of Studies of Water Regime and Drought Resistence of Plants from Timiryazev to Our Days*, Izd. AN SSSR, Moscow-Leningrad, 48 pp. (R)

Manabe, S. and Bryan, K.: 1969, 'Climate Calculation with a Combined Ocean-Atmosphere Model', *J. Atm. Sci.* **26**(4), 786—789.

Manabe, S. and Hahn, D. G.: 1977, 'Simulation of the Tropical Climate of an Ice Age', *J. Geophys. Res.* **82**(27), 3889—3911.

Manabe, S. and Stouffer, R. J.: 1980, 'Sensitivity of a Global Climate Model to an Increase of $CO_2$ Concentration in the Atmosphere', *J. Geophys. Res.* **85**(C10), pp. 5529—5553.

Manabe, S. and Wetherald, R. T.: 1967, Thermal Equilibrium of the Atmosphere with a Given Distribution of Relative Humidity', *J. Atm. Sci.* **24**(3), 241—259.

Manabe, S. and Wetherald, R. T.: 1975, 'The Effects of Doubling the $CO_2$ Concentration on the Climate of a General Circulation Model', *J. Atm. Sci.* **32**(1), 3—15.

Manabe, S. and Wetherald, R. T.: 1980, 'On the Horizontal Distribution of Climate Change Resulting from an Increase in $CO_2$ Content of the Atmosphere', *J. Atm. Sci.* **37**, 99—118.

Marchuk, G. I.: 1979, 'Modelling of Climatic Change and the Problem of Long-Term Weather Forecasts', *Meteorol. Gidrol.* **7**, 25—36. (R)

Margulis, L. and Lovelock, J. E.: 1974, 'Biological Modulation of the Earth's Atmosphere', *Icarus* **21**, 471—489.

Marchunova, M. S.: 1980, *The Conditions of the Formation and Characteristics of the Radiative Climate of the Antarctic*, Gidrometeoizdat, Leningrad, 214 pp. (R)

Martin, P. S.: 1958, 'Pleistocene Ecology and Biogeography of North America', *Zoogeography: Amer. Assoc. Sci. Publ.* **51**, 375—420.

Martin, P. S.: 1966, 'Africa and Pleistocene Overkill', *Nature* **212**(5060), 339—342.

Martin, P. S.: 1967, 'Prehistoric Overkill', in Martin, P. S. and Wright, H. E. (eds.), Yale University Press, N.Y.-L., pp. 75–120.

Martin, P. S.: 1973, 'The Discovery of America', *Science* **179**, 969—974.

Mayr, E.: 1963, *Animal Species and Evolution*, Belknap Press of Harvard Univ. Press, Cambridge, MA, 797 pp.

Mayr, E.: 1976, *Evolution and the Diversity of Life. Selected Essays*, Belknap Press of Harvard Univ. Press, Cambridge, MA–London, 721 pp.

Menzhulin, G. V.: 1976, 'The Influence of Climatic Change on the Productivity of Agricultural Plants', *Proc. Main Geoph. Observ.* **65**, 41—48. (R)

Menzhulin, G. V. and Savvateev, S. P.: 1980, 'The Influence of Present Climate Change on Agricultural Productivity, in *The Problems of Atmospheric Carbon Dioxide*, Gidrometeoizdat, Leningrad, pp. 186—197. (R)

McGhee, G. R.: 1982, 'The Fransian-Famennian Extinction Event: A Preliminary Analysis of Appalachian Marine Ecosystems', in *Geological Implications of Impacts of Large Asteroids and Comets on the Earth*. The Geological Society of America, special paper 190, pp. 491—500.

McGowen, C.: 1979, 'Selection Pressure for High Body Temperatures: Implication for Dinosaurs', *Paleobiology* **5**(3), 285—295.

McKenzie, D. P. and Parker, R. L.: 1967, 'The North Pacific: An Example of Tectonics on a Sphere', *Nature* **216**(5122), 1276—1280.

Meadows, D. H. *et al.*: 1972, *Limits to Growth*, Universe Book, NY, 205 pp.

Mercer, J. H.: 1978, 'West Antarctic Ice Sheet and $Co_2$ Greenhouse Effect — A Threat of Disaster, *Nature* **271**, 321—325.

Mesarovic, M. and Pestel, E.: 1975, *Mankind at the Turning Point*, Hutchinson, London, 210 pp.

Milankovich, M.: 1920, *Théorie mathématique des phénomènes thermiques produits par la radiation solaire*, Paris, 339 pp.

Milankovich, M.: 1930, 'Mathematische Klimalehre und Astronomische Theorie der Klimaschwankungen', in *Handbuch der Klimatologie, Bd 1*, Teil A, Berlin, 176b S.

Milankovich, M.: 1941, *Kanon der Erdbestrahlung und seine Anwendung am Eiszeitproblem*, Royal Serbian, Academy, v. 133, 633 pp.

Miles, M. K. and Gildersleeves, P. B.: 1977, 'A Statistical Study of the Likely Causative Factors in the Climatic Fluctuations of the Last 100 Years', *Met. Mag.* **106**(1263), 314—322.

Miller, J. A.: 1978, 'Warmed-Over Dinosaurs', *Science News* **113**(14), 218—220.

Mintz, Y.: 1965, 'Very Long-Term Global Integration of the Primitive Equation of Atmospheric Motion', WMO Technical Note, 66.

Mitchell, J. M.: 1961, 'Recent Secular Changes of Global Temperature', *Annals New York Acad. Sci.* **95**(1), 235—250.

Mitchell, J. M.: 1963, 'On the World-Wide Pattern of Secular Temperature Change', Changes of Climate, *Arid Zone Research* **20**, UNESCO Paris, 161—181.

Mitchell, J. M.: 1971, 'Air Pollution and Climatic Change', 64th annual meeting (Nov. 28, 1970), Am. Inst. Chem Engineers, San Francisco.

*Modelling of the Processes of Global Development*, 1979, Proc. VNIIST, 8, Moscow, 123 pp.

Moiseev, N. N.: 1979, 'System Analysis of the Dynamic Processes of the Biosphere', *Inf. Bull. Acad. Sci. USSR* **1**, 97—108. (R)

Mokhov, I. I.: 1981, 'On the $CO_2$ Effects on the Thermal Regime of the Earth's Climatic System', *Meteorol. Gidrol,* **4**, 24—34. (R)
Möller, F.: 1963, 'On the Influence of Changes in the $CO_2$ Concentration in Air on the Radiation Balance of the Earth's Surface and on the Climate', *J. Geophys. Res.* **68**(13), 3877—3886.
Monsi, M. und Saeki, T.: 1953, 'Über den Lichtfactor in den Pflanzengesellschaften und seine Bedeutung für die Stoffproduktion', *Japan. Botany* **14**(1), 22—52.
Monteith, J. L.: 1965, 'Light Distribution and Photosynthesis in Field Crops', *Ann. Bot.* **29**(113), 17—37.
Monteith, J. L., Szeicz, G. and Yabuki, K.: 1964, 'Crop Photosynthesis and the Flux of Carbon Dioxide Below the Canopy', *J. Appl. Ecol.* **1**, 321—337.
Morgan, W. J.: 1968, 'Rises, Tranches, Great Faults and Crustal Blocks', *J. Geophys. Res.* **73**, 1959—1982.
Müller, D.: 1960, 'Kreislauf des Kohlenstoffes', in *Handbuch der Pflanzenphysiologie* **12**(2), Berlin, pp. 934—948.
Newell, N. D.: 1963, 'Crises in the History of Life', *Scientific America* **208**(2), 77—92.
Newell, N. D.: 1967, 'Revolutions in the History of Life', in *Uniformity and Simplicity*, N.Y., Geol. Soc. Amer. Spec. pap. 89, pp. 63—91.
Newell, N. D.: 1980, 'Now-Asteroid Caused Extinctions', *Science News* **117**(2), 22.
Newman, M. J. and Rood, R. T.: 1977, 'Implications of Solar Evolution for the Earth's Early Atmosphere', *Science* **198**, 1035—1037.
'New Species of Man: Ancestors from "Afar"', *Science News* **115**(1), 36—37.
Nichiporovich, A. A.: 1973, 'Photosynthetic Activity and Primary Production of Phytocenosis at the Current Stage of the Evolution of the Biosphere', in *The Problems of Biocenology*, Nauka, Moscow, pp. 157—173. (R)
Odum, E. P.: 1971, *Fundamentals of Ecology*, 3d ed., Saunders, Philadelphia, 574 pp.
Odum, H. T.: 1971, *Environment, Power and Soceity*, Wiley, Interscience, N. Y.—London—Sydney—Toronto, 336 pp.
Oke, T. R.: 1978, *Boundary Layer Climates*, Methuen, Halsted Pres., John Wiley, London, N.Y.
O'Keefe, J. D. and Ahrens, T. J.: 1982, 'Impact Mechanics of Cretaceous-Tertiary Extinction Bolide', *Nature* **298**, 123—127.
Oldekop, E. M.: 1911, 'On the Evaporation from the Surface of River Basins', *Proc. Met. Observ. Yuriev Univ.* **4** 209 pp. (R)
Oliver, R. C.: 1976, 'On the Response of Hemispheric Mean Temperature to Stratospheric Dust: An Empirical Approach, *J. Appl. Met.* **15**(9), 333—350.
Olson, S. L. and James, H. F.: 1982, 'Fossil Birds from the Hawaiian Islands: Evidence for Wholesale Extinction by Man before Western Contact', *Science* **217** (4560), 633–635.
Oort, A. H. and Vonder Haar, T. H.: 1976, 'On the Observed Annual Cycle in the Ocean-Atmosphere Heat Balance over the Northern Hemisphere, *J. Phys. Oceanogr.* **6**(6), 781—800.
Oparin, A. I.: 1924, *The Origin of Life*, Moskovsky rabochiy, Moscow, 70 pp. (R)
Oparin, A. I.: 1957, *The Emergence of Life on the Earth*, 3d ed, Izd. AN SSSR, Moscow, 458 pp. (R)
Ostrom, J. H.: 1969, 'Terrestrial Vertebrates as Indicators of Mesozoic Climates', Proc. North. Amer. Paleont. Conv, pp. 347—376.
Palmer, A. R.: 1982, 'Biomere Boundaries: A Possible Test for Extraterrestrial Perturbations of the Biosphere', in *Geological Implications of Impacts of Large Asteroids and Comets on the Earth*, The Geological Society of America, special paper 190, pp. 469—475.
Pearson, R.: 1978, *Climate and Evolution*, Academic Press, London, New York, San Francisco, 274 pp.
Penman, H. L.: 1948, 'Natural Evaporation from Open Water, Bare Soil and Grass', *Proc. Roy. Soc.*, Ser. A. **193**, 120–145.

Pilbeam, D.: 1982, 'New Hominoid Skull Material from the Miocene of Pakistan', *Nature* **295**, 232—234.

Pitrat, C. W.: 1970, 'Phytoplankton and the Late Paleozoic Wave of Extinction', *Palaeogeogr., Palaeoclimatol., Palaeoecol.* **8**, 49—66.

Pivovarova, Z. I.: 1977, *The Radiative Features of the Climate of the USSR*, Gidrometeoizdat, Leningrad, 355 pp. (R)

Plass, G. H.: 1956, 'The Carbon Dioxide Theory of Climatic Change', *Tellus* **8**(2), 140—154.

Pollack, J. B.: 1982, 'Solar, Astronomical and Atmospheric Effects on Climate', in *Climate in Earth History*, National Academy Press, Washington, D.C., pp. 68—76.

Pollack, J. B., Toon, O. B. and Sagan, C.: 1975, 'The Effect of Volcanic Activity on Climate', *Proc. WMO/IAMAP Symp. on Long-Term Climatic Fluctuations*, WMO, Geneva, No. 421, pp. 279—285.

Pollack, J. B. *et al.*: 1976, 'Volcanic Explosions and Climatic Change: A Theoretical Assessment', J. B. Pollack, O. B. Toon, C. Sagan *et al., J. Geophys. Res.* **81**(6), 1071–1083.

Pollard, D.: 1978, 'An Investigation of the Astonomical Theory of the Ice Ages using a Simple Climate-Ice Sheet Model', *Nature* **272**, 233—235.

Poluektov R. A. (ed.): 1974, *Dynamic Theory of Biological Populations*, Nauka. Moscow, 455 pp. (R).

Polyakov, B. V.: 1946, *Hydrological Analysis and Calculations*, Gidrometeoizdat. Leningrad, 480 pp. (R).

Rabinowitch E. I. 1951, *Photosynthesis and Related Processes*, vol. 2, pt. 1, Interscience Publ., New York, pp. 603—1208.

Rakipova, L. R.: 1962, 'Changes in Climate by Forcing the Arctic Ice', *Meteorol. Gidrol.* **9**, 28—30 (R).

Remanathan, V. *et al.*: 1979, Increased Amospheric $CO_2$: Zonal and Seasonal Estimates of the Effect on the Radiation Energy Balance and Surface Temperature', *J. Geophys. Res.* **84**(C8), 4949—4956.

Raschke, E., Möller, F., and Bandeen, W.: 1968, 'The Radiation Balance of the Earth-Atmosphere System over both Polar Regions obtained from Radiation Measurements of the Nimbus II Meteorological Satellite', *Sveriges Met. och. Hydrol. Inst. Modellanden*, Ser. B, No. 28, 104 S.

Raschke, E. *et al.*: 1973, 'The Radiation Balance of the Earth-Atmosphere System from Nimbus III Radiation Measurements', NASA Technical Note, D—7242, National Aeronautics and Space Administration, Washington, 73 pp.

Rauner, Yu. L.: 1979, 'The Synchronism of Droughts in the Grain Regions of the Northern Hemisphere', *Bull. Acad. Sci. USSR, Ser.: Geogr.* **1**, 5—19 (R).

Raup, D. M.: 1979, 'Size of the Permo-Triassic Bottleneck and its Evolutionary Implications, *Science* **206**, 217—218.

Raup, D. M. and Sepkosky, J. J.: 1982, 'Mass Extinctions in the Marine Fossil Record', *Science* **215**(4539), 1501—1503.

Raup, D. M. and Stanley, S. M.: 1971, *Principles of Paleontology*, W. H. Freeman, San Francisco, 331 pp.

Riqlet, A.: 1969, 'L'histologie osseuse envisagée comme indicateur de la physiologie thérmique chez les tétrapodes fossiles', *C.R. Acad. Sci. Paris* **268**, 782—785.

Robinson, P. L.: 1971, 'A Problem of Faunal Replacement on Permo-Triassic Continents', *Palaeontology* **14**(I), 131—152.

Robock, A.: 1978, 'Internally and Externally Caused Climate Change', *J. Atm. Sci.* **35**(6), 1111—1122.

Rodin, L. Ye. and Bazilevich, N. I.: 1965, *The Dynamics of Organic Matter and Biological Cycle of Ash — Like Elements and Nitrogen in the Basic Types of the Earth's Vegetation*, Nauka, Moscow-Leningrad, 253 pp. (R)

Ronov, A. B.: 1959, 'On the Post-Cambrian Geochemical History of the Atmosphere and Hydrosphere', *Geochimiya* **5**, 397—409 (R)

Ronov, A. B.: 1964, 'General Tendencies in the Evolution of the Composition of the Earth's Crust, the Ocean and the Atmosphere', *Geochimiya* **8**, 715—743 (R)

Ronov, A. B.: 1976, 'Volcanism, Accumulation of Carbon, Life', *Geochimiya* **8**, 1252—1277 (R)

Ronov, A. B.: 1980, *Sedimentary Envelope of the Earth*, Nauka, Moscow, 79 pp. (R)

Ronov, A. B.: 1982, 'The Global, Balance of Carbon in the Neogene', *Geochimiya* **7**, 920—932 (R)

Ross, Yu. K.: 1964, 'On the Mathematical Theory of Photosynthesis of Vegetation Cover', *Rep. Acad. Sci. USSR* **157**(5), 1239—1242 (R)

Rotty, R. M.: 1979, 'Present and Future Production of $CO_2$ from Fossil Fuels. A Global Appraisal', CONF—770385. Workshop on the global effects of carbon dioxide from fossil fuels, pp. 36—43.

Rubinshtein, Ye. S., and Polozova, L. G.: 1966, *Modern Changes in Climate*, Gidrometeoizdat, Leningrad, 268 pp. (R)

Russell, D. A.: 1979, 'The Enigma of the Extinction of the Dinosaurs', *Annual Review of Earth and Planetary Science* **7**, 163—182.

Rutten, M. G.: 1971, *The Origin of Life by Natural Causes*, Elsevier, Amsterdam, 420 pp.

Sagan, C.: 1974, *The Cosmic Connection*, Caronet Books, London 274 p.

Sagan, C. and Mullen, G.: 1972, 'Earth and Mars: Evolution of Atmospheres and Surface Temperature', *Science* **177**, 52—56.

Sagan, C., Toon, O. B. and Pollack, J. B.: 1979, 'Human Impact on Climate of Global Significance Since the Domestication of Fire, *Science* **204**(4425), 1363—1368.

Sanderson, R. M.: 1975, 'Changes in the Area of Arctic Sea Ice 1966 to 1974', *Met. Mag.* **104**(1240), 313—323.

Savin, S. M.: 1977, 'The History of the Earth's Surface Temperature During the Past 100 Million Years', *Ann. Rev. Earth Planet. Sci.* **5**, 319—355.

Savinov, S. I.: 1913, 'The Greatest Values of Solar Radiation Tensity by Observations in Pavlovsk in 1892', *Bull. Acad. Sci.*, ser. 6, **7**(12), 707—720 (R)

Schneider, S. H.: 1972, 'Cloudiness as a Global Climatic Feedback Mechanism: The Effects on the Radiation Balance and Surface Temperature of Variations in Cloudiness', *J. Atm. Sci.* **29**, 1413—1422.

Schneider, S. H. and Thompson, S. L.: 1979, 'Ice Ages and Orbital Variations: Some Simple Theory and Modelling', *Quater. Res.* **12**, 188—203.

Schneider, S. H. and Thompson, S. L.: 1981, 'Atmospheric $CO_2$ and Climate: Importance of the Transient Response', *J. Geophys. Res.* **86**(C4), 3135—3147.

Schopf, T. J. M: 1980, *Paleoceanography*, Harvard Univ. Press, Cambridge, Mass., 341 pp.

Schopf, T. J. M.: 1981, 'Punctuated Equilibrium and Evolutionary Stasis', *Paleobiology,* **7**(2), 156—166.

Schreiber, P.: 1904, 'Über die Beziehungen zwischen dem Niederschlag und der Wasserführung der Flüße in Mitteleuropa', *Z. Meteorol.* **21**, Pt. 10.

Schwarzbach, M.: 1950, *Das Klima der Vorzeit. Eine Einführung in die Paläoklimatologie*, Enke, Stuttgart, 211 S.

Schwarzbach, M.: 1968, 'Neue Eiszeithypothesen', *Eiszeitalter und Gegenwart*, Bd **19**, S. 250—261.

Schwarzschild, M.: 1958, *Structure and Evolution of the Stars*, Princeton Univ. Press, Princeton, N.Y., 296 pp.

Sellers, W. D.: 1969, 'A Global Climatic Model Based on the Energy Balance of the Earth-Atmosphere System', *J. Appl. Met.* **8**(3), 392—400.

Severtsev, A. N.: 1925, *The Principal Trends of Evolutionary Process* (3rd edn), Izd. MGU, Moscow, 1967, 202 pp. (R).

Severtsev, A. N.: 1939, *Morphological Features of Evolution*, Izd. AN SSSR, Moscow, Leningrad, 610 pp.

Severtsev, S. A.: 1941, *The Dynamics of Population and Adaptive Evolution of Animals*, Izd. AN SSSR, Moscow, 316 pp. (R).
Schackleton, N. J. and Kennet, J. P.: 1975, 'Paleotemperature History of the Cenozoic and the Initiation of Antarctic Glaciation', *Initial Reports of DSDP* **29**, U.S. Government Printing Office, Washington, D.C., 743—755.
Sharaf, Sh. G. and Budnikova, N. A.: 1969, 'Secular Changes in the Earth's Orbital Elements and Astronomical Theory of Climatic Fluctuations', *Proc. Inst. Theoret. Astr.* **14**, 48—85 (R)
Shklovsky, I. S.: 1963, *Universe, Life, Intelligence*, lst ed (5th ed., 1980), Nauka, Moscow, 352 pp. (R)
Shklovsky, I. S.: 1976, 'On the Possible Uniqueness of Intelligent Life in the Universe', *Voprosy filos.* **9**, 80—93 (R)
Shmalgauzen, I. I.: 1940, *The Ways and Laws of the Evolutionary Process*, Izd. AN SSSR, Moscow-Leningrad, 231 pp. (R)
Shmalgauzen, I. I.: 1946, *The Factors of Evolution*, Izd. AN SSSR, Moscow-Leningrad, 396 pp. (R)
Shuleikin, V. V.: 1941, *Physics of the Sea*, Izd. AN SSSR, Moscow-Leningrad, 834 pp. (R)
Sinitsyn, V. N.: 1965, *Ancient Climates of Eurasia, P. 1*, Izd. LGU, Leningrad, 267 pp. (R)
Sinitsyn, V. M.: 1966, *Ancient Climates of Eurasia, P. 2*, Izd. LGU, Leningrad, 166 pp. (R)
Sinitsyn, V. M.: 1967, *Introduction to Paleoclimatology*, Nedra, Leningrad, 232 pp. (R)
Sinitsyn, V. M.: 1970, *Ancient Climates of Eurasia, P. 3*, Izd. LGU, Leningrad, 133 pp. (R)
Sinitsyn, V. M.: 1976, Climate of Laterites and Bauxites, Nedra, Leningrad, 152 pp. (R)
Simon, C.: 1981, 'Clues in the Clay', *Science News* **120**, 314—315.
Simon, C.: 1982 'Deep Crust Hints Meteoritic Impact', *Science News* **121**(5), 69.
Simpson, G. G.: 1944, *Tempo and Mode in Evolution*, Columbia Univ. Press, New York. 237 pp.
Simpson, G. G.: 1961, *Life of the Past. An Introduction to Paleontology*, Yale University Press, New Haven, 198 pp.
Simpson, G. G.: 1969, *Biology and Man*, Harcourt, Brace and World, New York, 175 pp.
Simpson, G. G.: 1978, *Concession to the Improbable: An Unconventional Autobiography*, Yale Univ. Press, New Haven and London, 291 pp.
Simpson, G. G.: 1982, Personal commmunication.
Smagorinsky, J.: 1963, 'General Circulation Experiments with the Primitive Equations. 1: The Basis Experiment', *Monthly Weather Rev.* **91**, 99—164.
Smagorinsky, J., Manabe S., and Holloway, I. L.: 1965, 'Numerical Results from Nine-Level General Circulation Model of the Atmosphere', *Monthly Weather Rev.* **93**, 727—768.
Smit, J. and Hertogen, J.: 1980, 'An Extraterrestrial Event at the Cretaceous—Tertiary Boundary', *Nature* **285**, 198—200.
Sochava, A. V.: 1979, 'Changes in the Composition of the Earth's Atmosphere and the Appearance of Multicellular Animals', in *Paleontology of the Pre-Cambrian and Early Cambrian*, Nauka, Leningrad, pp. 255—265 (R).
Sochava, A. V. and Glikman, L. S.: 1973, 'Cyclic Changes in the Free Oxygen Content of the Atmosphere and Evolution', in *Papers of Evolutionary Seminar*, Vladivostok, pp. 1, 68—87 (R).
Soolov, B. S.: 1975, 'Organic World of the Earth on the Way to Phanerozoic Differentiation', Rep. at Jubilee Session of Ac. Sc. USSR, Izd. VINITI, Moscow, 20 pp. (R).
Sokolov, B. S.: 1981, 'Paleontology, Geology and Evolution of the Biosphere', in *Problems of Evolution of Geological Processes*, Nauka, Novosibirsk, pp. 158—167 (R).
Sokolovsky, D. L.: 1936, 'The Relationship between Run-Off and Precipitation Under Different Geographical Conditions', *Meteorol. Gidrol.* **6**, 56—58 (R)
Stanley, S. M.: 1979, *Macroevolution*, W. H. Freeman, San Francisco, 332 pp.
Stebbins, G. L. and Ayala, F. J.: 1981, 'Is a New Evolutionary Synthesis Necessary?' *Science*

**213**(4511), 967—971.

Stephens, G. L., Cambell, G. G., and Vonder Haar, T. H.: 1981, 'Earth Radiation Budget', *J. Geophys. Res.* **86**(C10), 9739—9760.

Suarez, M. J. and Held, I. M.: 1976, 'Modeling Climatic Response to Orbital Parameter Variations', *Nature* **263**, 46—47.

Sukavhev, V. N.: 1945, 'Biogeocenology and Phytocenology', *Rep. AN SSSR* **47**(6), 447—449 (R)

Sukachev, V. N.: 1964, 'Basic Concepts of Forest Biocenology', in *Fundamentals of Forest Biocenology*, Nauka, Moscow, 5—49 (R)

Svirezhev, Yu. M. and Logofet, D. O.: 1978, *Resistence of Biological Communities*, Nauka, Moscow, 352 pp. (R)

*System Modelling of Processes of Global Development*, 1980, Proc. VNIIST, 14, Moscow, 141 pp. (R)

Takhtadzhyan, A. L.: 1980, 'The Order: Flowering Plants, or Angiosperms', in *The Life of Plants* **5**(1), Prosveshcheniye, Moscow, 7—114 (R)

Tatarinov, L. P.: 1972, 'Ecological Factors of the Origin of Amphibians', in *The Problems of Evolution*, 2, Nauka, Novosibirsk, pp. 144—153 (R)

Tatarinov, L. P.: 1976, *Morphological Evolution of Teriodonts and General Problems of Phylogenetics*, Nauka, Moscow, 258 pp. (R)

Taylor, B. L., Gal-Chen, T., and Schneider, S. H.: 1980, 'Volcanic Eruptions and Long-Term Temperature Records: An Empirical Search for Cause and Effect', *Quart. J. Roy. Met. Soc.* **106**, 175—199.

Technical Report by the WMO Executive Committee Panel of Experts on Climatic Change, 1977, *WMO Bulletin* **XXVI**(1), 50—55.

'The Appeal of the All-Union Conference of Scientists for the Salvation of Mankind from Nuclear Threat, for Disarmament and Peace. 1983', *Inf. Bull. Acad. Sci. USSR* **9**, 123—124 (R).

*The Development and Changes in the Organic World at the Boundary of the Mesozoic and Cenozoic: The Vertebrates* (V. N. Shimansky, A. N. Soloviev, and P. G. Danilchenko *et al.*), Nauka, Moscow, 1978, 136 pp. (R).

*The Global Problems of Present Epoch* (ed. N. N. Inozemtsev), Moscow, 1981, 285 pp. (R).

*The Life and Letters of Charles Darwin*, vol. 3, Murray, London, 1887.

*The Problem of the Search for Extraterrestrial Civilizations* (eds V. S. Troitsky and N. S. Kardashev), Nauka, Moscow, 1981, 263 pp. (R).

'The Quest for a Catastrophe', 1980, *Science News* **118**(9), 134.

*The World Water Balance and Water Resources of the Earth*, Gidrometeoizdat, Leningrad, 1974, 638 pp. (R).

Thomas, R. H. *et al.*: 1979, 'Effect of Climatic Warming on the West Antarctic Ice Sheet', *Nature* **277**, 355—358.

Tikhonov, A. N. *et al.*: 1969, 'On the Evolution of Fusion Zones in the Course of the Earth's Thermic History', *Rep. Acad. Sci. USSR* **188**(2), 342—344 (R).

Timofeev-Resovsky, N. V., Vorontsov, N. N. and Yablokov, A. V.: 1977, *A Brief Essay on the Theory of Evolution*, 2nd edn, Nauka, Moscow, 301 pp. (R).

Tooming, Kh. G.: 'Mathematical Modelling of the Structure and Production Process of Phytocenosis', *Zhurnal obshchey biologii* **35**(2), 181—195 (R)

Toon, O. B. *et al.*: 1982, 'Evolution of an Impact-Generated Dust Cloud and its Effects on the Atmosphere', in *Geological Implications of Impacts of Large Asteroids and Comets on the Earth*, The Geological Society of America, special paper 190, pp. 187—200.

Tsitsenko, G. V.: 1963, 'The Methods of Calculating Temperature of the Body Surface of Man Using the Heat Balance Equation', *Proc. Main Geoph. Observ.* **139**, 108—114 (R)

Tyndall, I.: 1861, 'On the Absorption and Radiation of Heat by Gases and Vapours and on the Physical Connection of Radiation Absorption and Conduction', *Phil. Mag.* **22**(144). pp. 167—194, 273—285.

Uchijima, Z.: 1966, 'An Improvement of Semi-Empirical Method of Evaluating the Total Photosynthesis of a Plant Community', *J. Agric. Met.* **22**(1).

*Uniformity and Simplicity.* 1967, *A Symposium on the Principle of the Uniformity of Nature* (ed.) C. C. Albritton, Jr, special papers, Geol. Soc. Amer., NY. 99 pp.

Urey, H. C.: 1973, 'Cometary Collisions and Geological Periods', Nature **242**, 32—33.

Valentine, J. W.: 1968, 'Climatic Regulation of Species Diversification and Extinction', *Geol. Soc. Amer. Bull.* **79**, 273—276.

Van Loon, H. and Williams, J.: 1976, 1977, 'The Connection between Trends of Mean Temperature and Circulation at the Surface', P. 1, Winter, *Monthly Weather Rev.* (1976) **104**(4), 365—380; P. 2, Summer, *Monthly Weather Rev.* (1976) **104**(8), 1003—1011; P. 3 Spring and Autumn, *Monthly Weather Rev.* (1976) **104**(12), 1593—1596; P. 4. Comparison of the Surface Changes in the Northern Hemisphere with the Upper Air and with the Antarctic in Winter, *Monthly Weather Rev.* (1977) **105**(5), 636—647.

Van Valen, L.: 1969, 'Evolution of Communities and Late Pleistocene Extinctions', in *Proc. of the North American Paleontological Convention*, Part E, pp. 469—485.

Van Valen, L.: 1971, 'The History and Stability of Atmospheric Oxygen', *Science* **171** 439—443.

Velichko, A. A.: 1973, *Natural Process in the Pleistocene*, Nauka, Moscow, 256 pp. (R)

Vernadsky, V. I.: 1926, *The Biosphere*, Nauchno-techn. Izd., Leningrad, 146 pp. (R)

Vernekar, A. D.: 1972, 'Long-Period Global Variations of Incoming Solar Radiation', *Met. Mag.* **12**(34), 21 pp.

Veselovsky, K. S.: 1857, *On the Climate of Russia*, Izd. Imper. Ak. Nauk., St. Petersburg, 326 pp. (R)

Vinnikov, K. Ya. and Groisman, P. Ya.: 1981, 'The Empirical Analysis of $CO_2$ Effects on Present Changes in the Mean Annual Surface Air Temperature of the Northern Hemisphere', *Meteorol. Gidrol.* **11**, 30—43.

Vinnikov, K. Ya. and Groisman, P. Ya.: 1982, 'Empirical Study of Climate's Sensitivity', *Bull. Acad. Sci. USSR, Phys. Atmos. Ocea.* **18**(11) 1159—1169 (R)

Vinnikov, K. Ya. *et al.*: 1980, 'Modern Climatic Changes in the Northern Hemisphere', *Meteorol. Gidrol.* **6**, 5—17 (R)

Vinogradov, A. P.: 1972, 'Variations in the Atmosphere Under Human Activities', *Geochimiya* **1**, 3—10 (R)

Volobuev, V. R.: 1953, *Soils and Climate*, Izd. Ak. Nauk Az. SSR, Baku, 320 pp. (R)

Vonder Haar, T. H. and Suomi, V. E.: 1969, 'Satellite Observations of the Earth's Radiation Budget', *Science* **163**, 667—669.

Vysotsky, G. N.: 1905, *The Steppes of European Russia*, Comprehensive Encyclopaedia of Russian Agriculture and Contiguous Sciences **9** Izd. A. F. Dervien, St. Petersburg (R)

Walker, J. C. G., Hays, P. B. and Kasting, J. F.: 1981, 'A Negative Feedback Mechanism for the Long-Term Stabilization of Earth's Surface Temperature', *J. Geoph. Res.* **86** 9776—9782.

Wallace, A. R.: 1864, 'Origin of Human Races and the Antiquity of Man Deduced from Natural Selection', *Anthrop. Rev.* **2**, CLVIII 3

Wallace, A. R.: 1869, 'Geological Climates and Origin of Species', *Quart. Rev.*, 126 pp.

Watt, K.: 1968, *Ecology and Resource Management*, McGraw-Hill, New York, 450 pp.

Wetherald, R. T. and Manabe, S.: 1975, 'The Effect of Changing the Solar Constant on the Climate of a General Circulation Model, *J. Atm. Sci.* **32**, 2044—2059.

Wetherald, R. T. and Manabe, S.: 1980, 'Cloud Cover and Climate Sensitivity', *J. Atm. Sci.* **37**(7), 1485—1510.

Wetherald, R. T. and Manabe S.: 1981, 'Influence of Seasonal Variation upon the Sensitivity of a Model Climate', *J. Geophys Res.* **86**(C2) 1194—1204.

Wetherill, G. W.: 1978, 'Accumulation of the Terrestrial Planets', in *Protostars and Planets: Studies of Star Formation and of the Origin of the Solar System* (ed. T. Gehrels). Univ. of Arizona Press, Tucson, Arizona, pp. 565—598.

Whittaker, R. H.: 1975, *Communities and Ecosystems* (2nd edn), Macmillan, London, 162 pp.
Willett, H.: 1950, 'On the Present Climatic Variation', *Centenary Proc. Roy. Met. Soc.* pp. 195—205.
Willett, H.: 1974, 'Do Recent Climatic Fluctuations Portend an Imminent Ice Age?', *Geofisica Int.* **14**(4), 265—295.
Williamson, P. G.: 1981, 'Paleontological Documentation of Speciation in Cenozoic Molluscs from Tuzkana Basin', *Nature* **293**(5832), 437—443.
Willson, R. C. *et al.*: 1981, 'Observations of Solar Irradiance Variability', *Science* **211**, 700—702.
Winslow, C.-E. A., Herrington, L. P., and Gagge, A. P.: 1937a, 'Physiological Reactions of the Human Body to Varying Environmental Temperatures', *Amer. J. Physiol.* **120**(1), 1—22.
Winslow, C.-E. A., Herrington, L. P., and Gagge, A. P.: 1937b, 'Physiological Reactions of the Human Body to Various Atmospheric Humidities', *Amer. J. Physiol.* **120**(2), 288—299.
'WMO Statement on Climatic Change', 1976: *WMO Bull.* **XXV**(3), 211—212.
Yanshin, A. L.: 1961, 'On the Depth of Saliferous Basins and Some Problems of the Formation of Massive Saline Deposits', *Geol. Geophys.* **1**, 3—15 (R).
Yanshin, A. L.: 1981, 'Methodological Significance of Vernadsky's Teaching about the Biosphere and its Tranformation into the Noosphere', in *Methodology of Science and Scientific Progress*, Nauka, Novosibirsk, pp. 194—204 (R).
Yaroshevsky, V. A.: 1968, *Weather and Fine-Wool Sheep-Breeding*, Gidrometeoizdat, Leningrad, 203 pp. (R).
Yasamanov, N. A.: 1980. 'On the Problem of the Evolution of Temperature Regime in the Phanerozoic', *Rep. Acad. Sci. USSR* **249**(6), 1427—1429. (R)
Yefimova, N. A.: 1965, 'The Distribution of Photosynthetically Active Radiation on the Territory of the Soviet Union', *Proc. Main Geoph. Observ.* **179**, 118—130 (R)
Yefimova, N. A.: 1968, 'The Main Feature of Meteorological Regime on the Fields of Winter Wheat and Winter Eye', *Proc. Main Geoph. Observ.* **229**, 23—26.
Yefimova, N. A.: 1969, 'Geographical Distribution of Sums of Photo-Synthetically Active Radiation', in *General Theoretical Problems of Biological Productivity*, Nauka, Leningrad, 160—163 (R)
Yefimova, N. A.: 1976, 'The Maps of Productivity of Vegetation Cover on the Continents', *Botanichesky zhurnal* **61**(12), 1641—1657 (R)
Yefimova, N. A.: 1977, *Radiative Factors on the Productivity of Plant Cover*, Gidrometeoizdat, Leningrad, 216 pp. (R)
Yefimova, N. A. and Tsitsenko, G. V.: 1963, 'The Comparison of Experimental Method and that of Calculation for Determining the Temperature of the Body Surface', *Proc. Main Geoph. Observ.* **139**, 115—121 (R)
Zagladin, V. V. and Frolov, I. T.: 1981, *The Global Problems of Our Age*, Mezhdunarodnye otnosheniya, Moscow, 238 pp. (R)
Zakharov, V. F. and Strokina, L. A.: 1978, 'The Present Changes in the Ice Cover of the Arctic Ocean', *Meteorol. Gidrol.* **7**, 35—43 (R)
Zeuner, F.: 1959, *The Pleistocene Period. Its Climate, Chronology and Faunal Successions*, Hutchinson, London, 449 pp.
Zubakov, V. A. and Borzenkova, I. I.: 1983, *Paleoclimates of the Late Cenozoic*. Gidrometeoizdat, Leningrad, 216 pp. (R)
Zubenok, L. I.: 1956, 'The Water Balance of the Continents and Oceans', *Rep. Acad. Sci. USSR* **108**(5), 829—832 (R)
Zubenok, L. I.: 1963, 'The Influence of Temperature Anomalies on the Arctic Ice Cover', *Meteorol. Gidrol.* **6**, 25—30 (R)
Zubenok, L. I.: 1968, 'On the Determination of Total Evaporation Over Separate Years', *Proc. Main Geoph. Observ.* **233**, 101—109 (R)
Zubov, N. N.: 1945, *The Arctic Ice*, Moscow, 357 pp. (R)
Ångström, A.: 1969, 'Apparent Solar Constant Variations and Their Relation to the Variability of Atmospheric Transmission', *Tellus* **21**(2), p. 205—218.

# SUBJECT INDEX